Ergebnisse der Mathematik und ihrer Grenzgebiete

3. Folge · Band 20

A Series of Modern Surveys in Mathematics

A. I. Kostrikin

Around Burnside

Translated from the English by James Wiegold

Springer-Verlag Berlin Heidelberg GmbH

A. I. Kostrikin

Professor, V. A. Steklov Institute
Vavilov Street 42
II 7966, GSP-I, Moscow
USSR

Dr. James Wiegold
Professor, School of Mathematics
University of Wales
College of Cardiff
Senghenydd Road
Cardiff CF2 4AG, Wales, U.K.

Original Russian language edition, entitled „ВОКРУГ БЕРНСАЙДА", published by МОСКВА «НАУКА».

Mathematics Subject Classification (1980): 11F41, 11G18, 11I20

ISBN 978-3-642-74326-9 ISBN 978-3-642-74324-5 (eBook)
DOI 10.1007/978-3-642-74324-5

Library of Congress Cataloging-in-Publication Data
Kostrikin, A. I. (Alekseĭ Ivanovich)
[Vokrug Bernsaĭda. English] Around Burnside / A. I. Kostrikin. p. cm.
(Ergebnisse der Mathematik und ihrer Grenzgebiete ; 3. Folge, Bd. 20)
Translation of: Vokrug Bernsaĭda.
ISBN 978-3-642-74326-9
1. Burnside problem. I. Title. II. Series.
QA171.K67413 1990 512'.2--dc20 89-26244

Originally published by Springer-Verlag Berlin Heidelberg New York 1990
Softcover reprint of the hardcover 1st edition 1990

Typesetting: Macmillan India Ltd., India.

2141/3020-543210 – Printed on acid-free paper

In Memory of my Parents
Kostrikin, Ivan Vukolovich (1887–1968)
Kostrikin, Evdokiya Stepanovna (1887–1971)

Translator's Preface

Perhaps it is not inappropriate for me to begin with the comment that this book has been an interesting challenge to the translator. It is most unusual, in a text of this type, in that the style is racy, with many literary allusions and witticisms: not the easiest to translate, but a source of inspiration to continue through material that could daunt by its combinatorial complexity. Moreover, there have been many changes to the text during the translating period, reflecting the ferment that the subject of the restricted Burnside problem is passing through at present.

I concur with Professor Kostrikin's "Note in Proof", where he describes the book as fortunate. I would put it slightly differently: its appearance has surely been partly instrumental in inspiring much endeavour, including such things as the paper of A. I. Adian and A. A. Razborov producing the first published recursive upper bound for the order of the universal finite group $B(d,p)$ of prime exponent (the English version contains a different treatment of this result, due to E. I. Zel'manov); M. R. Vaughan-Lee's new approach to the subject; and finally, the crowning achievement of Zel'manov in establishing RBP for all prime-power exponents, thereby (via the classification theorem for finite simple groups and Hall-Higman) settling it for all exponents.

The book is encyclopaedic in its coverage of facts and problems on RBP, and will continue to have an important influence in the area.

I am very grateful to Professor Kostrikin for his enormous help with difficulties in the translation, and to Springer-Verlag for their unfailing courtesy at all times.

Cardiff, November 1989 James Wiegold

A Note on the English Edition

It has come about that one of the famous problems of the remarkable British mathematician William Burnside has taken deep root in Russian soil. Problems of Burnside type have become singularly popular in Moscow and Novosibirsk. At times, it seems that they occupy **too** prominent a position in the work of Russian algebraists; and it is of course advisable for them to share their knowledge with Western colleagues.

The restricted Burnside problem is fundamental in this regard, and it occupies a central position in my book. The solutions to date, while far from definitive[1]), are nonetheless pretty impressive; they are due in about equal measure to German, British and Russian mathematicians. Right up to the present time, even at the stage of preparing the English version, I have had to make insertions and additions reflecting the latest results. The emendations do not affect the first five chapters, and what they amount to is this. The text of the first section of Chap. 6 has been rewritten, and a totally new proof has been furnished for Theorem 6.4.1. I hope that the additional commentaries to Chap. 6 and 7 will be found of interest. As regards Appendix I, which is devoted to the question of effective bounds, it pursues what is in reality a purely academic aim.

I look upon the English translation of this book (with its somewhat **recherché** title "Around Burnside"), so effectively carried out by Professor James Wiegold, as the final act at the end of a very long trek. It will appear in the "Ergebnisse" series, which commands world-wide respect. It remains only for me to express my heartfelt gratitude to Springer-Verlag for taking the initiative that led to the translation in the first place.

Moscow, 12 February 1989 A. I. Kostrikin

[1]) [EDITORS NOTE] This was written by the author before the latest developments mentioned in the translator's preface were known.

Preface

„. . . Ревность к истине, меня вдохновляющая, не дает мне успокоения: чтоб не смотрелась ложь перед истиной и не хвалилась тьма перед светом.“[1]

Maksim Grek

I first solved the restricted Burnside problem (RBP) for prime exponent p a long time ago, in 1958, that is, I showed that there is a largest finite d-generator group satisfying the relation $x^p = 1$ identically, its order depending only on d and p. It should be said that, even now, after a lapse of a quarter of a century, the theorem is still far from being a triviality. Furthermore, there has not grown out of it a solution of RBP for arbitrary prime-power exponents p^k, as was originally expected. On the other hand, the method of sandwiches, which lies as the very heart of the proof (it received this name more recently), has found unexpected applications in other disciplines. This gave a convenient motive to: "Regarder vingt-sept ans en arrière, se retourner pour voir où les yeux se perdent, et toutes les choses et tous les gens depuis ont vécu leur destinée individuelle dans le sort collectif. Les mots n'ont plus le même sens, tout demande explication sans fin."[2]

The reason for reverting to concerns of the past in this way is the need to make available to experts in the area a text that is easily checked and does not pretend to the deceptive brevity of the original paper. All errors to be found there have been subjected to impartial analysis, and, it is hoped, have been eliminated. The reader will see that the recent improvements mentioned in Chap. 5 provide a somewhat shorter proof of RBP. However, the most important applications, those in the theory of finite-dimensional simple Lie algebras over algebraically closed fields of characteristic $p > 0$, and to the global nilpotency problem for n-Engel Lie algebras over fields of characteristic zero, rest on the original 1958 approach.

For these reasons, the exposition contains some duplication. The numerous commentaries are intended to assist the reader in choosing short-cuts to pre-assigned goals. The first four chapters constitute the kernel of the book, and we must add to what has already been said that the main canvas of the arguments from the 1958 paper are scrupulously reproduced in this section. This has been done quite consciously, for reasons that are not exclusively mathematical in nature. On the other hand, the proof of a theorem about sandwiches, so important for Chap. 6, is obtained rather naturally along the way. No proof different in principle has yet been suggested.

[1] Rough translation: ". . . Zeal for the truth, which is my inspiration, gives me no calm: that lies should not be valued above the truth, and that darkness should not swagger before the light."
[2] Louis Aragon, Henri Matisse, roman, vol. I, p. 15, Gallimard, 1971.

What has stimulated me to write this short monograph is the realization that background activity around Burnside has been on the increase in recent years. Its title reflects the content very accurately. Lie algebras occupy the pre-eminent position, whereas the main result associated with the name of Burnside is group-theoretical. It is largely thanks to the restricted version that his remarkable problem has kept its vitality. My original intention had been to present also the group-theoretical part; however, the appearance of the beautiful book by Huppert and Blackburn has made this less crucial. The classical papers of W. Magnus, H. Zassenhaus and O. Grün, and the later ones of I. N. Sanov, M. Lazard, G. Higman and P. Hall, and M. Hall Jr. presented a clear snapshot of the connection between nilpotent groups and Lie algebras. Recent investigations by other authors have answered more subtle questions about this correspondence. As has been observed already, the present work pursues a more modest aim, namely that of giving a thorough exposition of the proof of the main theorem about sandwiches and its consequences, embracing the Lie algebras that either satisfy an Engel condition or are finite-dimensional. In a recent book, Yu. A. Bakhturin has developed general methods in the theory of varieties of Lie algebras. With that in mind, it must be said that I am dealing here with a situation which is very concrete but nevertheless rich in content.

Work on the book has proceeded with the cheerful accompaniment of the new results of E. I. Zel'manov, and very recent results of E. I. Khukhro, M. R. Vaughan-Lee and G. E. Wall. The author has had no opportunity to understand these last-mentioned developments in detail, and they are just briefly reviewed in Chap. 7.

I would like to draw attention at this point to an important fact: the lemmas, propositions and theorems cited in the main text practically never mention proper names. All detailed references to authorship, circumstances of derivation, limits of applicability and significance of this or that result are relegated to the commentaries that appear at the ends of chapters. The plan of the book will be clear from Chap. 1, and any additional comment would be superfluous. The traditional adage "Feci quod potui, faciant meliora potentes" does not free me from acknowledging responsibility for possible inaccuracies.

The circle of people who have expressed interest in the appearance of this book, and consequently unreservedly deserve my sincere thanks, has turned out to be much wider than might have been expected. For instance, V. A. Ufnarovskii, who had never previously been occupied with problems of Burnside type, has read through the first four chapters with great care and clear understanding, and has made a number of pertinent comments.

The skilled and conscientious work of the editor, Yu. A. Bakhturin, calls for especial appreciation; he has made improvements to the text, and eliminated practically all inaccuracies pointed out by various readers. I express my thanks to all of these, including the referees.

I note here with pleasure the invariably benign influence of I. R. Shafarevich on my scientific outlook, and thank him for his mathematical friendship. Unfortunately, this influence does not extend to style nor to manner of exposition, not even in an opus like this, despite the fact that he, my Teacher, was at its very source.

A. I. Kostrikin

Contents

Chapter 1

Introduction

We present below, as far as possible in a form not too over-burdened with technical details, approaches to the two central themes of the book, with motivations, and also the main results in the order that they are proved in subsequent chapters. Definitions and auxiliary results are given which the reader is advised to become familiar with before embarking on a systematic study of the remainder of the text. In other words, our immediate aim is to create some perspective.

§ 1. Historical Survey

It was in 1902 that W. Burnside[1] [36] posed the problem of the local finiteness of groups in which all elements have finite orders. This problem went on to acquire the status of *the Burnside problem* on periodic groups. The negative answer was obtained as late as 1964 by E. S. Golod [54, 55], and was based on a universal construction due to E. S. Golod and I. R. Shafarevich. Later, S. V. Aleshin [14], R. I. Grigorchuck [58] and V. I. Sushchanskij [248] produced a whole sequence of negative examples. Grigorchuk's construction is one of the simplest, and is very easy to control. Its later development (see [59]) led, in particular, to groups of intermediate growth, that is, growth that is not polynomial and not exponential, in both the periodic and torsion-free cases (thus solving Milnor's problem).

Burnside himself turned particular attention in [36] to the problem of the local finiteness of groups with the identity $x^n = 1$. The group

$$B(d, n) = \mathscr{F} / \mathscr{F}^n, \qquad d > 1 ,$$

obtained by factoring the free group $\mathscr{F} = \mathscr{F}(d)$ on d generators by the normal subgroup $\mathscr{F}^n$ generated by the n-th powers of all elements of $\mathscr{F}$ is nowadays called the *free Burnside group of* exponent (or period) n. Its finiteness had already been established at various times for $n = 2$ (the trivial case), $n = 3$ (Burnside), $n = 4$ (Burnside for $d = 2$; I. N. Sanov [227] for arbitrary d), $n = 6$ (M. Hall Jr. [85]), but was put in doubt for $n \geqslant 72$ by P. S. Novikov in his article [206], which appeared five years before the above-mentioned papers by Golod.

A proof that $B(d, n)$, $d \geqslant 2$, is infinite for odd exponent $n \geqslant 4381$ was given by P. S. Novikov and S. I. Adian in [207], and a proof for odd $n \geqslant 665$ in Adian's

[1] A sketch biography of Burnside is included: see p. 197.

book [2]. A considerably more accessible, geometrically visual version of the proof for odd $n > 10^{10}$ was proposed by A. Yu. Ol'shanskij in [210] (see also [212]). A little later [211] he constructed, for every sufficiently large prime p, an infinite p-group in which every proper subgroup is of order p, using the geometrical method that he had by then perfected. This is the strongest form of the negative answer to Burnside's question, in that it implies the existence of a boundless archipelago of finitely generated periodic groups satisfying the identity $x^p = 1$, as far removed in their properties from finite groups as it is possible to be. The finite groups that interested Burnside in the first place (and which interest us equally) occupy a very modest position in the variety defined by the identity $x^p = 1$, populated as it is by infinite monsters of various types.

It cannot be said that such a state of affairs was wholly unexpected by specialists. It is entirely conceivable that $B(d, n)$ could turn out to be infinite but to have only finitely many non-isomorphic finite factor-groups. By an elementary theorem of Poincaré, this would imply the existence of a largest finite group $B_0(d, n)$ having all finite groups on d generators and satisfying the identity $x^n = 1$ as homomorphic images. A naïve, but more graphic formulation: there is just one cyclic group of order n; how many finite 2-generator groups of exponent n is it possible to devise?

The problem of the existence of the universal finite group $B_0(d, n)$, which was called, quite logically, *the restricted Burnside problem* (RBP) by W. Magnus [175] in 1950, gained recognition in the early forties after the appearance of papers of Magnus [171–174], O. Grün [64], H. Zassenhaus [278] and R. Baer [24]. Indeed, it was then that the group-theoretical problem of the existence of $B_0(d, p)$ for prime exponent $n = p$ was reduced to that of the local nilpotency of a Lie algebra L over $\mathbb{Z}_p$ satisfying the identity

$$[uv^{p-1}] = [\ldots[[u, v], v], \ldots, v] = 0$$

or, as we prefer to say, the *Engel condition* E_{p-1}. Many paths have led to the deduction of this relation; for example, the Campbell–Hausdorff formula was used by Sanov [232], and the associated Lie algebra $L(B(d, p))$ by G. Higman [108], *etc* (see also Chap. 7). As became clear later (G. E. Wall [268–270], E. I. Khukhro [127–129]), the linear problem about Lie algebras is coarser. However, the idealized expression of the group operations of multiplication and commutation in the language of Lie algebras has made it possible to separate the hypothetical group $B_0(d, p)$ completely from its inconvenient partner $B(d, p)$, and to carry out an independent investigation. Incidentally, the long delay caused by this detachment from the more obviously attractive apparatus of Lie algebras occurred partly because it was only in the thirties that abstract Lie algebras themselves were extracted from the infinitesimal bosom of the theory of Lie groups as an object of independent study. Results not so delayed are: the existence of $B_0(2, 5)$ was established in 1955 [131]; of $B_0(d, 5)$ in 1956 ([132], [108]), and of $B_0(d, p)$ in 1958 [136].

Similar attempts to express the existence of the universal finite group $B_0(d, p^\alpha)$ in the language of ring theory when $\alpha > 1$ have not yet met with success even in the cases $p = 2$, $\alpha = 3$ and $p = 3$, $\alpha = 2$. There is a fairly extensive literature in this direction, and it is constantly being enlarged, but all the really important discoveries are still in the future (see §4 of Chap. 7). Meanwhile the Hall-Higman theorem [91] and the classification of finite simple groups together show that the problem of the existence of $B_0(d, n)$ can be reduced to the prime-power exponent case: $n = p^\alpha$. The greatest success in this context is the solution of RBP for exponents of the form $n = 2^\alpha p_1 p_2 \ldots p_m$, where $\alpha = 0, 1$ or 2, m is any natural number and the p_i $1 \leqslant i \leqslant m$, are pairwise different odd primes (see Chap. 8 of [116], for example).

To summarise what has been touched on so far:

i) the Burnside problem for periodic groups (with elements of unbounded orders);

ii) the Burnside problem for periodic groups satisfying the identity $x^n = 1$;

iii) the restricted Burnside problem for finite groups satisfying the identity $x^n = 1$.

Each of the three variants is of independent interest, and usually comes preheated with additional statements of adjacent group-theoretical problems. As the authors of [176] have remarked, a detailed discussion, in precise technical language, of the abutting variants ii) and iii) was carried out by Grün in his 1940 text [64]. Mentioned consecutively there were: a) the problem of the local nilpotency of a Lie ring satisfying the Engel condition E_{p-1} (in other terminology, and with reference to earlier papers by Magnus and Zassenhaus); b) possible corollaries of the positive and negative solutions of the Burnside problem for groups with the identity $x^p = 1$; c) a statement of RBP (without introducing the term) for prime exponent. Finally, there is the following statement: "This fact, whose truth needs to be established, and the corresponding proof for prime-powers p^i, is the most important problem in the theory of finite p-groups. We are still very far from a solution of this problem". It is interesting that Grün, when discussing the natural problems floating around in the group-theoretical world, should calmly envisage the possibility that there exists a finitely generated infinite group with the identity $x^p = 1$ and having no normal subgroups of finite index. This same point of view is reflected in [232].

Sanov and Magnus had remarked in the fifties on the absence of effective approaches to solving the Burnside problem, and they suggested that attention should be focussed on the much narrower (but still interesting) class of objects, namely the *finite* groups satisfying the identity $x^n = 1$. They were fired with the wish to make a little progress, the wish being based on the emergence of the new linear methods in group theory. Magnus [175] proposed the term "restricted Burnside problem" (RBP), which has become accepted and is sometimes expressed by the more roundabout form of words "What has come to be known as the restricted Burnside problem". M. R. Vaughan–Lee even chooses the symbol $R(d, n)$ to denote the hypothetical universal finite group ($B_0(d, n)$ in [176] and in this book), which is sometimes denoted by $\bar{B}(d, n)$ in other papers. The literal translation of the English term RBP into Russian as "ogranichennaya problema Bernsaida" introduces an ambiguity, evoking an association with variant ii), that is, with the Burnside problem as it relates to *all* groups of some finite exponent, not just the finite ones. As we have mentioned already, this problem has been solved in the negative in general ([207, 210]), so that the usage by Sanov [232], after the appearance of Magnus's paper [175], of the term ОПБ must be regarded as very successful. On the other hand, the term "weakened Burnside problem" occasionally suggested by some Russian authors for ОПБ(RBP) has not been accepted into the literature in English. Taking the conservative point of view in relation to established terms that cause no misunderstanding, we regard RBP and ОПБ as mathematical (though not linguistic) synonyms, this being the best way of reflecting the essence of the matter. However, there is no real depth in this terminological discussion, since it will always be clear to experts what is being discussed.

Finally, we must recall, in the nature of things, and also because of the light hand of A. G. Kurosh [152] and other algebraists, many statements of problems *of Burnside type* have arisen. A desire to cover them all would take us off into too-distant regions. Thus we have deliberately chosen a fairly circumscribed neighbourhood, and our excursions "around Burnside" are strictly limited by the scope of the theories of finite groups and Lie algebras. But even within these limits, preference has naturally been given to the author's own investigations on RBP for prime exponent (the first four chapters), and thereafter to the complementary results of Yu. P. Razmyslov (see § 3 of Chap. 6) and E. I. Zel'manov (the major part of Chap. 5, and § 4 of Chap. 6).

The reader will find a fairly thorough exposition of Lie ring methods in the theory of nilpotent groups in the book [116] of B. Huppert and N. Blackburn already mentioned. It is accompanied by the solution of some concrete problems. All facts that have a direct relation to RBP for prime exponent are contained in our Chap. 7, together with proofs. With regret, we have omitted the proof of the reduction theorem of Hall and Higman, since it rests heavily on a development of the theory of p-soluble groups. Results on the Burnside problem, and many remarks of a historical nature, are to be found in the book [176] by Magnus and Chandler, in various books on combinatorial group theory and in the short survey [201] by M. F. Newman. The last-mentioned has an extensive bibliography (see [202]). Supplemented by articles of recent years, it has become our literature list. We have not attempted to group articles according to any scheme. But when you acquaint yourself with the literature – however sketchily – you discover with surprise just how thorny the path to a solution of even comparatively special problems really was. For example, the beautiful result of Razmyslov [221, 222] established the end of a whole scientific thrust, represented by some tens of articles and called the problem of Hall and Higman on groups of exponent 4.

§ 2. Engel Lie Algebras: Definitions and Examples

2.1. Let L be any Lie algebra over a field F, with multiplication $(x, y) \mapsto [x, y]$, which is also written in the form

$$[x, y] = x(\operatorname{ad} y) .$$

Clearly,

$$\mathrm{ad}(\alpha y + \alpha' y') = \alpha\,\mathrm{ad}\,y + \alpha'\,\mathrm{ad}\,y'; \qquad \alpha, \alpha' \in F\,.$$

By the Jacobi identity

$$[[x, y], z] + [[y, z], x] + [[z, x], y] = 0\,,$$

the map[1] $\mathrm{ad}: y \to \mathrm{ad}\,y$, associating with each element $y \in L$ its *adjoint endomorphism* $\mathrm{ad}\,y$, gives the so-called *adjoint representation* of L:

$$\mathrm{ad}[x, y] = [\mathrm{ad}\,x, \mathrm{ad}\,y]\,.$$

Here $[x, y]$ is the abstract Lie bracket, and at the same time

$$[\mathrm{ad}\,x, \mathrm{ad}\,y] = \mathrm{ad}\,x \cdot \mathrm{ad}\,y - \mathrm{ad}\,y \cdot \mathrm{ad}\,x$$

is the commutator of linear operators. By definition, $\mathrm{Ker\,ad} = Z(L)$ is the centre of the Lie algebra L. It also follows from the Jacobi identity that

$$[x, y]\,\mathrm{ad}\,z = [x, y\,\mathrm{ad}\,z] + [x\,\mathrm{ad}\,z, y]\,,$$

that is, $\mathrm{ad}\,z$ is a derivation. It is called an *inner derivation* of L. Thus, when $Z(L) = 0$, the epimorphism $\mathrm{ad}: L \to \mathrm{Im\,ad}$ establishes an isomorphism between L and its algebra of inner derivations. We shall make wide use of this fact.

2.2. The algebra $\mathrm{ad}\,L$ generates an associative algebra $A(L)$ over F whose elements are F-linear combinations of all products

$$\mathrm{ad}\,x_1 \cdot \mathrm{ad}\,x_2 \ldots \mathrm{ad}\,x_m\,,$$

where the dot means the usual composition of endomorphisms. We note that if $w(\mathrm{ad}\,x_1, \ldots, \mathrm{ad}\,x_m)$ is an element of $A(L)$, then $zw(\mathrm{ad}\,x_1, \ldots, \mathrm{ad}\,x_m) = 0$ for every $z \in L$ if and only if $w(\mathrm{ad}\,x_1, \ldots, \mathrm{ad}\,x_m) = 0$.

If U and V are F-subspaces of L, then as usual the subspace spanned by all products $[u, v]$ with $u \in U$, $v \in V$, is denoted by the symbol $[U, V]$. The subspace U is said to be an *ideal* of L whenever $[U, L] \subseteq U$. The *lower central series* of L is the chain of ideals

$$L = L^1 \supseteq L^2 \supseteq \cdots \supseteq L^m \supseteq \cdots,$$

where $L^m = [L^{m-1}, L]$ for $m \geqslant 2$. It follows from the Jacobi identity that $[L^{m-s}, L^s] \subseteq L^m$, and also that a product of arbitrary elements

$$w_1 \in L^{k_1}, \ldots, w_s \in L^{k_s}\,, \qquad k_1 + \cdots + k_s = m + 1\,,$$

with any arrangement of the brackets [,], can be expressed as a linear combination of *left-normed* products

$$\begin{aligned}[x_0 x_1 x_2 \ldots x_m] &= [\ldots[[x_0, x_1], x_2], \ldots, x_m] \\ &= x_0\,\mathrm{ad}\,x_1\,\mathrm{ad}\,x_2 \ldots \mathrm{ad}\,x_m\,.\end{aligned}$$

---

[1] Our notation differs in sign from the conventional $\mathrm{ad}\,y: x \to [y, x]$.

A Lie algebra L is *abelian* if $L^2 = 0$, that is, $[x, y] = 0$ for all $x, y \in L$. The following definition introduces a more general concept, one that is important for our purposes:

2.3. Definition. We agree to say that a Lie algebra L is *nilpotent of class m* if $L^m \neq 0$ but $L^{m+1} = 0$; in other words, if $\operatorname{ad} x_1 \cdot \operatorname{ad} x_2 \ldots \cdot \operatorname{ad} x_m = 0$ for all $x_i \in L$, $i = 1, 2, \ldots, m$. An algebra L is said to be *locally nilpotent* if every finitely generated subalgebra is nilpotent.

In what follows, the notation $[x_0 x_1 x_2 \ldots x_m] = [x_0, x_1, x_2, \ldots, x_m]$ is used without further stipulation. In the case where all x_i for $i > 0$ are the same, we write

$$[yx^n] = y(\operatorname{ad} x)^n .$$

2.4. Definition. A non-zero element $x \in L$ is said to be an *Engel element* (or a *nil-element*) if there exists a smallest natural number $n(x)$, is called the *index* of x, such that $(\operatorname{ad} x)^{n(x)} = 0$. When $n(x) = 1$, we arrive at the concept of a *central* element.

A Lie algebra L in which every element is Engel is said to be an *Engel* Lie algebra (or a *nil-algebra*). If $n(x) \leqslant n$ for all $x \in L$, we speak of an n-Engel Lie algebra, or *an algebra with the condition* E_n. It is normally assumed that $n = n(a)$ for some $a \in L$.

Thus, by definition, a Lie algebra L with E_n satisfies the identity $[yx^n] = 0$, also written in the form $(\operatorname{ad} x)^n = 0$. It should be noted that this relation cannot be considered as an identity for $A(L)$, in the accepted sense of the term: $[\operatorname{ad} x, \operatorname{ad} y]^n = 0$, but this does not mean that $(\operatorname{ad} x \cdot \operatorname{ad} y)^n = 0$! This is the source of the difficulty in studying Lie algebras with additional identities (see [26]), in particular, Lie algebras with E_n. The following question is fundamental in connection with Engel algebras:

2.5. The Local Nilpotency Problem: *is every Lie algebra with* E_n *locally nilpotent*?

Golod [54, 55] has given the negative answer for the analogous problem for nil-algebras with unbounded indices $n(x)$: a non-nilpotent nil-algebra on $d \geqslant 3$ generators is constructed there. No other constructions of this type have been suggested as yet.

It would appear that the local nilpotency problem for Lie algebras with E_n ought to have a positive solution, as in the finite-dimensional case, where Engel's classical theorem gives the answer. Our aim is to confirm this belief by means of a theorem valid for all n not exceeding the characteristic p of the ground field F. However, when the index n is significantly bigger than p ($n \gg p$), the problem remains open. Besides, at the moment it is not clear what group-theoretical corollaries would follow from a complete solution to Problem 2.5, in addition to RBP, that is. This fact holds back the investigation of the general case to some extent (see, however, § 4 of Chap. 7). As a rule, we shall assume that $n \leqslant p$ in what follows. It is always implicit that n is arbitrary if the characteristic of F is zero.

2.6. What approaches to the solution of Problem 2.5 can be suggested? Using purely combinatorial tools within the free Lie algebra $\mathscr{L}$ on d generators

$x_1, \ldots, x_d$, we would have to find expressions of the form

$$[x_{i_0} x_{i_1} \ldots x_{i_m}] = \sum_k [u_k v_k^n]$$

for all left-normed commutators $[x_{i_0} x_{i_1} \ldots x_{i_m}]$ of length $m+1$ greater than some fixed natural number m_0. The elements u_k, v_k of L, which depend in a somewhat complicated and not fully determinate fashion on the given sequence of indices $i_0, \ldots, i_m$, must, nevertheless, carry auxiliary information about the mysterious number m_0; and the existence of m_0 itself has to be established. Up to now, this has been achieved explicitly only for $n \leqslant 4$. Even for $d = 2$, $x_1 = x$, $x_2 = y$, there are no heuristic considerations of any sort that make it possible to obtain, for example, a formula of the following special type:

$$[xy](\operatorname{ad} x \cdot \operatorname{ad} y)^s = [xyxy \ldots xy] = \sum_k [u_k v_k^n] .$$

Rather, we should expect the existence of a universal formula

$$[x_0 x_1 x_2 \ldots x_m] = \sum [u_k v_k^n]$$

in the free Lie algebra $\mathscr{L}$ with countably many independent generators $x_0, x_1, \ldots$. For small n, this is indeed the case.

2.7. Example. Let L be a Lie algebra with E_2. To economise on space, we shall rid ourselves temporarily of the prefix ad, and use capital letters X, Y, sometimes with indices, instead of $\operatorname{ad} x$, $\operatorname{ad} y, \ldots$. Either of the two fairly cumbersome formulae

$$\begin{aligned} 4XYZ = {} & 3((X + [YZ])^2 - X^2 - [YZ]^2) + ((Y + [XZ])^2 \\ & - Y^2 - [XZ]^2) - ((Z + [XY])^2 - Z^2 - [XY]^2) \\ & + 2X((Y+Z)^2 - Y^2 - Z^2) - 2Y((X+Z)^2 - X^2 - Z^2) \\ & + 2Z((X+Y)^2 - X^2 - Y^2) ; \\ 4XYZ = {} & -((X + [YZ])^2 - X^2 - [YZ]^2) + ((Y + [XZ])^2 - Y^2 \\ & - [XZ]^2) + 3((Z + [XY])^2 - Z^2 - [XY]^2) \\ & + 2((Y+Z)^2 - Y^2 - Z^2)X - 2((X+Z)^2 - X^2 - Z^2)Y \\ & + 2((X+Y)^2 - X^2 - Y^2)Z , \end{aligned}$$

valid in the free Lie algebra $\mathscr{L}$ (more exactly, for $x, y, z \in \mathscr{L}$) shows that our algebra L is nilpotent for $p > 2$: $L^4 = 0$. We shall convince ourselves next that it is far more practical to operate within the Engel algebra L itself.

Since $X^2 = 0$, $Y^2 = 0$, and $(X+Y)^2 = 0$, we have $XY + YX = 0$. Using this relation, with Y replaced by Z, we get

$$\begin{aligned} z(XY - 2YX) &= [zxy] - 2[zyx] = -[z[yx]] - [zyx] \\ &= [yxz] + [yzx] = y(XZ + ZX) = 0 \end{aligned}$$

for arbitrary $z \in L$, and this is equivalent to the relation $XY - 2YX = 0$, as was remarked in 2.2. For $p = 2$ and $p > 3$, we get at once from the two relations

$$XY + YX = 0, \qquad XY - 2YX = 0$$

that $XY = 0$, that is, $L^3 = 0$. It should be noted that we have used here only an effect connected with the Jacobi identity, and that XY has not been written as a sum of squares of elements of ad $\mathscr{L}$!

For $p = 3$ we have achieved nothing as yet, but if the relation

$$X_1[X_2X_3] + [X_2X_3]X_1 = 0$$

is rewritten in expanded form

$$X_1X_2X_3 - X_1X_3X_2 + X_2X_3X_1 - X_3X_2X_1 = 0$$

and we note that $X_{\pi 1}X_{\pi 2}X_{\pi 3} = (\operatorname{sgn}\pi)X_1X_2X_3$ for every permutation π in the symmetric group S_3, we will get $4X_1X_2X_3 = 0$. Thus, if L satisfies E_2, then

$$L^m = 0, \qquad \text{where } m = \begin{cases} 3 & \text{if } p \neq 3\,, \\ 4 & \text{if } p = 3\,. \end{cases}$$

The small nuance (characteristic $p = 2$ has turned out to be "better' than $p = 3$) in this case is not reflected in the qualitative picture: L turns out to be not only locally nilpotent, but even nilpotent in the sense of Definition 2.3. The suspicion arises that the following will perhaps have a positive solution:

2.8. Nilpotency Problem: *is every Lie algebra with* E_n *nilpotent*?

Alas, the answer to this question is in the negative [21, 220]. The rather delicate construction of examples of non-nilpotent Lie algebras with E_n, $n = p - 2$, and thus for the case $n = p - 1$, which is the most important from the group-theoretical point of view, will be expounded in Chap. 6. For $n > p$, there is a much simpler example due to P. M. Cohn (see 6.2.1). One can get some feel for how powerful the influence of the characteristic of the field is on the nature of the answer by considering the complexity of the following example:

2.9. Example. Let L be a Lie algebra with E_3, $p \geqslant 3$. As in the case $n = 2$, we see that the following relations are consequences of the fact that $X^3 = 0$:

$$YX^2 + XYX + X^2Y = 0\,, \qquad (\alpha_1)$$

$$XY^2 + YXY + Y^2X = 0\,, \qquad (\alpha_2)$$

Replacing Y by Z in (α_1), we get

$$\begin{aligned} 0 &= y(X^2Z + XZX + ZX^2) = [yx^2z] + [yxzx] + [yzx^2] \\ &= -z([YX^2] + [YX]X + YX^2)\,, \end{aligned}$$

whence, since z is arbitrary,

$$[YX^2] + [YX]X + YX^2 = 0\,,$$

or, as is the same thing,

$$3YX^2 - 3XYX + X^2Y = 0\,. \qquad (\beta_1)$$

Interchanging X and Y here, we get

$$3XY^2 - 3YXY + Y^2X = 0\,. \qquad (\beta_2)$$

Multiplying relations (α_i) and (β_i), $i = 1, 2$, on the left and right by X and Y, we arrive at the following system of equations (in which some summands have been shifted around):

$$(\alpha_1)\,Y\colon\ YX^2Y + XYXY + X^2Y^2 = 0\,,$$

$$(\alpha_2)X\colon XY^2X + YXYX + Y^2X^2 = 0\,.$$

$$Y(\alpha_1)\colon\ YX^2Y + YXYX + Y^2X^2 = 0\,,$$

$$Y(\beta_1)\colon\ YX^2Y - 3YXYX + 3Y^2X^2 = 0\,,$$

$$X(\beta_2)\colon XY^2X - 3XYXY + 3X^2Y^2 = 0\,,$$

$$(\beta_1)\,Y\colon 3YX^2Y - 3XYXY + X^2Y^2 = 0\,.$$

It follows from this system that

$$XY^2X = X^2Y^2,\quad YX^2Y = X^2Y^2,\quad 3Y^2X^2 = -2X^2Y^2\,,$$

$$3XYXY = 4X^2Y^2,\quad 3YXYX = -X^2Y^2,\quad 5X^2Y^2 = 0\,.$$

For $p > 5$, we find that

$$X^2Y^2 = 0\,.$$

For $p = 3$ it is clear that the stronger relation $X^2Y = 0$ (see (β_1)) holds. But we know from the formulae at the beginning of 2.7 that, for $p > 2$,

$$X_1X_2X_3 = \sum_i A_iU_i^2, \qquad X_4X_5X_6 = \sum_i V_j^2B_j\,,$$

where A_i and B_j are either elements of F or linear forms in $X_1, \ldots, X_6$. Thus

$$X_1X_2X_3X_4X_5X_6 = \sum_{i,j} A_iU_i^2V_j^2B_j\,,$$

and since $U_i^2V_j^2 = 0$ by what has been proved already, we can deduce the nilpotency of L for $p \geqslant 3$, $p \neq 5$: $L^7 = 0$. It is known that the nilpotency class is actually less (see 8.1); but far more fundamental is the question of the nilpotency of L when $p = 5$. In this case the above system of six relations have the equation $Y^2X^2 = X^2Y^2$ as a consequence (in fact $YZ^2 = X^2Y$). Thus

$$X_1^2X_2^2 \ldots X_m^2 = X_{\pi 1}^2X_{\pi 2}^2 \ldots X_{\pi m}^2\,, \qquad \pi \in S_m\,,$$

for arbitrary m: however, we do not get an identity of the form $X_1^2X_2^2 \ldots X_m^2 = 0$ (from which nilpotency would follow), even if we are more inventive in the treatment of identities. This will be proved in Chap. 6. Even more so, an algebra with E_4 need not be nilpotent if $p = 5$, a fact that has a direct relevance to RBP for exponent 5.

Meanwhile, if our algebra L with E_3 is generated by d elements $x_1, \ldots, x_d$, say, then every commutator $[x_{i_1} x_{i_2} \ldots x_{i_m}]$ of length $m \geqslant 2d + 1$ will contain at least three occurrences of one and the same generator. But, by using the identity $YX^2 = X^2Y$ and the formulae introduced at the very beginning of 2.7, every product of the form $X_1 X_i \ldots X_j X_1 X_k \ldots X_l X_1$ can be transformed without difficulty into a sum of products with the elements X_1 collected together, that is, products that contain X_1^3, and are thus zero. Therefore, $L^{2d+1} = 0$.

It can be readily imagined that this sort of complexity in combinatorial manipulations increases swiftly with increasing n. Although the answer to question 2.8 is negative, as has been remarked already, there is a general theorem due to Zel'manov [284] stating that Lie algebras with E_n over a field of characteristic zero (and also characteristic $p \gg n$) are globally nilpotent (that is, nilpotent in the sense of Definition 2.3). The proof of this theorem is set forth in Chap. 6.

§ 3. The Locally Nilpotent Radical

The direct approach to the solution of the Problem 2.5 on the local nilpotency of Lie algebras with E_n, including those with $n < p$, is beset by extremely cumbersome calculations even in the very simplest cases (see Examples 2.7 and 2.9). The free two-generator Lie algebra with E_{p-1} over $\mathbb{Z}_p$ is certainly inaccessible to hand calculation (for instance, the class is about 30 when $p = 7$), and, as far as we know, nobody has yet used a computer to achieve this aim. Thus, the approach outlined below based on a consideration of the locally nilpotent radical is still the only one available. It is absolutely ineffective for determining upper bounds on nilpotency classes[1], but it is nevertheless totally natural to use it.

3.1. Lemma. *Let L be a Lie algebra over a field F, generated by a locally nilpotent ideal M and a nil-element x. Then L is locally nilpotent.*

Proof. By assumption, $(\operatorname{ad} x)^n = 0$ for some $n > 0$. Since M is an ideal of L, and L is generated by $\{M, x\}$, we have

$$L = \langle x \rangle + \sum_{j=0}^{n-1} M (\operatorname{ad} x)^j .$$

To check that L is locally nilpotent, we shall show that every finite subset of L generates a nilpotent subalgebra (see Definition 2.3). Every element a of L is of the form $a = \lambda x + \sum_{i,j} x_i (\operatorname{ad} x)^j$, with $x_i \in M$. Therefore, it is enough to consider the subalgebra T of L generated by elements $x_1, \ldots, x_r$ of M and x. Such a T contains the subalgebra N generated by the elements $x_i (\operatorname{ad} x)^j$, $1 \leqslant i \leqslant r$, $0 \leqslant j \leqslant n - 1$.

[1] Unfortunately, the bound recently obtained in terms of primitive recursive functions (see Appendix I) is completely unrealistic.

Since they all lie in the locally nilpotent ideal M, N is nilpotent and, clearly, $N \operatorname{ad} x \subset N$, so that T is generated by $\{N, x\}$.

As is well known [31] and follows easily from the definitions, T is nilpotent if and only if it has a finite central series

$$T = T_1 \supset T_2 \supset \cdots \supset T_m \supset T_{m+1} = 0$$

of ideals T_i; this means that $[T, T_i] \subseteq T_{i+1}$; $i = 1, \ldots, m$.

We consider the series in T with general term $N^k(\operatorname{ad} x)^j + N^{k+1}$:

$$T \supseteq N \supseteq N \operatorname{ad} x + N^2 \supseteq N(\operatorname{ad} x)^2 + N^2 \supseteq \cdots \supseteq N(\operatorname{ad} x)^{n-1} + N^2$$
$$\supseteq N^2 \supseteq N^2 \operatorname{ad} x + N^3 \supseteq \cdots$$

Clearly, the length of this series is not more than $ns + 1$, where s is the nilpotency class of N; the fact that it is a central series follows from the inclusion

$$[N^k(\operatorname{ad} x)^j + N^{k+1}, T] \subseteq N^k(\operatorname{ad} x)^{j+1} + N^{k+1},$$

which needs to be checked for all cases $1 \leqslant k \leqslant s, 0 \leqslant j \leqslant n-1$. Since N is an ideal of T, each of its powers N^{j+1} is an ideal of T (easy induction on j), and thus it is enough to show that

$$[N^k(\operatorname{ad} x)^j, T] \subseteq N^k(\operatorname{ad} x)^{j+1} + N^{k+1}.$$

But every element of T is of the form $a + \lambda x$, $a \in N$, $\lambda \in F$, so that the verification can be done in less than no time. If $u \in N^k$, then

$$[u(\operatorname{ad} x)^j, a + \lambda x] = \lambda u(\operatorname{ad} x)^{j+1} + [u(\operatorname{ad} x)^j, a],$$
$$\lambda u(\operatorname{ad} x)^{j+1} \in N^k(\operatorname{ad} x)^{j+1}, \qquad u(\operatorname{ad} x)^j \in N^k, \qquad [u(\operatorname{ad} x)^j, a] \in N^{k+1}.$$

This proves that T is nilpotent, and thus that L is locally nilpotent. □

3.2. Proposition. *Every Engel Lie algebra L has a unique maximal locally nilpotent ideal $R = R(L)$; the factor-algebra L/R has no non-zero locally nilpotent ideals, that is, $R(L/R) = 0$.*

Proof. Arguing by contradiction, we assume that N/R is a locally nilpotent ideal of L/R, where R is some fixed maximal locally nilpotent ideal of L; the existence is guaranteed by Zorn's lemma. Consider a finitely generated subalgebra M of N. The factor-algebra $(M + R)/R \cong M/M \cap R$ is nilpotent and finitely generated, so that there is a series

$$M \cap R = M_0 \subset M_1 \subset \cdots \subset M_t = M$$

of ideals of M such that each term M_i is generated by the preceding term M_{i-1} and a single element a_i (one of the generators of M). By assumption, $(\operatorname{ad} a_i)^{n(a_i)} = 0$, $1 \leqslant i \leqslant t$. Since $M \cap R$ is locally nilpotent, t applications of Lemma 3.1 show that M is locally nilpotent, and therefore that it is even nilpotent, being finitely generated. This all means that N is a locally nilpotent ideal. However, N contains R and R is maximal, and we arrive at the conclusion that $N = R$.

3.3. We shall not need to do it, but it can be proved that every Lie algebra L (not just the Engel Lie algebras, as in Proposition 3.2) has a locally nilpotent radical $R(L)$, namely the sum of its locally nilpotent ideals. However, the more important radical property $R(L/R(L)) = 0$ is not always satisfied, as is shown by the canonical example of a two-dimensional non-abelian Lie algebra. The Engel condition in Proposition 3.2 is therefore essential.

§ 4. Basic Conventions. Elementary Combinatorics

4.1. The fact that every Lie algebra L satisfying E_n is its own locally nilpotent radical $R(L)$ would imply the positive solution to Problem 2.5. Assuming that $L \neq R(L)$ and going over to $\bar{L} = L/R(L)$, we obviously get a Lie algebra with the same Engel condition E_n which is finitely generated if the original algebra L is. In this situation, Proposition 3.2 yields the important property that $R(\bar{L}) = 0$.

Thus, arguing by contradiction, we can assume from the outset that our Lie algebra with E_n answering Problem 2.5 negatively has no non-zero locally nilpotent ideal. In particular, $Z(L) = 0$.

4.2. Since it has zero centre, L can be identified with its algebra $\mathrm{ad}\,L$ of inner derivations (Remark 2.1). Having made this identification, we can suppose that L is contained as a set in the associative algebra $A(L)$, and that the Lie multiplication can be written as $[x, y] = xy - yx$. This fact leads to significant economies, and makes the identities intuitively easier to comprehend. After the discussion in Examples 2.7 and 2.9, such a step seems particularly natural and efficient.

However, we must observe great caution in the interpretation of symbols, especially in regard to the agreement made at the end of 2.3 to omit brackets and commas in left-normed commutators. As has been said already, an equation $u_1 u_2 \dots u_{m-1} u_m = 0$ in $A(L)$, with $u_i \in L$, is equivalent to the identity $[xu_1 u_2 \dots u_{m-1} u_m] = 0$ in $x \in L$. But $[xu_1 u_2 \dots u_{m-1} u_m] = [xu_1 u_2 \dots u_{m-1}]u_m - u_m[xu_1 u_2 \dots u_{m-1}]$. Continuing this expansion of the commutator $[xu_1 u_2 \dots u_m]$ as a linear combination of associative products, we come to a relation of the form

$$\sum (-1)^k u_{i_1} \dots u_{i_k} x u_{j_1} \dots u_{j_{m-k}} = 0 ,$$

which, generally speaking, is not obtained from the original equality $u_1 u_2 \dots u_{m-1} u_m = 0$ by formally multiplying on the left and right by elements of $A(L)$. We shall make constant use of this *principle of double interpretation of relations in L and A(L)*.

Another remark on this plane. Assume that the equation $u_1 u_2 \dots u_s y v_1 \dots v_t = 0$, with u_i, y, $v_j \in L$, is an identity in y. Then we have the identity $[xu_1 \dots u_s y v_1 \dots v_t] = 0$ in x and y in L. However, since $[xu_1 \dots u_s y v_1 \dots v_t] = -[y[xu_1 \dots u_s]v_1 \dots v_t]$, we get the relation $[xu_1 \dots u_s]v_1 \dots v_t = 0$ in

$A(L)$, which is an identity in x. For $s > 0$, this gives something additional to what was used in 2.7 and 2.9. Besides, the effect is achieved on replacing x by a sum of Lie products depending on the u_i and v_j. This is more easily perceived in each concrete situation.

4.3. A few words now on notation and terminology. As a rule, elements of the field F are denoted by Greek letters, elements of L by lower-case Latin letters, and elements of $A(L)$ by upper-case Latin letters and polynomials in them. Practically all the discussions are appropriate for rings with the obvious restrictions on their additive groups; however, the extra generality that would be achieved is negligible (at least where applications to groups of prime exponent are concerned), and caution about scalar multipliers produces additional trouble.

We shall not use any special terminology or symbolism, with the exception of that already mentioned. Lemmas, propositions and theorems are referred to outside the chapters where they appear by a full triple $\alpha.\beta.\gamma$, which points to Chapter α, item $\beta.\gamma$ within section β. When reference is made within Chapter α, the numeral denoting the assertion is abbreviated to $\beta.\gamma$. Relations and identities are given consecutive numerals $(\nu.\mu)$ inside each chapter ν.

Further, suppose that we have a relation or identity $Aw(x_1, \ldots, x_m)B = 0$ in $A(L)$, obtained from $Aw'(x_1, \ldots, x_m)B = 0$ as the result of the explicit application of the identity with numeral $(\nu \cdot \mu)$ to the segment $w'(x_1, \ldots x_m)$. For ease of visualisation and also to economise on space, this process of transforming words is expressed schematically by a line under $w'(x_1, \ldots, x_m)$, with the numeral $(\nu \cdot \mu)$ written under the line:

$$0 = \underset{(\nu\cdot\mu)}{\underline{Aw'(x_1, \ldots, x_m)B}} = Aw(x_1, \ldots, x_m)B .$$

Sometimes underlining without the numeral $(\nu \cdot \mu)$ of the identity being used is sufficiently informative, that is, when it is perfectly obvious which transformation is being applied to $w'(x_1, \ldots, x_m)$.

We shall not misuse the implication sign $\Rightarrow$ nor the equivalence sign $\Leftrightarrow$, though they do appear in the text. The end of the proof of every assertion is marked by the symbol $\square$.

4.4. We shall need some formal identities, whose origin goes back at least to the article [118]. By definition,

$$(u_1 + u_2 + \cdots + u_s)^r = \sum_{j_1 + \cdots j_s = r} \begin{Bmatrix} u_1 & u_2 & \ldots & u_s \\ j_1 & j_2 & \ldots & j_s \end{Bmatrix},$$

where the symbol enclosed in curly brackets denotes the sum of all associative monomials of degree j_k in u_k, $1 \leqslant k \leqslant s$. For example, when $u_1 = u$, $u_2 = v$, the sum indicated contains a summand of the form

$$\begin{Bmatrix} u & v \\ 1 & r-1 \end{Bmatrix} = \begin{Bmatrix} u_1 & u_2 & u_3 \ldots u_s \\ 1 & r-1 & 0 \ldots 0 \end{Bmatrix} = \sum_{k=0}^{r-1} v^k u v^{r-1-k} .$$

With this notation, we have

$$\begin{Bmatrix} u_1 \dots u_r \\ 1 \dots 1 \end{Bmatrix} = (u_1 + \dots + u_r)^r - \sum_{i=1}^{r} (u_1 + \dots + \hat{u}_i + \dots + u_r)^r$$

$$+ \sum_{1 \leqslant i < j \leqslant r} (\dots + \hat{u}_i + \dots + \hat{u}_j + \dots)^r$$

$$- \dots (-1)^{r-1} \sum_{i=1}^{r} u_i^r \tag{1.1}$$

(as always, the symbol ^ placed over a letter means that it is deleted). In particular, if the set $\{u_1, \dots, u_r\}$ is partitioned into s families of $j_1, j_2, \dots, j_s$ equal letters, it follows from (1.1) that

$$j_1! \dots j_s! \begin{Bmatrix} u_1 \dots u_s \\ j_1 \dots j_s \end{Bmatrix} = \sum_i L_i(u_1, \dots, u_s)^r, \qquad j_1 + \cdots + j_s = r.$$

Here the $L_i(u_1, \dots, u_s)$ are linear forms in $u_1, \dots, u_s$ which can be determined without difficulty from relations (1.1). When $\operatorname{char} F = 0$ or $r < p = \operatorname{char} F$, we deduce that

$$\begin{Bmatrix} u_1 \dots u_s \\ j_1 \dots j_s \end{Bmatrix} = \sum \alpha_i L_i(u_1, \dots, u_s)^r, j_1 + \cdots + j_s = r. \tag{1.2}$$

A simple induction on r and s gives the formulae:

$$\begin{Bmatrix} u & v \\ 1 & r-1 \end{Bmatrix} = \sum_{k=0}^{r-1} \binom{r}{k+1} v^{r-1-k} [uv^k]; \tag{1.3}$$

$$uv^s - v^s u = \sum_{k=1}^{s} \binom{s}{k} v^{s-k} [uv^k]; \tag{1.4}$$

$$[uv^s] = \sum_{k=0}^{s} (-1)^k \binom{s}{k} v^k u v^{s-k}. \tag{1.5}$$

4.5. Suppose now that L is a Lie algebra with E_n, $n < p$, embedded in the associative algebra $A(L)$; that is, $u^n = 0$ and $[vu^n] = 0$ for all $u, v \in L$. By identity (1.1),

$$\sum_{\pi \in S_n} u_{\pi 1} u_{\pi 2} \dots u_{\pi n} = \begin{Bmatrix} u_1 & u_2 \dots u_n \\ 1 & 1 \dots 1 \end{Bmatrix} = 0. \tag{1.6}$$

More generally, in accordance with identity (1.2),

$$\begin{Bmatrix} u_1 \dots u_s \\ j_1 \dots j_s \end{Bmatrix} = 0, \qquad j_1 + \cdots + j_s = n. \tag{1.7}$$

In particular,

$$\sum_{k=0}^{n-1} v^k u v^{n-1-k} = \begin{Bmatrix} u & v \\ 1 & n-1 \end{Bmatrix} = 0. \tag{1.8}$$

At this point we recall the *double interpretation* principle from 4.2. Replacing u by w in (1.8) and using formula (1.5), we have

$$\begin{aligned}
0 &= \sum_{k=0}^{n-1} [uv^k wv^{n-1-k}] = -\sum_{k=0}^{n-1} [w[uv^k]v^{n-1-k}] \\
&= -\sum_{k=0}^{n-1}\sum_{i=0}^{k} (-1)^i \binom{k}{i} [wv^i uv^{n-1-i}] \\
&= \sum_{i=0}^{n-1}\left(\sum_{k=i}^{n-1}\binom{k}{i}\right)(-1)^{i+1}[wv^i uv^{n-1-i}] \\
&= \sum_{i=0}^{n-1}(-1)^{i+1}\binom{n}{i+1}[wv^i uv^{n-1-i}] .
\end{aligned}$$

Since w is arbitrary, we get from this that

$$\sum_{i=0}^{n-1}(-1)^{i+1}\binom{n}{i+1} v^i uv^{n-1-i} = 0 . \tag{1.9}$$

Multiplication of the left-hand side of (1.9) by v yields $[uv^n]$, since $v^n = 0$; and this is obviously zero. Thus, the element of novelty in identity (1.9) is comparatively small, and indeed it disappears when $n = p - 1$ since $(-1)^{i+1}\binom{p-1}{i+1} \equiv 1 \pmod p$ and (1.9) is the same as (1.8). However, for $p \gg n$, (1.9) has a certain effect, and the time will come for us to use it.

Since $\binom{p}{k} \equiv 0 \pmod p$ for $1 \leqslant k \leqslant p-1$, when $r = p$ identity (1.3) becomes

$$[uv^{p-1}] = \begin{Bmatrix} u & v \\ 1 & p-1 \end{Bmatrix}, \tag{1.10}$$

which of course also follows from (1.5) with $s = p - 1$. Finally, we get the identity

$$uv^p - v^p u = [uv^p]$$

from (1.4) or (1.5) with $s = p$; this is normally used to define Lie p-algebras (restricted Lie algebras). Identities (1.6)–(1.9) are said to be obtained as a result of *linearizing* the original identity $u^n = 0$.

4.6. Proposition. *Let L be a Lie algebra with E_n over a field F of characteristic $p > n$ (or characteristic 0). Then the following assertions hold:*

(i) *Every product in the associative algebra $A(L)$ of elements $x_i \in L$ of length $m \geqslant n$ can be written in the form*

$$x_1 x_2 \dots x_{m-1} x_m = \sum_i w_{i,1} w_{i,2} \dots w_{i,n-1}, \qquad w_{i,k} \in L ,$$

where $\sum_{k=1}^{n-1} \deg w_{i,k} = m$ for each i (here $\deg w_{i,k}$ means the degree of $w_{i,k}$ as a multilinear commutator in $x_1, x_2, \dots, x_m$).

(ii) *Every element A of $A(L)$ takes the form*

$$A = \sum \alpha_i a_i^{n-1} + \sum \beta_i b_i^{n-2} + \cdots + \sum \gamma_i c_i^2 + \sum d_i ,$$
$$\alpha_i, \beta_i, \ldots \gamma_i \in F; \quad a_i, b_i, \ldots, c_i, d_i \in L .$$

For $m < p$, the following relation holds in every Lie algebra:

$$x_1 \ldots x_m = \sum \alpha_i a_i^m + \sum \beta_i b_i^{m-1} + \cdots + \sum d_i .$$

Proof. Since $uv = vu + [u, v]$ for all $u, v \in L$,

$$n!\, x_1 x_2 \ldots x_n = \begin{Bmatrix} x_1 & x_2 \ldots x_n \\ 1 & 1 \ldots 1 \end{Bmatrix} + P_n ,$$

where P_n is a linear combination of products

$$u_1 u_2 \ldots u_{n-1} = x_{i_1} \ldots x_{i_{k-1}} [x_{i_k}, x_{i_{k+1}}] \ldots x_{i_n}$$

of length $n - 1$ and total degree n in $x_1, \ldots, x_n$. By assumption $n! \neq 0$ in F, while $\begin{Bmatrix} x_1\, x_2 \ldots x_n \\ 1 \;\; 1 \ldots 1 \end{Bmatrix} = 0$ by (1.6). Thus, $x_1 x_2 \ldots x_n$ has a linear expression in terms of $u_1 u_2 \ldots u_{n-1}$. For the same reason, $u_1 u_2 \ldots u_{n-1} x_{n+1}$ (and therefore also $x_1 x_2 \ldots x_n x_{n+1}$) has a linear expression in terms of elements of the form $v_1 v_2 \ldots v_{n-1}$, where $v_s = x_i$, $[x_i, x_j]$ or $[x_i, x_j, x_k]$, and $\sum_s \deg v_s = n + 1$. Continuing in this way, we arrive at an expression like that mentioned in (i).

Assertion (ii) is a direct consequence of (i) and the formula

$$r!\, w_1 w_2 \ldots w_r = \begin{Bmatrix} w_1 & w_2 \ldots w_r \\ 1 & 1 \ldots 1 \end{Bmatrix} + P_r , \qquad 1 \leqslant r \leqslant n - 1 ,$$

which we have used already during the proof of (i) for $r = n$. It remains only to apply (1.1) several times, for the various values of r.

The expression for $x_1 \ldots x_m$ with $m < p$ is obtained in a completely analogous way. □

4.7. Corollary. *Let X be a generating set for an associative algebra A over a field F of characteristic $p \geqslant 0$, and $[X]$ the Lie subalgebra generated by X with respect to the operation $[x, y] = xy - yx$. Suppose further that A satisfies the identity $u^n = 0$ in elements $u \in [X]$, and that $n < p$ (with n arbitrary if $p = 0$). Then A is nilpotent (in the sense of associative algebras) if $[X]$ is nilpotent. The converse is trivial.*

The *proof* is an immediate consequence of Proposition 4.6(i), but we need to clarify something. In the present situation, the identity $u^n = 0$ does not usually imply that $[X]$ satisfies E_n. The distinction between the pairs $([X], A)$ and $(L, A(L))$, where L satisfies E_n, is roughly the same as that between a universal enveloping algebra and the associative algebra of linear operators corresponding to a linear representation of a Lie algebra. True, by formula (1.5) we have

$$[uv^{2n-2}] = \sum_{i=0}^{2n-2} (-1)^i \binom{2n-2}{i} v^i u v^{2n-2-i}$$
$$= (-1)^{n-1} \binom{2n-2}{n-1} v^{n-1} u v^{n-1} = 0\,.$$

Since $u^n = 0$, we get $\sum_i v^i u v^{n-1-i} = \left\{\begin{matrix} u & v \\ 1 & n-1 \end{matrix}\right\} = 0$ (see (1.2)). Thus

$$v^{n-1} u v^{n-1} = v^{n-1}\left(\sum_i v^i u v^{n-1-i}\right) = 0\,.$$

Therefore, the Lie algebra $[X]$ satisfies E_k with $k \leqslant 2n-2$. For $p > 0$, we can do it differently: $u^n = 0 \Rightarrow u^p = 0 \Rightarrow (u + \lambda v)^p = 0 \Rightarrow \sum_i v^i u v^{p-1-i} = 0 \Rightarrow [uv^{p-1}] = 0$ (see (1.10)), that is, in this case $[X]$ satisfies E_k for some $k \leqslant \min\{2n-2, p-1\}$.

However, in proving assertion 4.6(i), we have proceeded exclusively from the identity $u^n = 0$, without worrying about its provenance. Our conclusions are therefore correct. □

It must be added that this corollary is not essential for our exposition.

Remark. We have convinced ourselves that a Lie algebra L over a field F of characteristic $p > 0$ satisfying E_p and embedded in $A(L)$ also satisfies E_{p-1}. Therefore, any statement about nilpotency or local nilpotency valid for $n = p - 1$ remains in force when $n = p$.

§ 5. The Method of Sandwiches

5.1. For a fixed value of the characteristic p of the ground field F, the proof that a Lie algebra L with E_n, $n < p$, is locally nilpotent proceeds by the natural induction on n. The basis of the induction is contained in Examples 2.7 and 2.9. According to the induction hypothesis, for every finitely generated subalgebra $W \subset L$ there is an index m (a natural number) such that every element $w \in W^m$ has an expression of the form

$$w = \sum_i [u_i v_i^{n-1}]\,, \qquad u_i, v_i \in W\,.$$

Moreover, by Proposition 3.2 and the basic agreement in 4.1, we can assume that $L \neq 0$, $R(L) = 0$. The desired contradiction is achieved if we can construct an abelian ideal $J \neq 0$ in L, for example; that is,

$$J^2 = [J, J] = 0\,, \qquad [J, L] \subseteq J\,.$$

However, it is not necessary to assume that $R(L) = 0$ in order to be able to establish the existence of an abelian ideal $J \neq 0$. This fact strengthens the main theorem. However, we do assume that $L \subset A(L)$.

For every element $c \neq 0$ in J, the set $[cL^\infty]$ of all linear combinations of commutators $[cx_0x_1 \dots x_{k-1}x_k]$, $x_i \in L$, is contained in J and clearly comprises an abelian ideal of L; it is called a principal ideal. Without loss of generality, we can therefore state at the very outset our aim of finding an abelian ideal J of L generated by some element $c \neq 0$.

5.2. By the Jacobi identity, we have

$$[v_1v_2 \dots v_s[cu_1u_2 \dots u_m]]$$

$$= \sum_{t;\, i_1, \dots, i_t} (-1)^t [cv_1 \dots v_s u_{i_1} \dots u_{i_t} c u_{j_1} \dots u_{j_{m-t}}]\,,$$

so that

$$[J, J] = 0 \Leftrightarrow [cx_0x_1 \dots x_{k-1}x_k c] = 0, \qquad k = 0, 1, \dots;\, x_i \in L\,.$$

The question is really about constructing at least one element $c \neq 0$ with the required property. Since $[cx_0x_1 \dots x_kc] = -[x_0cx_1 \dots x_kc]$, the fact that $x_0 \in L$ is arbitrary allows us to express the property as a system of identities:

$$cx_1x_2 \dots x_{k-1}x_kc = 0\,, \qquad k = 0, 1, \dots;\, x_i \in L\,. \tag{1.11}$$

Definition. A non-zero element c of L satisfying the system of identities[1] (1.11) is called a *sandwich* of L. More exactly, we shall speak of a sandwich $c \in L$ of *thickness* r if identities (1.11) are satisfied for $k = 0, 1, \dots, r$ but

$$ce_1e_2 \dots e_re_{r+1}c \neq 0$$

for some elements $e_1, e_2, \dots, e_r, e_{r+1}$ of L.

For $r \leqslant 1$ we speak of a *thin* sandwich, and a *thick* sandwich when $r \geqslant 2$.

This concept is introduced in order that the construction of a sufficiently thick sandwich generating an abelian ideal can be divided into comparatively elementary stages.

5.3. We note that Proposition 4.6(ii) allows us to restate the definition of sandwich of thickness r (Definition 5.2) as follows when $p > r$:

$$cu^kc = 0 \quad \text{for} \quad k = 0, 1, \dots, r \text{ and all } u \in L\,,$$

$$ce^{r+1}c \neq 0 \text{ for at least one element } e \in L\,. \tag{1.12}$$

By Proposition 4.6(ii) again, we may assume that the ideal $J = [cL^\infty]$ is spanned by elements

$$c, [cu], [cu^2], \dots, [cu^{n-1}]\,, \qquad u \in L\,.$$

The definition that we have given of sandwich of thickness $r < p - 1$ makes sense in arbitrary Lie algebras, not only for those satisfying E_n.

---

[1] Not to be confused with the formal concept of identity in the variety-theoretical sense (see [26]).

Proposition. *The thickness of a sandwich is always odd.*

Proof. If $cu^k c = 0$ for $k = 0, 1, \ldots, 2m$, then $[cu^{2m+1}c] = -[u\underline{cu^{2m}c}] = 0$. Furthermore, by (1.5) we have

$$[cu^{2m+1}c] = [cu^{2m+1}]c - c[cu^{2m+1}] = (cu^{2m+1}c + \cdots)$$
$$-((-1)^{2m+1}cu^{2m+1}c + \cdots) = 2cu^{2m+1}c + \ldots,$$

where the dots denote terms that contain monomials $cu^k c$, $k \leqslant 2m$, and are therefore zero. Thus, $cu^{2m+1}c = 0$ is an identity in u, and has been obtained "free of charge", so to speak. □

Every sandwich c satisfies the relation $c^2 = 0$, so that the Proposition gives that $cuc = 0$ identically in u. For a thin sandwich, nothing more is required. Thus, the concepts of *thin sandwich, element of nil-index* 2, *and absolute zero divisor* are the same for Lie algebras. At the same time, thick sandwiches are new constructive objects.

5.4. Definition. The subalgebra $\mathfrak{C} \leqslant L$ of L generated by all the sandwiches will be called the *sandwich subalgebra* of L (or a *sandwich algebra* if $\mathfrak{C} = L$).

For convenience, we introduce the following notation:

$\mathfrak{C}_m^* = \mathfrak{C}_m^*(L)$ is the set of all sandwiches of thickness $r = 2m - 1$ (see (1.12)),

$\mathfrak{C}^* = \mathfrak{C}^*(L) = \bigcup_{m=1}^{\infty} \mathfrak{C}_m^*$ is the set of all sandwiches.

Proposition. *The set $\mathfrak{C}^*$ is multiplicatively closed (closed under commutation, that is, weakly closed in the sense of* [120], *Chap.* 3). *In other words, the sandwich subalgebra is the linear space over F spanned by $\mathfrak{C}^*$: $\mathfrak{C} \doteqdot \langle \mathfrak{C}^* \rangle_F$.*

The *proof* is almost obvious, since

$$c_1^2 = 0,$$
$$c_2^2 = 0 \Rightarrow [c_1 c_2]^2 = \underline{c_1 c_2 c_1}\, c_2 + c_2 \underline{c_1 c_2 c_1} - c_1 \underline{c_2^2} c_1 - c_2 \underline{c_1^2} c_2 = 0,$$

that is, $c_1, c_2 \in \mathfrak{C}^* \Rightarrow [c_1, c_2] \in \mathfrak{C}^*$. □

Sandwich subalgebras are of interest in their own right, and a section in Chapter 3 is devoted to them.

5.5. The method of sandwiches is not just a matter of the above definitions; rather, it is the realisation that it has turned out to be a very fruitful idea for constructing elements of $\mathfrak{C}_{m+1}^*$ from $\mathfrak{C}_m^*$. It is proved during the following three chapters that the transition from $\mathfrak{C}_m^*$ to $\mathfrak{C}_{m+1}^*$ is possible provided that $2m + 5 \leqslant p$ (and is always possible if char $F = 0$). Since $n < p$ and therefore $2\left[\dfrac{n-1}{2}\right] + 3 \leqslant p$, after a finite calculation we get an element $c \in \mathfrak{C}_{m_0}^*$, $m_0 = \left[\dfrac{n-1}{2}\right]$ that generates an abelian ideal $[cL^\infty]$ of L, or, as is the same thing, is a representative of the thick sandwiches:

$$cu^k c = 0\,, \qquad k = 0, 1, \ldots, n-1; \qquad u \in L\,.$$

Generally speaking, if we start immediately from the definition of $\mathfrak{C}^*_{mo}$, these identities are valid for $k \leqslant n-3$ only. However, by (1.7) and (1.8) we have

$$cu^{n-2}c = \begin{Bmatrix} c & u \\ 2 & n-2 \end{Bmatrix} - \sum_{\substack{0 \leqslant j+k \leqslant n-2 \\ k \leqslant n-3}} u^j cu^k cu^{n-2-j-k} = 0\,,$$

$$cu^{n-1}c = c\begin{Bmatrix} c & u \\ 1 & n-1 \end{Bmatrix} - \sum_{0 \leqslant k \leqslant n-2} cu^k cu^{n-1-k} = 0\,,$$

so that the descent to thick sandwiches is not too onerous, given the condition $2m + 5 \leqslant p$ mentioned above. A further fact is that, for $p = 5$, the conditions $2m + 5 \leqslant p$, $m \geqslant 1$ are not satisfiable. However, thick sandwiches are unnecessary for $p = 5$, as is attested by the following example.

5.6. Example. Let L be a Lie algebra with E_4, and suppose that char $F = p \geqslant 5$. We prove that L is locally nilpotent by establishing the existence of a non-zero abelian ideal.

Firstly, there is an element $b = [at^3] \neq 0$, otherwise L would satisfy E_3, the case investigated in Example 2.9. We have

$$b^3 = \sum_{i+j+k+l=9} \alpha_{ijkl} t^i at^j at^k at^l\,.$$

Using the relation $at^3 = -tat^2 - t^2at - t^3a$ (see (1.8)), and the corollary $t^3at^3 = 0$, we can reduce the expression for b^3 to the form $b^3 = \alpha t^3at^2at^2at^2$. But $0 = t^3[ta^2]t^3 = -2t^3atat^3 = 2t^3at^2at^2$, so that $b^3 = 0$. In passing, we prove the following lemma, which will be useful later:

Lemma. *Let L be any Lie algebra over F,* char $F > 3$, *suppose that L is embedded in $A(L)$, and let $b \neq 0$ be a nil-element of index 3. The following relations hold:*

$$bub^2 = b^2ub\,, \qquad b^2ub^2 = 0\,, \tag{1.13}$$

$$[ub^2]^2 = b^2u^2b^2\,. \tag{1.14}$$

Proof. Since $b^3 = 0$, relation (1.5) gives that $-3bub^2 + 3b^2ub = [ub^3] = 0$, which reduces to the first of relations (1.13) since char $F > 3$. Multiplying it on the left by b, we get the second relation in (1.13). Finally, by (1.13) we have $[ub^2]^2 = (ub^2 - 2bub + b^2u)^2 = -2bubub^2 + 4bu\underline{b^2ub} - 2\underline{b^2ub}ub + b^2u^2b^2 = -2bubub^2 + 4bubub^2 - 2bu\underline{b^2ub} + b^2u^2b^2 = b^2u^2b^2$, which is relation (1.14). □

Returning to our example, we note that either $b^2 = 0$, or else there exists an element $c = [fb^2] \neq 0$. Using (1.14) and (1.7) for $s = j_1 = j_2 = 2$ in the latter case, we find that $c^2 = [fb^2]^2 = b^2\underline{f^2b^2} = -\underline{b^2fbfb} = 0$, since now alongside relations (1.13) we have a further one:

$$bfb^2 + b^2fb = \begin{Bmatrix} b & f \\ 3 & 1 \end{Bmatrix} = 0 .$$

By the arguments of 5.5, we can add $cu^2c = 0$, $cu^3c = 0$ to the relations $c^2 = 0$, $cuc = 0$. Using (1.7) once more, we get

$$cu^2c = \begin{Bmatrix} c & u \\ 2 & 2 \end{Bmatrix} - \underline{c^2}u^2 - u^2\underline{c^2} - u\underline{cuc} - \underline{cuc}u - u\underline{c^2}u = 0 ,$$

while Proposition 5.3 gives $cu^3c = 0$. □

It is interesting to note that the elementary calculations used here to lead so quickly to our goal for $n = 4$ (whereas this case remained open for such a long time) contain the germ of the general scheme of the arguments that follow.

5.7. Proposition. *A finitely generated sandwich Lie algebra M embedded in $A(M)$ is nilpotent if and only if the associative algebra $A(M)$ is nilpotent. The characteristic of the ground field F is assumed to be not 2.*

Proof. Suppose that M is generated by a finite set $S = \{x_1, \ldots, x_d\}$ of sandwiches. It is clear that the nilpotency of $A(M)$ implies that of M, since every commutator $a \in M$ of length N in $x_1, \ldots, x_d$ can be written as a linear combination of associative monomials of length N in $A(M)$, so that $A(M)^N = 0 \Rightarrow M^N = 0$.

It is intuitively clear, but less obvious, that an implication of the type $M^{m+1} = 0 \Rightarrow A(M)^N = 0$ (with N depending on m) holds. We shall argue by contradiction. Consider an associative monomial

$$0 \neq x_{\alpha_1} x_{\alpha_2} \ldots x_{\alpha_N} \in A(M), \qquad \alpha_i \in \{1, \ldots, d\} ,$$

whose length N is so great that one of the generators, x_1 say, occurs a pre-assigned number N' of times in it. Since $x_1^2 = 0$ and $x_1 u x_1 = 0$, we have $k \geqslant 2, l \geqslant 2, \ldots$ in the monomials

$$x_1 x_{i_1} \ldots x_{i_k} x_1 x_{j_1} \ldots x_{j_l} x_1 \ldots .$$

Moreover, any attempt to bring two occurrences of x_1 closer together leads to an expression of the form

$$x_1 x_{i_1} \ldots x_{i_k} x_1 = \Sigma \pm x_{i'_1} \ldots x_{i'_s} [x_1 x_{i''_1} \ldots x_{i''_t}] x_1 ,$$

where $t \geqslant 2$ also. Without loss of generality, we may assume at the outset that we have a product

$$y_1 y_2 \ldots y_{N'} \neq 0$$

in $A(M)$, with $y_1 = [x_1 x_{i_1} \ldots x_{i_k}]$, $y_2 = [x_1 x_{j_1} \ldots x_{j_l}]$. The lengths of these commutators is bounded above by some number m, since M is nilpotent, and thus only a finite number $d' < d^m$ of them can be different; thus in fact we have a product

$$y_{s_1} y_{s_2} \dots y_{s_{N'}} \neq 0\,, \qquad 1 \leqslant s_i \leqslant d'$$

of elements of the set $S' = \{y_1, \dots, y_{d'}\}$.

Since the length N' can be as large as we please, some commutator, y_1 shall we say, occurs a pre-assigned number N'' of times, that is, we have a monomial of the form

$$y_1 y_{\alpha_1} \dots y_{\alpha_r} y_1 y_{\beta_1} \dots y_{\beta_s} y_1 \dots y_1 y_{\gamma_1} \dots y_{\gamma_t} y_1 \neq 0\,.$$

We note that $y_1^2 = 0, \dots, y_{d'}^2 = 0$, so that $r \geqslant 2$, $s \geqslant 2, \dots, t \geqslant 2$. As in the first case, any attempt to draw the two occurrences of y_1 together leads us (after a change of notation) to a product

$$z_1 z_2 \dots x_{N''} \neq 0$$

of elements

$$z_1 = [y_1 y_{\alpha_1} \dots y_{\alpha_r}], \qquad z_2 = [y_1 y_{\beta_1} \dots y_{\beta_s}], \dots, z_{N''} = [y_1 y_{\gamma_1} \dots y_{\gamma_t}]$$

of the nilpotent algebra M, which therefore have their indices $r, s, \dots$ bounded above by some absolute constant (and below by 2). This time we have a product

$$z_{t_1} z_{t_2} \dots z_{t_{N''}} \neq 0\,,$$

of elements of the set $S'' = \{z_1, \dots, z_{d''}\}$, where $d'' < d^m$ as before.

We can repeat this process unboundedly often, thus increasing the length N of the original monomial. On the other hand, on calculating the lengths $m_1, m_2, \dots$ with respect to $x_1, \dots, x_d$ of first generation commutators $y_1, y_2, \dots$, then the second generation $z_1, z_2, \dots$ *etc*, we get that

$$m_1 \geqslant 3, \qquad m_2 \geqslant 3^2, \dots, m_q \geqslant 3^q, \dots\,.$$

For sufficiently large q, we have $m_q > m$. However, all such commutators in M are zero. The contradiction thus achieved shows that there exists an N depending on m and such that $A(M)^N = 0$.

The condition $\operatorname{char} F = p \neq 2$ is necessary only in order for the concept of sandwich to make sense. □

This simple property of sandwich algebras, although it is not necessary for what follows, is interesting in itself and allows us to introduce a small improvement into the exposition. In particular, it makes the following definition a sensible one:

5.8. Definition. Let L be any Lie algebra embedded in $A(L)$, and $S = \{x_1, \dots, x_d\}$ a finite set of sandwiches of L generating a nilpotent subalgebra. We shall call an element

$$\tilde{a} = [a x_{i_1} \dots x_{i_s}]\,,$$

where $i_1, \dots, i_s$ is a fixed succession of numbers from $\{1, \dots, d\}$, an *S-continuation* (or, more graphically, an *$\{x_1, \dots, x_d\}$-continuation*) of the non-zero element

a of L if $\tilde{a} \neq 0$ but $[ax_{j_1} \dots x_{j_s} x_{j_{s+1}}] = 0$ for all choices of $j_1, j_2, \dots, j_s, j_{s+1}$ from $\{1, 2, \dots, d\}$.

We do not exclude the possibility that $\tilde{a} = a$, but it is far more important that $\tilde{a}$ always exists; this is guaranteed by Proposition 5.7 as applied to the subalgebra of $A(L)$ generated by the sandwiches $x_1, \dots, x_d$. The notation $\tilde{a}$ in no way reflects the variety of possible continuations, but normally we are satisfied with an arbitrary choice of one of them.

Let $f_1, \dots, f_m$ be commutators of the same general structure as the product $x_{i_1} \dots x_{i_s}$ (they are therefore sandwiches). In particular,

$$\sum_{i=1}^{m} \deg_S f_i(x_1, \dots, x_d) = s .$$

We shall call any element $[af_1 \dots f_m] \neq 0$ with the smallest possible index m, $1 \leqslant m \leqslant s$, a *subtwist* of the S-continuation $\tilde{a}$.

We shall inevitably arrive at a subtwist of $\tilde{a}$ if we rearrange the components x_{i_k} in $\tilde{a}$ in all possible ways; starting at any given instant, the commutators can be rearranged at will. Thus, by definition,

$$[af_1 \dots f_m] = [af_{\pi 1} \dots f_{\pi m}] \neq 0 , \qquad \pi \in S_m ,$$

$$[af_1 \dots f_m x_i] = 0 , \qquad 1 \leqslant i \leqslant d .$$

This last relation is perhaps not all that clear, but it becomes so if every commutator f_k is expanded in the associative algebra $A(L)$, and one notices that the expression so obtained is a linear combination of elements of type $\tilde{a}$.

§ 6. Filtrations in Lie Algebras

At the beginning of this century, È. Cartan classified the infinite-dimensional transitive complex simple Lie algebras. In so doing he effectively used the concept of filtration, an idea that proved later to be a convenient tool for studying finite-dimensional simple Lie algebras over algebraically closed fields F of finite characteristic. Since the problem of giving a complete classification of such Lie algebras is still open, the problems posed in this section retain their urgency.

6.1. Definition. Let L be a Lie algebra and L_0 any maximal subalgebra. By the *filtration of L through L_0* we understand the sequence of subalgebras

$$L = L_{-1} \supset L_0 \supset L_1 \supset \cdots \supset L_s \supset L_{s+1} \supset \dots , \tag{1.15}$$

where

$$L_{i+1} = \{x \in L_i \mid [x, L] \subseteq L_i\} , \qquad i = 0, 1, \dots .$$

For $i + j \geqslant -1$, we have

$$[L_i, L_j] \subset L_{i+j} . \tag{1.16}$$

For, by definition $[L_i, L] \subset L_{i-1}$, while if (1.16) holds for $i+j \leqslant k-1$, the Jacobi identity gives

$$[[L_i, L_j], L] \subset [[L_i, L], L_j] + [L_i, [L_j, L]] \subset [L_{i-1}, L_j] + [L_i, L_{j-1}] \\ \subset L_{i+j-1}$$

for $i+j=k$, and this inclusion is equivalent to (1.16). In particular, $[L_0, L_0] \subset L_0$ (this is simply the statement that L_0 is a subalgebra) and $[L_i, L_0] \subset L_i$, that is, L_i is an ideal of L_0. The following cases are possible.

a) $L_{s+1} = L_s \neq 0$ for some $s \geqslant 0$. It then follows from the definition of L_{s+1} that L_s is an ideal of L contained in the maximal subalgebra L_0.

b) $\bigcap_{i=0}^{\infty} L_i = 0$. Then, either (1.15) is an infinite sequence (a filtration of infinite length), or $L_{s+1} = 0$, $L_s \neq 0$ for some s, and the filtration is of length s by definition. We say that the filtration is *trivial* if $s=0$, *short* if $s=1$ and *long* if $s \geqslant 2$.

Case a) is of no interest for us, since it cannot arise in a simple Lie algebra. In the situation studied by Cartan, L_0 has finite codimension. In that case it can be shown that $\dim L_i/L_{i+1} < \infty$ for all i, so that the filtration through L_0 has infinite length. When F is of zero characteristic and L_0 is a maximal subalgebra of smallest codimension in a simple Lie algebra L, it follows that either L has a short filtration, in which case $\dim L < \infty$; or else $L_2 \neq 0$, in which case $L_s \neq 0$ for every non-negative s.

Suppose now that L is a finite-dimensional simple Lie algebra over an algebraically closed field of characteristic $p > 5$. We wish to translate the above general remarks on filtrations into the language of sandwiches as applied to L. As before, we assume that L is embedded in $A(L)$.

6.2. Proposition. *The following assertions hold*:

(i) *Every Lie algebra L with a filtration of length $s \geqslant 2$ contains a sandwich of thickness $r \geqslant s-2$.*

(ii) *Every Lie algebra with a sandwich of thickness r has a filtration of length $s \geqslant r+1$.*

Therefore, L has a long filtration if and only if it has non-zero sandwich algebra $\mathfrak{C} = \mathfrak{C}(L)$.

Proof. (i) Let

$$L = L_{-1} \supset L_0 \supset L_1 \supset \cdots \supset L_s \supset 0\,, \qquad s \geqslant 2\,,$$

be a filtration in L. By (1.16), for every non-zero element $c \in L_s$ and all $x_i \in L$, we get that $[cx_0x_1 \ldots x_kc] \in L_{2s-1-k} = 0$, $k \leqslant s-2$; that is, $cx_1 \ldots x_kc = 0$. By Definition 5.2, this means that c is a sandwich of thickness $r \geqslant s-2$.

(ii) Conversely, assume that c is a sandwich of thickness $r \geqslant 1$ in L. We choose any maximal subalgebra L_0 of L containing $\operatorname{Ker}(\operatorname{ad} c)$. Note that

$$L_0 \supseteq \operatorname{Ker}(\operatorname{ad} c) \supseteq \sum_{k=0}^{r+1} [cL^k]\,,$$

where $[cL^k]$ is the linear space spanned by the commutators $[cx_1x_2 \dots x_k]$, $x_i \in L$. We see from Definition 6.1 that the term L_1 of the filtration through L_0 always contains the subspace $\sum_{k=0}^{r+1}[cL^k]$. More generally, $L_i \supset \sum_{k=0}^{r+1-i}[cL^k]$. In particular, $L_{r+1} \supset \langle c \rangle_F \neq 0$, that is, $s \geqslant r+1$.

Our discussion is based on the fact that all the inclusions

$$\sum_{k=0}^{l+1}[cL^k] \supset \sum_{k=0}^{l}[cL^k]$$

are proper. Indeed, if this were false for some l, then the principal ideal $J = [cL^\infty]$ would be just $\sum_{k=0}^{l}[cL^k]$. However, since L is simple, $J = L$, so that $J \subset \operatorname{Ker}(\operatorname{ad} c)$ if $l \leqslant r$; that is, we get the contradiction that $[c, L] = 0$.

Finally, it follows from (i) and (ii) that $\mathfrak{C}(L) \neq 0 \Leftrightarrow s \geqslant 2$. □

6.3. Finite-dimensional simple Lie algebras with non-zero sandwich subalgebras, and thus with long filtrations (the *strongly degenerate algebras*), are very similar to algebras of Cartan type, which in turn are analogues of the infinite-dimensional Cartan algebras in characteristic zero. Algebras of Cartan type have non-trivial deformations; the fundamental conjecture extant at the end of the sixties predicts that the class of abstract strongly degenerate algebras will be the same as the class of all deformations of algebras of Cartan type.

Non-classical simple Lie algebras exist only when $\operatorname{char} F = p > 0$, and their study was initiated later than in the classical case. Nevertheless, there is a fifty-year history of "modular" Lie algebras. A typical early example is the *Zassenhaus algebra*

$$W_1(m) = \left\langle e_i \mid -1 \leqslant i \leqslant p^m - 2, [e_i, e_j] = \left\{\binom{i+j+1}{j} - \binom{i+j+1}{i}\right\} e_{i+j} \right\rangle.$$

It contains a sandwich of thickness $p^m - 4$, namely $c = e_{p^m-2}$ (here p is an odd prime and m any natural number). Investigations carried out in the USSR, the USA and the FRG have led to the discovery of a number of the springs in the classification mechanism. In this book, strongly degenerate Lie algebras are encountered exclusively in the context of sandwiches.

Why does the existence of a filtration of length 2 in L make it possible to construct a filtration of length $s \geqslant p - 3$?

It seems that the answer to this question will have to be obtained using linear algebraical methods not resting on the combinatorial considerations developed in connection with the idea of thick sandwich. However, nobody has done it yet.

On the other hand, why not try to construct filtrations of high length in Lie algebras with E_n directly, avoiding sandwiches and using Proposition 6.2 only at the very end? This is a purely rhetorical question. Sandwiches first came to light about 30 years ago, and have not changed; rather, they have found new applications.

§ 7. Main Results and Structure of Proofs

The following results were first obtained at the end of the fifties, and they form the kernel of the book.

7.1. Main Theorem. *Let L be a Lie algebra with zero centre over a field F of characteristic $p > 5$, containing at least one thin sandwich and belonging to one of two types:*

1) *L satisfies* E_n *and $n < p$;*
2) $\dim_F L < \infty$; *L is simple when $p = 7$.*

Then L has a sandwich of thickness $r \geqslant p - 4$ (of arbitrary thickness in the zero characteristic case).

The *proof* is executed in § 4 of Chap. 2, and the two following chapters.

7.2. Corollary. *A finite-dimensional simple Lie algebra over a field F of characteristic $p > 3$ and containing at least one sandwich has a filtration of length $s \geqslant p - 3$.*

The *proof* follows without difficulty from the main theorem (Theorem 7.1) and Proposition 6.2. □

7.3. Theorem. *Every non-zero Lie algebra with* E_n, *$n < p$ (n arbitrary in the zero characteristic case) contains a non-zero abelian ideal.*

The *proof* is obtained from the results of § 2 in Chap. 2, the main Theorem 7.1, and the arguments in 5.5. □

7.4. Theorem. *Every Lie algebra L with* E_n, *$n < p =$* $\operatorname{char} F$ *(n arbitrary in the zero characteristic case) is locally nilpotent.*

The *proof* is a combination of Theorem 7.3 and the arguments in 5.1. □

7.5. Corollary. *The class $\mathfrak{K}_p$ of finite groups on a given number d of generators satisfying the identical relation $x^p = 1$ contains only finitely many non-isomorphic groups. In other words, every group in $\mathfrak{K}_p$ is a factor-group of a largest group $B_0(d, p)$ in $\mathfrak{K}_p$ (this is the positive solution of the restricted Burnside problem for prime exponent p).*

The *proof* is the known reduction of 7.5 to the assertion of Theorem 7.4 (see Chapter 7). □

Remark. A shorter path through the proof of Theorem 7.4, and thus to a solution of RBP for prime exponent p, is to go through §§ 3, 4, 5 of Chap. 1, §§ 1, 2 of Chap. 2, §§ 3, 5 and the beginning of § 4 of Chap. 3, and §§ 1, 2 of Chap. 5.

7.6. Important results that are connected ideologically with the theorems just mentioned are to be found in Chaps. 5–7. They are due to several authors and relate to problems of Burnside type in Lie algebras (E. I. Zel'manov and Yu. P.

Razmyslov) as well as to the precise connection between finite groups and Lie algebras (Higman and Hall, G. E. Wall, E. I. Khukhro, M. R. Vaughan-Lee). An account of these results, which were obtained by various methods, will be supplied together with a fairly detailed commentary in the appropriate places.

7.7. The expert in group theory who is interested only in Corollary 7.5 can scan the book by taking advantage of the pleasant path (see the remark in 7.5) avoiding many of the details of the proof of Theorem 7.1. However, as we have mentioned already, that theorem has a wider significance. Therefore, it is apposite to sketch at this point a general outline of the arguments, so as to be of assistance to those readers who find themselves obliged to apply the full strength of the method of sandwiches.

First Step: descent to sandwiches, that is, a proof that the set $\mathfrak{C}_1^*(L)$ is non-empty in every Lie algebra L with E_n, $n < p$ (Theorem 2.2.1.). No commentary is necessary here.

Second Step: investigation of the various methods of producing sandwiches of thickness m, $1 \leqslant m \leqslant (p\text{-}3)/2$. The fairly simple Lemma 2.4.2 asserts that the map

$$\Phi_a : c \mapsto [ca^{2m+1}c]$$

takes the set $\mathfrak{C}_m = \mathfrak{C}_m^* \cup \{0\}$ to itself. For sufficiently large r, it is expected that the element

$$c_r = \Phi_{a_r}(\Phi_{a_{r-1}} \ldots (\Phi_{a_2}(\Phi_{a_1}(c))) \ldots), \qquad r = 1, 2, \ldots, \tag{1.17}$$

lies in $\mathfrak{C}_{m+k}^* \subset \mathfrak{C}_m^*$ for $k \geqslant 1$. Theorem 2.4.6 states that this is indeed true, even for $r \leqslant 2$ provided that $m \geqslant 2$. The proof is obtained through direct calculations, which it would be desirable to eliminate, for example by increasing the parameter r or by considering other ways of producing sandwiches (we are forced to use such in the case $m = 1$). Thus, the possession of a single sandwich of thickness $m \geqslant 2$ enables us to get sandwiches of whatever thickness is necessary, and by what we saw in 5.5, this leads us to our goal.

Third and Decisive Step: transition from $\mathfrak{C}_1^*$ to $\mathfrak{C}_2^*$, which is the most difficult. Probably, when $m = 1$ it can be proved directly that the chain (1.17) of elements c_r breaks off for sufficiently large $r_0 : c_{r_0} = 0$. However, as in the original paper [137], we choose here the roundabout path that uses, in addition to (1.17), thin sandwiches of the form

$$c_{(r)} = [ca_0^3ca_1^2ca_2^2c \ldots ca_r^2c] \tag{1.18}$$

(see §§ 1, 2 of Chap. 3), and—perhaps more essentially—a study of arbitrary sandwich algebras. We prove: if $\{c_1, c_2, \ldots, c_d\}$ is any finite set of thin sandwiches in a Lie algebra with E_n, then all commutators in them of large enough length l are zero:

$$[c_{i_1} c_{i_2} \ldots c_{i_{l-1}} c_{i_l}] = 0, \qquad 1 \leqslant i_k \leqslant d\,. \tag{1.19}$$

We prove later that the condition E_n is unnecessary here, although this is of no importance for Theorem 7.1. The proof of relations like (1.19) (the local nilpotency of sandwich algebras with E_n) can be carried out using one of the simplified schemes (see Variants 1 and 2 in § 1 of Chap. 5). True, if he prefers to behave differently, and struggles through the labyrinth of the arguments in § 4 of Chap. 3, the reader will overcome relatively easily the barrier on the path to "thick pairs of thin sandwiches" in Chap. 4. These pairs are an effective instrument in the closing stages of the proof. We note that a direct proof that a chain $\{c_{(r)}\}$ of sandwiches of the form (1.18) breaks off appears to be a problem of the same sort of difficulty to the case of a chain like (1.17).

7.8. Some Technical Remarks. The method of double interpretation of relations in L and in $A(L)$ (see 4.2) and the linearization of identities with the implicit use of relations from 4.4–4.6 is an elementary apparatus; temporary difficulties may arise if it is not used fully. We consider two almost random examples, which really will be encountered below.

a) Let L be a Lie algebra embedded in $A(L)$, and suppose that $c \in \mathfrak{C}_1^*$; that is, $c \neq 0$, $c^2 = 0$, $cxc = 0$ for all $x \in L$. If it suddenly happens that $[cx^3c] = 0$ for all $x \in L$, then $[cxyzc] = 0$ as well (linearization !), and this means that $[xcyzc] = 0$, that is, $cyzc = 0$. On the other hand, after expansion of the commutator $[cx^3c] = 2cx^3c + \sum_{k \leqslant 2} \ldots cx^kc \ldots$, the given Lie identity acquires the second interpretation $cx^3c = 0$. Linearizing once more, we arrive at the identity $cxyzc = 0$. Taken together, all this means that $c \in \mathfrak{C}_2^*$.

b) To deduce relation (2.32) below for $c \in \mathfrak{C}_m^*$ and $c_0 = [ca^{2m+1}c]$, we use (2.29) (in fact, (2.30) as well) to get the identity

$$c_0 u^{2m} aca^{2m} c = c_0 au^{2m} ca^{2m} c ,$$

which expresses shift to the left of a single element $a \in L$. But every shift is accompanied by the appearance of commutators in products like $c_0 u^k [au^l] u^{2m-l-k} ca^{2m} c$. Such products are zero for $l \geqslant 1$, as can be seen from linearizing the identities (2.29) and (2.30), or directly from (2.29) and (2.30) if one uses Proposition 4.6(ii) as applied to the element $u^k[au^l]u^{2m-l-k}$ of $A(L)$. If we were to go into details of this sort in every circumstance, we would run the risk of wallowing in trivialities. Although such trivialities do in fact arise, it is to be hoped that they do not constitute the whole set-up.

§ 8. Commentary

8.1. Supplement to 2.9. Here is a more economical argument in the case of a Lie algebra L with E_3 for $p > 5$. By 2.7, we have

$$[x_1 x_2 x_3] = \sum_i [t_i w_i^2] ,$$

so that it follows by the arguments of 2.9 that

$$[x_1 x_2 x_3 x_4 x_5 x_6] = \sum_{i,j} [t_i \underline{w_i^2 v_j^2} b_j] .$$

Thus $L^6 = 0$: a result obtained by Higgins [106].

8.2. A Direct Combinatorial Attack on the local nilpotency problem for Lie algebras with E_{p-1} over $\mathbb{Z}_p$ (which is what is really needed for RBP) has been attempted twice at a serious level. In 1953, the author solved RBP for exponent 5 (see [131]), by proving that the 2-generator free Lie algebra $L = L\,(2, 4)$ with E_4 over $\mathbb{Z}_5$ is nilpotent. It turned out that $L^{13} = 0$. More exactly, the following values were obtained for the dimensions r_k of the factors L^k/L^{k+1} over $\mathbb{Z}_5$:

k:	1	2	3	4	5	6	7	8	9	10	11	12	13 ,
r_k:	2	1	2	3	2	4	4	4	6	3	2	$\leqslant 1$	0 ,

so that $33 \leqslant \dim L \leqslant 34$. Subsequently (see [98], [151]) a computer was used to establish that in fact $\dim L = 34$, and that the order of $B_0(2, 5)$ is 5^{34}.

Independently of the author, Higman [108] solved RBP for exponent 5 and arbitrary generating number d; this was in 1956. He proved that there is a natural number N such that the d-generator free Lie algebra $L = L(d, 4)$ with E_4 over $\mathbb{Z}_5$ has nilpotency class at most dN. It was conjectured that $N \leqslant 9$ (clearly $N \geqslant 6$, as the detailed calculation for $d = 2$ shows), but there have been no further investigations in this direction. About the same time, the author solved RBP for $p = 5$ and 7, using simpler (but not directly combinatorial) methods.

8.3. The locally nilpotent radical in associative rings was first introduced by Levitzki and applied by him to solve a problem of Kurosh. If A is an associative algebra satisfying the identical relation $x^n = 0$, over a field of characteristic $p > n$, the necessity for the Levitzki radical is eliminated, since Higman [107] proved using combinatorial methods that $A^{f(n)} = 0$, where $f(n) \leqslant 2^n - 1$. Here $f(n) > (n/e)^2$ for sufficiently large n (e is the base of natural logarithms), and by now there have been improvements to the limits of the possible values for $f(n)$ (see the Commentary to Chap. 6).

As we have noted, the analogous result cannot be obtained for Lie algebras with E_n, $n<p$, whereas it is true for $p \gg n$. The locally nilpotent radical was first applied in studying Lie algebras with E_n in the author's paper [132]. Further information on the radical can be drawn from Lazard's lectures [157].

8.4. Since formula (1.1) is so actively used, it may make sense to indicate a short derivation of it. Set

$$S_k = \sum_{1 \leqslant i_1 < i_2 < \cdots < i_k \leqslant r} (u_{i_1} + u_{i_2} + \cdots + u_{i_k})^r$$

for the sum of the $\binom{r}{k}$ powers of linear expressions obtained by allowing $i_1, \ldots, i_k$

to run over all choices of k indices from the set $\{1, 2, \ldots, r\}$. Further, set

$$S = \sum_{k=1}^{r} (-1)^{r-k} S_k$$

for the alternating sum on the right hand side of (1.1.). It is clear that the symbol

$$\begin{Bmatrix} u_{i_1} & u_{i_2} & \ldots & u_{i_l} \\ j_1 & j_2 & \ldots & j_l \end{Bmatrix}, \quad j_1 + j_2 + \cdots + j_l = r; \quad j_1 > 0, \ldots, j_l > 0$$

(for $l = 2$ this is none other than $\begin{Bmatrix} u_1 & u_2 & \ldots & u_r \\ 1 & 1 & & 1 \end{Bmatrix}$, that is, the left-hand side of (1.1)) occurs in S_k with coefficient $\binom{r-l}{k-l}$, and in S with coefficient

$$\sum_{k=1}^{r} (-1)^{r-k} \binom{r-l}{k-l} = (-1)^{r-l} \sum_{t=0}^{r-l} (-1)^t \binom{r-l}{t} = \delta_{l,r},$$

that is, $S = \begin{Bmatrix} u_1 & u_2 & \ldots & u_r \\ 1 & 1 & \ldots & 1 \end{Bmatrix}$. □

8.5. The idea of sandwiches, which is very natural and simple, was formulated at the beginning of 1956, when it was first realized that the identity $[cu^3c]^2 = 0$ could perhaps be the indication that there is a universal construction for abelian ideals. Effectively, thick sandwiches were described in § 5 of [133]. The "culinary" terminology (thin sandwich, thick sandwich), having first been used internally, pressed itself on the author's consciousness and hit the literature in 1979, in [143]; however, it now seems too linear: any system of identities $cu_1 \ldots u_k c = 0$, $k = 0, 2, \ldots$ was called a sandwich, and the element c itself appeared in the role of the outside of the sandwich. This is perhaps more intuitive, but as an element of usage it is perhaps too long and inconvenient.

8.6. Although finite-dimensional simple Lie p-algebras have now been classified (see the short note [29]), there is still a boundless class of simple Lie algebras that are not p-algebras, so that all problems relating to filtrations maintain their urgency.

Chapter 2

The Descent to Sandwiches

We shall be embarking on the proof of the main result, but to begin with we can forget that, remembering only that our Lie algebra L with $Z(L) = 0$ is embedded in the associative algebra $A(L)$ and belongs to one of two types: 1) L is n-Engel: 2) L is finite-dimensional. The ground field F has finite characteristic $p > n$ or else is of zero characteristic, in which case we use the symbol $p = \infty$ for convenience. In this latter case all restrictions on n vanish.

§ 1. Descent to Nil-elements of Index 3

1.1. Lemma. *Let L be a completely arbitrary Lie algebra over a field F of characteristic p; v a nil-element of L of index m, where $4 \leqslant m \leqslant p-1$. Then $[u\,v^{m-1}]^{m-1} = 0$ for all $u \in L$, that is, L contains a nil-element of index not more than $m-1$.*

Proof. By formula (1.5) about expanding commutators we have

$$[uv^{m-1}]^{m-1} = \sum_{i_1,\ldots,i_m} \alpha_{i_1,\ldots,i_m} v^{i_1} uv^{i_2} uv^{i_3} \ldots uv^{i_m}.$$

We note here that if in the term

$$T = v^{i_1}u \ldots uv^{i_k}u \ldots uv^{i_m}, \quad i_1 + \cdots + i_m = (m-1)^2, \tag{2.1}$$

of this sum an "internal" index i_k (that is, one with $2 \leqslant k \leqslant m-1$) is zero, then the remaining indices are all $m-1$; since $m-1 \geqslant 3$, one of the following products occurs in T:

$$v^{m-1}uv^{m-1}u^2v^{m-1}, \quad v^{m-1}u^2v^{m-1}uv^{m-1}.$$

However, by the double interpretation principle (see 1.4.2),

$$v^m = 0 \Rightarrow [uv^m] = \sum_{i=1}^{m-1} (-1)^i \binom{m}{i} v^i uv^{m-i} = 0\,. \tag{2.2}$$

In particular

$$v^{m-2}[uv^m] = -mv^{m-1}uv^{m-1} = 0\,,$$

whence it follows, since $m < p$, that

$$v^m = 0 \Rightarrow v^{m-1}uv^{m-1} = 0\,, \tag{2.3}$$

and thus that $T = 0$.

Therefore $[uv^{m-1}]^{m-1}$ is a linear combination of terms T of the form (2.1), where in addition we have

$$i_2 \neq 0, \quad i_3 \neq 0, \ldots, i_{m-1} \neq 0; \quad i_k \leqslant m-1 \quad \text{for all } k\,.$$

Returning to (2.2), we note that vuv^{m-1} is a linear combination of $v^2uv^{m-2}, \ldots, v^{m-1}uv$. This allows us to rewrite a typical monomial T (see (2.1)) as a linear combination of similar monomials with $i_m \leqslant m-2$. We apply suitable transformations to these in turn, lowering the index i_{m-1} (by raising i_{m-2}), then lowering i_{m-2} by raising i_{m-3} *etc.* This wave of transformations ends with the penultimate (second from the left) index i_2. Finally, we arrive at monomials of the form (2.1) with indices

$$i_1 \leqslant m-1, \quad 0 < i_k \leqslant m-2, \quad k = 2, 3, \ldots, m\,.$$

A single strict inequality $i_k < m-2$ leads to a contradiction:

$$i_1 + i_2 + \cdots + i_m < (m-1) + (m-1)(m-2) = (m-1)^2\,.$$

Therefore, we have the unique possible variant

$$[uv^{m-1}]^{m-1} = \alpha v^{m-1}(uv^{m-2})^{m-1}\,.$$

We consider the linear homogeneous system

$$\underset{(2.2)}{\underline{v^{m-3}[uv]^m}} = \binom{m}{2} v^{m-1}uv^{m-2} - mv^{m-2}uv^{m-1} = 0\,,$$

$$(-1)^m[uv^m]v^{m-3} = -mv^{m-1}uv^{m-2} + \binom{m}{2} v^{m-2}uv^{m-1} = 0\,,$$

connecting the two monomials $v^{m-1}uv^{m-2}$ and $v^{m-2}uv^{m-1}$. The system has determinant

$$\Delta = \left(\frac{m(m-1)}{2}\right)^2 - m^2 = \frac{1}{4}m^2(m-3)(m+1) \neq 0\,,$$

if $4 \leqslant m < p-1$, so that $v^{m-1}uv^{m-2} = 0$ in this case, and thus $v^{m-1}(uv^{m-2})^{m-1} = 0$.

Suppose now that $m = p-1$. Replacing u in (2.2) by $[uv^{m-3}u] = [uv^{p-4}u]$ and using the fact that p is odd, we get that

$$v^{m-2}[uv^{m-3}uv^m] = -mv^{m-1}[uv^{m-3}u]v^{m-1} = -2mv^{m-1}uv^{m-3}uv^{m-1} = 0.$$

Therefore, it follows from the relation

$$0 = v^{m-1}uv^{m-4}[uv^m] = -\underline{mv^{m-1}uv^{m-3}uv^{m-1}} + \binom{m}{2}v^{m-1}uv^{m-2}uv^{m-2} - \binom{m}{3}\underset{(2.3)}{\underline{v^{m-1}uv^{m-1}uv^{m-3}}}$$

that $v^{m-1}uv^{m-2}uv^{m-2} = 0$, that is, $[uv^{m-1}]^{m-1} = 0$ in this case also. □

1.2. Proposition. *Every Lie algebra L with a nil-element of index $m \leqslant p-1$ has a nil-element of index* 3.
(Naturally, the trivial case of a Lie algebra with E_2 is excluded from the discussion.)

Proof. Application of the method of descent described in Lemma 1.1 to a nil-element of index m, $m \geqslant 4$, yields an element b of index 3 in a finite number of steps.

1.3. Theorem. *Every finite-dimensional Lie algebra L over an algebraically closed field F of characteristic $p > 5$ has a nil-element of index $m \leqslant p-1$ (and then, by Proposition* 1.2, *it has a nil-element of index* 3).

We omit the proof of this important theorem, which is due to A. A. Premet [216], and which I had stated earlier as a conjecture. We shall not need the result in what follows.

§ 2. Descent to Thin Sandwiches (General Case)

Every attempt to effect a universal method of descent from a nil-element of index 3 to thin sandwiches is doomed to failure, for an obvious reason. Namely, every finite-dimensional complex simple Lie algebra with non-degenerate Killing form $(x|y) = tr\ \mathrm{ad}\,x\ \mathrm{ad}\,y$ has arbitrarily many nil-elements of index 3 (for example, multiples of root elements). But there is no sandwich c in L since $(\mathrm{ad}\ c\ \mathrm{ad}\ x)^2 = 0$ (the analogue of the relation $\underline{cxc}x = 0$), and therefore $(c|x) = tr\ \mathrm{ad}\ c\ \mathrm{ad}\ x = 0$ for all $x \in L$, a contradiction to the non-degenerateness of $(x|y)$. The same argument goes over almost without change to the case of classical modular Lie algebras.

Therefore, the class of finite-dimensional algebras taken as a whole is impervious to a realization of our planned scheme:

$$\{\exists b \neq 0 | b^3 = 0\} \Rightarrow \{\exists c \neq 0 | c^2 = 0\},$$

although an analysis of the reasons hindering this aim is very instructive and permits us to glance afresh at the classical problem for Lie algebras. This, however, is not the place to include this type of problem. Let us return to Engel algebras.

2.1. Theorem. *Let L be a Lie algebra over a field F of characteristic p satisfying* E_n, $n<p$. *Then L has at least one sandwich.*

Of course, what one has in mind here is a thin sandwich—a nil-element $c \neq 0$ of index 2: $c^2 = 0$.

2.2. Plan of the Proof (the proof is begun in detail in 2.3). As a starting-point we take a nil-element $b \neq 0$ of index 3, whose existence was established in Proposition 1.2. Relations (1.13) and (1.14), which are identities in $u \in L$, suggest the idea of introducing the element

$$g_m = g_m(u) = [u[bu]^m b^2] , \tag{2.4}$$

depending on a natural parameter m. By (1.14), $g_0^2 = [ub^2]^2 = b^2u^2b^2$. Induction on m shows that

$$g_m^2 = b^2(u^2b^2)^{m+1} . \tag{2.5}$$

On the other hand, $g_{n-1}(u) = 0$ because of the condition E_n, so that

$$b^2(u^2b^2)^t = 0 \quad \text{for some fixed } t, \qquad 0 \leqslant t \leqslant n . \tag{2.6}$$

If $t = 0$, then $b^2 = 0$; if $t = 1$, $[ub^2]^2 = 0$ for all $u \in L$. This means that b, or $[ub^2]$ for suitable $u \in L$, is the sandwich that we are looking for.

For $t > 1$, it will be shown that there exists $f \in L$ such that the element $b_0 = g_m(f) \neq 0$ of nil-index $\leqslant 3$ satisfies the following relation for all $u \in L$:

$$b_0^2(u^2b_0^2)^s = 0; \quad s = [t/2] . \tag{2.7}$$

Descent over t leads us to an element $b_1 \neq 0$ such that $b_1^3 = 0$, $b_1^2u^2b_1^2 = 0$, where u is any element of L. When $b_1^2 \neq 0$, we have at our disposal an improved nil-element of index 3: $c = [ub_1^2] \neq 0$ for some $u \in L$, but $c^2 = b_1^2u^2b_1^2 = 0$. This is what we have been aiming at.

We shall now accomplish our plan and produce the missing details in the process.

2.3. First of all, to check the truth of (2.5) for arbitrary m, we establish a number of auxiliary identities. Replacing u in (1.13) by $[bu^3] = bu^3 - 3ubu^2 + 3u^2b^u - u^3b$, we get that

$$b^2ubu^2b^2 = b^2u^2bub^2 = bub^2u^2b^2 = b^2u^2b^2ub , \tag{2.8}$$

since $b^3 = 0$ and $p \geqslant 5$. Further, replacing u by $[bu^2]$ in (1.14), we get after some easy calculations that

$$\begin{aligned}
[[bu^2]b^2]^2 &= b^2[bu^2]^2b^2 = b^2(bu^2 - 2ubu + u^2b)^2b^2 \\
&= -\underbrace{2b^2ububu^2b^2}_{(1.13)} + \underbrace{4b^2ubu^2}_{(1.13)}\underline{bub^2} - 2b^2u^2\underbrace{bubub^2}_{(1.13)} + b^2u^2b^2u^2b^2 \\
&= -\underbrace{2bubub^2u^2b^2}_{(2.8)} + 4bub^2u^2b^2ub - \underbrace{2b^2u^2b^2ubub}_{(2.8)} + b^2u^2b^2u^2b^2 \\
&= b^2u^2b^2u^2b^2 ,
\end{aligned}$$

whence, since $[bu^2] = -[u[bu]]$, we get

$$[u[bu]b^2]^2 = b^2u^2b^2u^2b^2 . \tag{2.9}$$

Identities (1.14) and (2.9) are the required identity (2.5) for $m = 0$ and $m = 1$ respectively.

2.4. *The following recurrence relation holds for $m \geqslant 2$:*

$$g_m = [g_{m-2}u^2b^2] . \tag{2.10}$$

Proof. When written in the form

$$[bu]b^2 = -b^2[bu] ,$$

the first relation in (1.13) allows us to choose a somewhat different form for the element (2.4), namely: $g_m = (-1)^m[u[bu]^2b^2[bu]^{m-2}] = (-1)^m[g_2[bu]^{m-2}]$. But

$$g_2 = [u[bu]^2b^2] = -[bu^2[bu]b^2] = -[bu^2bub^2] + [bu^3\underline{b^3}]$$
$$= -\underbrace{[bu^2b\ ub^2]}_{\text{(Jacobi)}} = [ub^2u^2b^2] ,$$

so that

$$g_m(u) = (-1)^m[ub^2u^2b^2[bu]^{m-2}] . \tag{2.11}$$

Relations (2.8), supplemented by the two obvious relations

$$b^2u^2b^2[bu] = -b^2u^2b^2ub, \qquad [bu]b^2u^2b^2 = bub^2u^2b^2 ,$$

lead to skew-symmetry:

$$b^2u^2b^2[bu] = -[bu]b^2u^2b^2 ,$$

$(m-2)$-fold application of which to the expression (2.11) puts g_m into the required form: $g_m = (-1)^m(-1)^{m-2}[u[bu]^{m-2}b^2u^2b^2] = [g_{m-2}u^2b^2]$. □

2.5. A nil-element of index 3 reproduces elements like itself as follows:

$$[ub^2]^3 = \underbrace{[ub^2]^2[ub^2]}_{(1.14)} = \underbrace{b^2u^2b^2ub^2}_{(1.13)} = 0 .$$

This gives us the right to replace b in (1.14) by $[ub^2]$, and u by $[u[bu]^{m-2}]$:

$$[u[bu]^{m-2}[ub^2]^2]^2 = [ub^2]^2[u[bu]^{m-2}]^2[ub^2]^2 ,$$

or, by (1.14),

$$[u[bu]^{m-2}b^2u^2b^2]^2 = b^2u^2\underbrace{(b^2[u[bu]^{m-2}]^2b^2)}_{(1.14)}u^2b^2$$
$$= b^2u^2\underbrace{[u[bu]^{m-2}b^2]^2}_{(2.4)}u^2b^2 = b^2u^2g_{m-2}^2u^2b^2 .$$

Thus, by (2.10) we have a recurrence relation

$$(g_m(u))^2 = b^2u^2(g_{m-2}(u))^2u^2b^2 ,$$

that enables us to apply induction on m. As was remarked at the end of 2.3, we have a basis for induction: $g_0^2 = b^2u^2b^2$, $g_1^2 = b^2u^2b^2u^2b^2$. This proves identity (2.5). □

2.6. In accordance with the remark at the end of 2.4, $[bu]b^2 = -b^2[bu]$. Therefore

$$0 = [b[bu]^n] = [b[bu][bu]^{n-1}] = [ub^2[bu]^{n-1}] = (-1)^{n-1}[u[bu]^{n-1}b^2]$$
$$= (-1)^{n-1}g_{n-1}(u)\,,$$

so that $(g_{t-1}(u))^2 = 0$ for some index $t \leqslant n$ (which we take to be the smallest possible). Whether or not $g_{t-1}(u) = 0$ is completely immaterial. An obvious use of (2.5) leads us to identity (2.6).

We have already analyzed the cases $t = 0$, $t = 1$ in 2.2. For $t > 1$, it remains for us now to find an element b_0 satisfying identity (2.7). We choose $f \in L$ and $m \geqslant 0$ such that

$$g_{m+1}(f) = 0, \qquad g_m(f) \neq 0; \qquad \text{we set} \quad a = [f[bf]^m]\,. \tag{2.12}$$

Since $g_0(f) \neq 0$ for some f and $g_{n-1}(u) = 0$ for all $u \in L$, f and m of the required form do in fact exist. Clearly,

$$[ab^2]^3 = 0, \qquad [ab^2] = g_m(f) \neq 0\,.$$

This element $g_m(f)$ is a candidate for the role of b_0.

2.7. *The following relation holds:*

$$[ab^2ab] = 0\,. \tag{2.13}$$

Proof. By (2.11),

$$g_{m+2}(f) = (-1)^{m+2}[fb^2f^2b^2[bf]^m]\,,$$

and

$$g_{m+1}(f) = (-1)^{m+1}[fb^2f^2b^2[bf]^{m-1}]\,.$$

Since $g_{m+1}(f) = 0$ by choice (see (2.12)), we have also that

$$g_{m+2}(f) = -[g_{m+1}(f)\cdot[bf]] = 0\,. \tag{2.14}$$

Further,

$$[ab^2\underline{a}b] = \left[f[bf]^m b^2\left(\sum_{i=0}^{m}(-1)^i\binom{m}{i}[bf]^i f[bf]^{m-i}\right)b\right].$$

For terms with $i > 0$ we have

$$[f[bf]^m b^2[bf]\ldots] = -[f[bf]^{m+1}b^2\ldots] = -[g_{m+1}(f)\ldots] = 0\,,$$

and thus

$$[ab^2ab] = [f[bf]^m b^2 f[bf]^m b]\,.$$

Liberating one of the commutators $[bf]$ standing between b^2f and b on the right-hand side from its brackets, we get

$$[ab^2ab] = [f[bf]^m b^2 f(bf - fb)[bf]^{m-1}b]\,.$$

But $f(bf-fb) = -(bf-fb)f + bf^2 - f^2b$, $b^3 = 0$, so that

$$[ab^2ab] = -[f[bf]^m \underline{b^2[bf]}f\ldots] - [f[bf]^m b^2 f^2 b[bf]^{m-1}b]\,.$$

The first term is $[g_{m+1}(f)f\ldots] = 0$, so that

$$[ab^2ab] = -[f[bf]^m b^2 f^2 b[bf]^{m-1}b]\,.$$

We note at this point that

$$b[bf]^{m-1}b = b^2 A_{m-1}\,,$$

where $A_0 = 1$, and A_{m-1} is the term defined by the inductive relations

$$b[bf]^k b = b^2 f[bf]^{k-1}b - b\underline{fb[bf]^{k-1}b} = b^2(f[bf]^{k-1}b) - \underbrace{bfb^2}_{(1.13)}A_{k-1}\,.$$

Thus

$$A_k = f[bf]^{k-1}b - fbA_{k-1}\,.$$

Therefore

$$[ab^2ab] = -[\underline{f[bf]^m b^2}\,f^2b^2A_{m-1}]$$

$$= -[\underbrace{g_m(f)f^2b^2}_{(2.10)}A_{m-1}] = -[\underbrace{g_{m+2}(f)}_{(2.14)}A_{m-1}] = 0. \qquad \square$$

2.8. Setting $w = [ab^2ab]$, expanding the commutator in $A(L)$ and using (2.13), we get

$$0 = w = b^2a^2b + ba^2b^2 + 2abab^2 + 2b^2aba - 4babab\,.$$

Multiplication of w on the right by b gives

$$0 = wb = b^2a^2b^2 + 2b^2abab - \underbrace{4babab^2}_{(1.13)}$$

$$= b^2a^2b^2 + 2b^2abab - \underbrace{4bab^2ab}_{(1.13)} = b^2a^2b^2 - 2b^2abab\,,$$

that is,

$$b^2a^2b^2 = 2b^2abab\,. \tag{2.15}$$

Similarly,

$$0 = wab^2 = b^2a^2\underbrace{bab^2}_{(1.13)} + ba^2\underbrace{b^2ab^2}_{(1.13)} + 2aba\underbrace{b^2ab^2}_{(1.13)} + 2b^2\underbrace{aba^2b^2}_{(1.13)} - \underbrace{4bababab^2}_{(1.13)}$$

$$= b^2a^2b^2ab + \underbrace{2bab^2a^2b^2}_{(2.8)} - \underbrace{4b^2ababab}_{(2.15)}\,.$$

We thus get the relation

$$b^2a^2b^2ab = 0\,. \tag{2.16}$$

2.9. We have now reached the concluding stage in the proof of Theorem 2.1. Let h be any element of L. Repeated application of (1.13) to whb^2 gives

$$0 = whb^2 = b^2a^2\underline{bhb^2} + 2b^2abahb^2 - 4baba\underline{bhb^2}$$

$$= b^2a^2b^2hb + 2bab^2ahb^2 - \underset{(2.15)}{\underline{4b^2ababhb}} = 2bab^2ahb^2 - b^2a^2b^2hb\ ,$$

whence it follows that

$$0 = bawhb^2 = \underset{(1.13)}{\underline{2babab^2ahb^2}} - \underset{(2.8)}{\underline{bab^2a^2b^2hb}} = \underset{(2.15)}{\underline{2b^2ababahb^2}} - \underset{(2.16)}{\underline{b^2a^2b^2abhb}}\ ,$$

that is,

$$b^2a^2b^2ahb^2 = 0 = b^2a^2b^2hab^2 \tag{2.17}$$

(here we have used the identity $0 = \underset{(1.13)}{\underline{b^2[ah]b^2}} = b^2ahb^2 - b^2hab^2$) .

Similarly, by considering the equation $0 = b^2hwab$, we get that

$$b^2hab^2a^2b^2 = 0 = b^2ahb^2a^2b^2\ . \tag{2.18}$$

Now set

$$b_0 = g_m(f) = [ab^2] \neq 0\ .$$

We already know that $b_0^2 = [ab^2]^2 = b^2ab^2$, $b_0^3 = 0$. We introduce the monomial

$$H = b_0^2(h^2b_0^2)^s = b^2a^2b^2(h^2b^2a^2b^2)^s, \qquad s = [t/2]\ ,$$

into the discussion, where h is an arbitrary element of L. We use the fact that h can be anything, and include H in a system of monomials connected by a non-degenerate system of linear equations. To this end, for convenience we set

$$\delta = 2s + 1 - t = 0 \quad \text{or } 1\ ,$$

and, on setting $u = h + \lambda a$ in (2.6), where h is any element of L and λ runs over the prime subfield $F_0 \subset F$, we get a system

$$(b^2a^2)^\delta b^2\{(h + \lambda a)^2b^2\}^t = 0, \qquad \lambda \in F_0\ ,$$

that is,

$$(b^2a^2)^\delta b^2\{h^2b^2 + \lambda(ah + ha)b^2 + \lambda^2a^2b^2\}^{2s+1-\delta} = 0\ . \tag{2.19}$$

Multiplying the complete decomposition of the left-hand side together over all the powers of λ, we see that that the coefficient of λ^k will appear as a homogeneous element $H_{2\delta+k}$ of $A(L)$ of degree $2\delta + k$ in a and $2(2s + 1 - \delta) - k$ in h:

$$H_{2\delta} + \lambda H_{2\delta+1} + \cdots + \lambda^{2(s+1-\delta)}H_{2s+2} + \cdots = 0\ .$$

For $\lambda = 0$ we obtain $H_{2\delta} = 0$, which is clear anyway since

$$H_{2\delta} = \underset{(2.6)}{\underline{(b^2a^2)^\delta b^2(h^2b^2)^t}} = 0\ .$$

We claim that

$$H_{2s+2} = H \quad \text{and} \quad H_k = 0, \quad \text{if} \quad k > 2s + 2\ .$$

For definiteness, suppose that $\delta = 0$. We are interested in any summand T of H_{2s+2} of degree $2s$ in h. Clearly, the factor $(ah + hb)b^2$ of degree 1 in h occurs an even number of times in T, $2r$ times shall we say. This means that T has $s-r$ occurrences of h^2b^2 and $2s + 1 - (s - r) - 2r = s - r + 1$ occurrences of a^2b^2. But $T = 0$ if a^2b^2 or $(ah + ha)b^2$ stands in front of an a^2b^2, that is, if there is an occurrence of the factors

$$\underbrace{b^2a^2b^2a^2b^2 = 0}_{(2.18):h=a} \quad \text{or} \quad \underbrace{b^2(ah + ha)b^2a^2b^2 = 0}_{(2.18)} \,,$$

in T; since the number of factors h^2b^2 is strictly less than the number of a^2b^2, only summands of the form

$$T = b^2a^2b^2 \ldots h^2b^2a^2b^2 \ldots h^2b^2a^2b^2 \ldots h^2b^2a^2b^2$$

can make a non-zero contribution to H_{2s+2}, where the dots stand for products of factors $(ah + ha)b^2$. However, every occurrence of $(ah + ha)b^2$ is in fact excluded, since it implies the existence in T of a factor

$$\underbrace{b^2a^2b^2(ah + ha)b^2 = 0}_{(2.17)} \,.$$

We have come to the conclusion that $r = 0$ and that

$$H_{2s+2} = b^2a^2b^2(h^2b^2a^2b^2)^s = H \,.$$

For this same reason, all homogeneous components H_k, $k > 2s + 2$, are identically zero.

The case $\delta = 1$ is to be considered in a completely analogous way.

Thus, after cancellation by $\lambda \neq 0$, we arrive at a homogeneous system

$$H_{2\delta+1} + \lambda H_{2\delta+2} + \cdots + \lambda^{2(s-\delta)} H_{2s+1} + \lambda^{2(s-\delta)+1} H = 0 \,,$$

$$\lambda = 1, 2, \ldots, 2(s + 1 - \delta) \,,$$

with Vandermonde determinant

$$\Delta = \prod_{2(s+1-\delta) \geqslant i > j \geqslant 1} (i - j) \,.$$

The largest prime factor of Δ does not exceed

$$2(s + 1 - \delta) - 1 = 2s + 1 - 2\delta = t - \delta \leqslant t \leqslant n < p \,.$$

Thus $\Delta \not\equiv 0 \pmod{p}$, and

$$H = b_0^2(h^2b_0^2)^s = 0 \,.$$

This concludes the proof of Theorem 2.1. □

§ 3. Descent to Thin Sandwiches (the Case $p \gg n$)

If the ground field F has zero characteristic, or characteristic that is sufficiently large in comparison with n, the proof that a Lie algebra L with E_n has a sandwich is very much simpler. Before this assertion materializes, we look at the ideal $E_{n-1}(L)$

of L generated by all the elements $[uv^{n-1}]$, where $u, v \in L$. It is fairly clear from what was said in Chap. 1 that

$$E_{n-1}(L) = \langle [uv^{n-1}] | u, v \in L \rangle_F$$

is the linear span over F of the elements $[uv^{n-1}]$. This is graphically illustrated by the formula

$$[uv^{n-1}w] = [uwv^{n-1}] + \sum_i \alpha_i [uz_i^{n-1}],$$

which follows easily from (1.2):

$$v^{n-1}w - wv^{n-1} = \sum_{i=0}^{n-2} v^i [vw] v^{n-2-i}$$

$$= \begin{Bmatrix} [vw] & v \\ 1 & n-2 \end{Bmatrix} = \sum_i \alpha_i z_i^{n-1}, \qquad z_i \in L.$$

It turns out that $E_{n-1}(L)$ is a sandwich ideal for $p \gg n$. More exactly:

3.1. Theorem. *For every pair of elements u, v of an n-Engel Lie algebra L over a field F of characteristic $p > n + [n/2]$,*

$$[uv^{n-1}]^2 = 0.$$

Proof. We shall need determinants of the form

$$\Delta(\nu, \mu) = \begin{vmatrix} (-1)^\nu \binom{n}{\nu} \cdots & (-1)^{\nu+\mu} \binom{n}{\nu+\mu} \\ \cdots\cdots & \cdots\cdots \\ (-1)^{\nu+\mu} \binom{n}{\nu+\mu} \cdots & (-1)^{\nu+2\mu} \binom{n}{\nu+2\mu} \end{vmatrix}$$

$$= (-1)^{\nu(\mu+1)+\binom{\mu+1}{2}} \frac{\binom{n+\mu}{\mu+1}\binom{n+\mu-1}{\mu+1}\cdots\binom{n-\nu+1}{\mu+1}}{\binom{\nu+2\mu}{\mu+1}\binom{\nu+2\mu-1}{\mu+1}\cdots\binom{\mu+1}{\mu+1}}.$$

Clearly, $\Delta(\nu, \mu) \not\equiv 0 \pmod p$ if $p > n + \mu$.

By (1.5) we have

$$[uv^{n-1}]^2 = \sum_{i,j=0}^{n-1} (-1)^{i+j} \binom{n-1}{i}\binom{n-1}{j} v^i uv^{n-1-i+j} uv^{n-1-j}.$$

Therefore, the theorem follows from the equations

$$A_{i,j} = v^i uv^{n-1-i+j} uv^{n-1-j} = 0, \qquad 0 \leqslant j \leqslant i \leqslant n-1,$$

which we shall now verify. Since $v^{n-1}uv^{n-1} = 0$ (see (2.3)), $A_{0,0} = 0$. Assume we have proved that $A_{i,j} = 0$ for $i < k$ and all $j \leqslant i$; we shall prove it for $i = k$.

Multiplying identity (1.9) on the left by $v^k u v^{n-1-k-s}$ and on the right by v^s, we get the following system of identities (in which $j = i - s$):

$$\sum_{j=0}^{k} (-1)^{j+s+1} \binom{n}{j+s+1} A_{k,j} = 0, \quad 0 \leqslant s \leqslant n-1-k .$$

When $n - k \geqslant k + 1$, we get what is required from the first $k + 1$ equations of the system: $A_{k,0} = \cdots A_{k,k} = 0$. This is indeed the case since $\Delta(1, k) \neq 0$.

Suppose now that $n - k < k + 1$. We express $A_{k,0}, \ldots, A_{k,n-1-k}$ in terms of the $A_{k,r}$ with $r \geqslant n - k$, using the entire system of identities introduced above and the fact that $\Delta(1, n-1-k) \neq 0$. Further, multiplying (1.9) on the left by v^s and on the right by $v^{r-s} u v^{n-1-r}$, we get yet another system of identities, in which $j = i$:

$$\sum_{j=0}^{n-1} (-1)^{j+1} \binom{n}{j+1} A_{s+j,r} = 0, \quad s = 0, 1, \ldots, n-1-k ,$$

and which we can write in the form

$$\sum_{j=0}^{n-1-k} (-1)^{k+i+1-s} \binom{n}{k+i+1-s} A_{k+i,r} = 0, \quad s = n-k-1, \ldots, 1, 0 .$$

For each fixed r, $n - k \leqslant r \leqslant k$, we get a homogeneous linear system in the $n - k$ unknowns $A_{k,r}, A_{k+1,r}, \ldots, A_{n-1,r}$ with determinant $\Delta(2k + 2 - n, n - 1 - k) \neq 0$. In particular, we get the necessary equations $A_{k,r} = 0$, thus completing the inductive step.

All we have to do now is to observe that all determinants of the type $\Delta(v, \mu)$ encountered during the proof were such that μ does not exceed $[n/2]$, so that the prime $p > n + n/2$ cannot occur in their canonical decompositions. □

3.2. Corollary. *The following assertions are equivalent:*

(i) *Every Lie algebra L with* E_n *over a field F of characteristic $p > n + [n/2]$ is locally nilpotent.*

(ii) *Every sandwich algebra with* E_n *over a field F of characteristic $p > n + [n/2]$ is locally nilpotent.*

Proof. Implication (i) ⇒ (ii) is trivial. Assume now that (ii) holds. To prove (i), we can assume at the outset that L has trivial locally nilpotent radical: $R(L) = 0$. It follows from Theorem 3.1 that the ideal $E_{n-1}(L)$ introduced at the beginning of this section is a sandwich ideal. By (ii), it is locally nilpotent, and $E_{n-1}(L) \neq 0$, else L would satisfy E_{n-1}. This contradiction establishes implication (ii) ⇒ (i). □

§ 4. Descent from $\mathfrak{C}_2^*$ to $\mathfrak{C}_{(p-3)/2}^*$

In this section L is a completely arbitrary Lie algebra over a field F of characteristic p. The definition of the set $\mathfrak{C}_m^* = \mathfrak{C}_m^*(L)$ is given in 1.5.4.

4.1. Lemma. *If* $c \in \mathfrak{C}_m^*$, *then*

$$cx_1x_2 \ldots x_{2m}c = cx_{\pi(1)}x_{\pi(2)} \ldots x_{\pi(2m)}c$$

for every permutation $\pi \in S_{2m}$.

Proof. In fact,

$$cx_1 \ldots x_ix_{i+1} \ldots x_{2m}c = cx_1 \ldots x_{i+1}x_i \ldots x_{2m}c + cx_1 \ldots [x_ix_{i+1}] \ldots x_{2m}c .$$
$$= cx_1 \ldots x_{i+1}x_i \ldots x_{2m}c ,$$

since $cy_1 \ldots y_kc = 0$, $k \leqslant 2m - 1$, for any $y_i \in L$. □

4.2. Lemma. *Suppose that* $c \in \mathfrak{C}_m^*(L)$, $1 \leqslant m \leqslant \frac{p-3}{2}$. *If*

$$c_0 = [ca^{2m+1}c] \neq 0$$

for some $a \in L$, *then* $c_0 \in \mathfrak{C}_r^*(L)$ *for* $r \geqslant m$.

Proof. By assumption, $cu^kc = 0$ for $0 \leqslant k \leqslant 2m - 1$, or, as is the same thing (see 1.5.3), $cx_1x_2 \ldots x_kc = 0$. Thus, we have in associative form:

$$c_0 = [ca^{2m+1}c] = 2ca^{2m+1}c - (2m+1)aca^{2m}c - (2m+1)ca^{2m}ca , \qquad (2.20)$$

which means that the identities $c_0u^kc_0 = 0, 0 \leqslant k \leqslant 2m - 3$, are obvious corollaries of the definition of c. Furthermore, (2.20) gives

$$c_0u^{2m-2}c_0 = (2m+1)^2 \underbrace{ca^{2m}cau^{2m-2}aca^{2m}c}_{\text{(Lemma 4.1)}} = (2m+1)^2ca^{2m}ca^2u^{2m-2}ca^{2m}c .$$

But

$$\binom{2m+2}{2} ca^{2m}ca^2u^{2m-2}c = c[ca^{2m+2}]u^{2m-2}c = cx_1 \ldots x_{2m-1}c = 0 ,$$

where $x_1 = [ca^{2m+2}]$, $x_2 = \cdots = x_{2m-1} = u$. Since $2m + 2 < p$, we have $ca^{2m}u^{2m-2}c^2 = 0$ so that $c_0u^{2m-2}c_0 = 0$. The identity $c_0u^{2m-1}c_0 = 0$ is automatic (see Proposition 1.5.3). □

The very method of producing sandwiches given in the lemma makes them particularly effective for constructing various identities. The elementary results of this section serve as good illustrations of the technical device of piling up identical elements next to each other. One has simply to introduce the sandwich

$$c_1 = [c_0b^{2m+1}c_0] = [ca^{2m+1}cb^{2m+1}[ca^{2m+1}c]] , \qquad (2.21)$$

to cause the hope to emerge that we will be able to jump sandwiches from $\mathfrak{C}_m^*$ to $\mathfrak{C}_{m+1}^*$. We shall see that this hope is not groundless.

4.3. We amass a *bank of auxiliary identities. Suppose that* $c \in \mathfrak{C}_m^*$, $m \geqslant 2$. *Then*

$$cvu^{2m-1+s}c = \sum_{i=0}^{s} \alpha_i cu^{2m-1+s-i}vu^ic, \qquad \alpha_i \in F ; \qquad (2.22)$$

$$cu^{2m-1+s}vc = \sum_{i=0}^{s} \beta_i cu^i cu^{2m-1+s-i}c, \qquad \beta_i \in F\,; \tag{2.23}$$

$$A(u) = ca^{2m}ca^{2m-1}uc = 0; \qquad \bar{A}(u) = cua^{2m-1}ca^{2m}c = 0\,; \tag{2.24}$$

$$B_1(u) = ca^{2m}ca^{2m}uc = 0; \qquad \bar{B}_1(u) = cua^{2m}ca^{2m}c = 0\,; \tag{2.25}$$

$$B_2(u) = ca^{2m+1}ca^{2m-1}uc = 0; \qquad \bar{B}_2(u) = cua^{2m-1}ca^{2m+1}c = 0\,. \tag{2.26}$$

In what follows, the symbols $X = x_1x_2 \ldots x_{k-1}x_k$ and $\bar{X} = x_kx_{k-1} \ldots x_2x_1$ will often arise together; it should also be noted that the coefficients α_i, β_i in (2.22) and (2.23) depend on s.

Proof. By (1.14), we have

$$vu^{2m-1+s} = u^{2m-1+s}v + \sum_{i=1}^{2m-1+s} \binom{2m-1+s}{i} u^{2m-1+s-i}[vu^i]\,.$$

Therefore, since $c \in \mathfrak{C}_m^*$, we find that

$$cvu^{2m-1+s}c = cu^{2m-1+s}vc + \sum_{i=1}^{s} \binom{2m-1+s}{i} cu^{2m-1+s-i}[vu^i]c\,,$$

which is obviously a sharpened form of identity (2.22) (we get an explicit form of (2.22) by expanding commutators using (1.5)). The verification of identity (2.23) is completely analogous. The first of the identities in (2.24) is valid since

$$\binom{2m+2}{2} A(u) = c[ca^{2m+2}]a^{2m-3}uc = cx_1x_2 \ldots x_{2m-1}c = 0\,.$$

The symmetrical relation $\bar{A}(u) = 0$ is found from the same sort of arguments. Finally, identities (2.25) and (2.26) follow from the homogeneous linear system

$$\binom{2m+3}{3+\delta} B_1(u) - \binom{2m+3}{2+\delta} B_2(u) = (-1)^\delta ca^\delta[ca^{2m+3}]a^{2m-3-\delta}uc = 0\,,$$
$$\delta = 0, 1\,,$$

with determinant $\Delta = -\dfrac{(2m+4)(2m+3)^2(2m+2)^2(2m+1)}{144} \neq 0$ (and also from the symmetrical system). □

4.4. Lemma. *The identities* $c_0u^{2m}ca^{2m}c = 0$, $ca^{2m}cu^{2m}c_0 = 0$ *hold for all* $a, u \in L$, *where* $c_0 = [ca^{2m+1}c]$, $c \in \mathfrak{C}_m^*$, $m \geqslant 2$.

Proof. We set

$$[ca^{2m+1}c]u^{2m}ca^{2m}c = K_1 - K_2\,,$$

where $K_1 = [ca^{2m+1}]\, cu^{2m}ca^{2m}c$, $K_2 = c[ca^{2m+1}]u^{2m}ca^{2m}c$. Since $c \in \mathfrak{C}_m^*$, using (1.5) we get

$$cu^{2m}c = c[cu^{2m}] = [cu^{2m}]c\,.$$

Therefore

$$
\begin{aligned}
K_1 &= [ca^{2m+1}]\underbrace{c[cu^{2m}]a^{2m}c}_{(2.22)} \\
&= \alpha_0[ca^{2m+1}]ca^{2m}\underbrace{[cu^{2m}]c}_{(1.5)} + \alpha_1[ca^{2m+1}]ca^{2m-1}\underbrace{[cu^{2m}]ac}_{(1.5)} \\
&= \alpha_0\underbrace{[ca^{2m+1}]ca^{2m}}_{(1.5)}cu^{2m}c - 2m\alpha_1\underbrace{[ca^{2m+1}]ca^{2m-1}}_{(1.5)}ucu^{2m-1}ac \\
&= \alpha_0\underbrace{B_2(a)}_{(2.26)}u^{2m}c - (2m+1)\alpha_0\underbrace{aA(a)}_{(2.24)}u^{2m}c - 2m\alpha_1\underbrace{B_2(u)}_{(2.26)}u^{2m-1}ac \\
&\quad + 2m(2m+1)\alpha_1\underbrace{aA(u)}_{(2.24)}u^{2m-1}ac = 0 .
\end{aligned}
$$

Similarly,

$$
\begin{aligned}
K_2 &= \alpha_0\underbrace{cu^{2m}[ca^{2m+1}]ca^{2m}c}_{(1.5)} + \alpha_1 cu^{2m-1}\underbrace{[ca^{2m+1}]uca^{2m}c}_{(1.24)} = \alpha_0 cu^{2m}\underbrace{B_2(a)}_{(2.26)} \\
&\quad - (2m+1)\alpha_0 cu^{2m}\underbrace{aA(a)}_{(2.24)} - (2m+1)\alpha_1 cu^{2m-1}\underbrace{aca^{2m}uca^{2m}c}_{(2.23)} \\
&\quad + (2m+1)\binom{2m+1}{2}\alpha_1 cu^{2m-1}\underbrace{a^2ca^{2m-1}uca^{2m}c}_{(2.23)} = \alpha_1' cu^{2m-1}\underbrace{a\bar{B}_1(u)}_{(2.25)} \\
&\quad + \alpha_2' cu^{2m-1}\underbrace{a\bar{A}([au])}_{(2.24)} + \alpha_3' cu^{2m-1}\underbrace{a^2\bar{A}(u)}_{(2.24)} = 0 .
\end{aligned}
$$

We can safely omit the verification of the symmetrical relation $ca^{2m}cu^{2m}c_0 = 0$. □

4.5. Lemma. *Every element* $c_0 = [ca^{2m+1}c]$, $c \in \mathfrak{C}_m^*$, $m \geqslant 2$, *yields the identity*

$$T = c_0u^{2m}c_0v^{2m}c_0 = 0, \qquad u, v \in L .$$

Proof. By (2.20) we have

$$
\begin{aligned}
T &= 2c_0u^{2m}ca^{2m+1}cv^{2m}c_0 - (2m+1)c_0u^{2m}\underbrace{aca^{2m}cv^{2m}c_0}_{(\text{Lemma 4.4})} \\
&\quad - (2m+1)\underbrace{c_0u^{2m}ca^{2m}c}_{(\text{Lemma 4.4})}av^{2m}c_0 = 2c_0u^{2m}ca^{2m+1}cv^{2m}c_0 = 2Pv^{2m}c_0 ,
\end{aligned}
$$

where $P = c_0u^{2m}ca^{2m+1}c$. Using (2.20) again, we rewrite P in the form $P = P_1 - P_2 - (2m+1)aP_3$, where

$$
\begin{aligned}
P_1 &= ca^{2m+1}cu^{2m}ca^{2m+1}c , \\
P_2 &= c[ca^{2m+1}]u^{2m}ca^{2m+1}c , \\
P_3 &= ca^{2m}cu^{2m}ca^{2m+1}c .
\end{aligned}
$$

We transform P_3 first:

$$\begin{aligned}P_3 &= \underbrace{ca^{2m}[cu^{2m}]ca^{2m+1}c}_{(2.23)}\\ &= \beta_0 c[cu^{2m}]a^{2m}ca^{2m+1}c + \beta_1 ca[cu^{2m}]a^{2m-1}ca^{2m+1}c\\ &= \beta_0 cu^{2m}\underbrace{ca^{2m}ca^{2m+1}c}_{(2.26)} - 2m\beta_1 cau^{2m-1}\underbrace{cua^{2m-1}ca^{2m+1}c}_{(2.26)}\\ &= \beta_0 cu^{2m}\bar{B}_2(a) - 2m\beta_1 cau^{2m-1}\bar{B}_2(u) = 0\,.\end{aligned}$$

In investigating P_1 and P_2, we attempt to transfer as many occurrences of a as possible from the left-hand interval between adjacent occurrences of c to the middle portion of the monomial:

$$\begin{aligned}P_1 &= \underbrace{ca^{2m+1}[cu^{2m}]ca^{2m+1}c}_{(2.23)} = \beta_0' c[cu^{2m}]a^{2m+1}ca^{2m+1}c\\ &\quad + \beta_1' ca[cu^{2m}]a^{2m}ca^{2m+1}c + \beta_2' ca^2[cu^{2m}]a^{2m-1}ca^{2m+1}c\\ &= \beta_1'' cu^{2m}ca^{2m+1}ca^{2m+1}c + \beta_2'' cu^{2m-1}acua^{2m}ca^{2m+1}c\\ &\quad + \beta_3'' cu^{2m-2}a^2cu^2a^{2m-1}ca^{2m+1}c + \beta_4'' cau^{2m}\underbrace{ca^{2m}ca^{2m+1}c}_{(2.26)}\\ &\quad + \beta_5'' ca^2u^{2m-1}\underbrace{cua^{2m-1}ca^{2m+1}c}_{(2.26)}\,.\end{aligned}$$

Denoting the monomial with coefficient β_i'' by Q_i, we see that $Q_4 = Q_5 = 0$. Moreover, setting

$$Q = cu^{2m-2}a^3cu^2a^{2m-2}ca^{2m+1}c\,,$$

we take advantage of the relation

$$\begin{aligned}0 &= \underbrace{cu^{2m-2}[cu^2a^{2m+1}]}ca^{2m+1}c = cu^{2m-2}[cu^2]a^{2m+1}ca^{2m+1}c\\ &\quad - (2m+1)cu^{2m-2}a[cu^2]a^{2m}ca^{2m+1}c\\ &\quad + \binom{2m+1}{2} cu^{2m-2}a^2[cu^2]a^{2m-1}ca^{2m+1}c\\ &\quad - \binom{2m+1}{3} cu^{2m-2}a^3[cu^2]a^{2m-2}ca^{2m+1}c\\ &= Q_1 + 2(2m+1)\underbrace{cu^{2m-2}aucua^{2m}ca^{2m+1}c}_{(\text{Lemma } 4.1)}\\ &\quad - (2m+1)cu^{2m-2}au^2\underbrace{ca^{2m}ca^{2m+1}c}_{(2.26)}\\ &\quad + \binom{2m+1}{2}Q_3 - 2\binom{2m+1}{2} cu^{2m-2}a^2u\underbrace{cua^{2m-1}ca^{2m+1}c}_{(2.26)}\\ &\quad - \binom{2m+1}{3}Q = Q_1 + 2(2m+1)Q_2 + \binom{2m+1}{2}Q_3 - \binom{2m+1}{3}Q\,.\end{aligned}$$

By expressing Q_3 in terms of Q_1, Q_2 and Q, we can convince ourselves that $P_1 = \gamma Q + \gamma_1 Q_1 + \gamma_2 Q_2$. If we take into account the relation

$$0 = cu^{2m-2}a^3cu^2[ca^{2m+3}]a^{2m-4}c = \binom{2m+3}{5}Q - \binom{2m+3}{4}Q_0,$$

where

$$Q_0 = cu^{2m-2}a^3cu^2a^{2m-1}ca^{2m}c,$$

we get finally that $P_1 = \gamma_0 Q_0 + \gamma_1 Q_1 + \gamma_2 Q_2$.

The situation with P_2 is much the same, but simpler:

$$\begin{aligned} P_2 &= c\underbrace{[ca^{2m+1}]u^{2m}}_{(2.22)}ca^{2m+1}c = \alpha_0 cu^{2m}[ca^{2m+1}]ca^{2m+1}c \\ &\quad + \alpha_1 cu^{2m-1}[ca^{2m+1}]uca^{2m+1}c \\ &= \alpha_0 Q_1 - (2m+1)\alpha_0 cu^{2m}\underbrace{aca^{2m}ca^{2m+1}c}_{(2.26)} \\ &\quad - (2m+1)\alpha_1 cu^{2m-1}\underbrace{aca^{2m}u}_{(2.23)}ca^{2m+1}c \\ &\quad + \binom{2m+1}{2}\alpha_1 cu^{2m-1}\underbrace{a^2ca^{2m-1}uca^{2m+1}c}_{(2.26)} \\ &= \alpha_0 Q_1 - (2m+1)\alpha_1\beta_0 Q_2 - (2m+1)\alpha_1\beta_1 cu^{2m-1}\underline{\underline{acaua^{2m-1}ca^{2m+1}c}} \\ &= \delta_1 Q_1 + \delta_2 Q_2 . \end{aligned}$$

Since $P = P_1 - P_2$, the expressions obtained for the P_i show that $P = \lambda_0 Q_0 + \lambda_1 Q_1 + \lambda_2 Q_2$, or, to stress the dependence of $P = c_0 u^{2m} ca^{2m+1}c$ on the element u of L,

$$P(u) = \lambda_0 Q_0(u) + \lambda_1 Q_1(u) + \lambda_2 Q_2(u) .$$

It is clear that the very same arguments lead us to the symmetrical formula

$$\bar{P}(v) = ca^{2m+1}cv^{2m}c_0 = \mu_0 \bar{Q}_0(v) + \mu_1 \bar{Q}_1(v) + \mu_2 \bar{Q}_2(v) ,$$

where

$$\bar{Q}_0(v) = ca^{2m}ca^{2m-1}v^2ca^3v^{2m-2}c ,$$

$$\bar{Q}_1(v) = ca^{2m+1}ca^{2m+1}cv^{2m}c ,$$

$$\bar{Q}_2(v) = ca^{2m+1}ca^{2m}vcav^{2m-1}c.$$

Let us return to the original monomial T. On the one hand,

$$\begin{aligned} T = 2P(u)v^{2m}c_0 &= 2\lambda_0 cu^{2m-2}a^3c \ldots \underbrace{ca^{2m}cv^{2m}c_0}_{\text{(Lemma 4.4)}} \\ &\quad + 2\lambda_1 c \ldots ca^{2m+1}\underline{ca^{2m+1}cv^{2m}c_0} + 2\lambda_2 c \ldots cua^{2m}\underline{ca^{2m+1}cv^{2m}c_0} \\ &= 2\lambda_1 c \ldots ca^{2m+1}\bar{P}(v) + 2\lambda_2 c \ldots cua^{2m}\bar{P}(v) . \end{aligned}$$

On the other hand, using the expression for $\bar{P}(v)$ in terms of the $\bar{Q}_i(v)$ together with the equation

$$cua^{2m}\bar{Q}_0(v) = \underbrace{cua^{2m}ca^{2m}c}_{(2.25)} \ldots = 0\,,$$

we conclude that T is a linear combination of elements of the form

$$R = c \ldots cxa^{2m}ca^{2m+1}ca^{2m}yc \ldots c\,,$$

where $x = a$ or u, $y = a$ or v. It is impossible to shift the segment $ca^{2m+1}c$ in the factor $R' = cxa^{2m}ca^{2m+1}ca^{2m}yc$ without producing zero. Indeed, if $i > 0$, we have

$$\underbrace{cxa^{2m-i}ca^{2m+1}ca^{2m+i}yc}_{(2.26)} = 0, \qquad cxa^{2m+i}\underbrace{ca^{2m+1}ca^{2m-i}yc}_{(2.26)} = 0\,.$$

By similar considerations,

$$cxa^{2m+1}\underbrace{ca^{2m}ca^{2m}yc}_{(2.25)} = 0, \qquad cxa^{2m+2}\underbrace{ca^{2m}ca^{2m-1}yc}_{(2.24)} = 0\,,$$

$$\underbrace{cxa^{2m}ca^{2m}}_{(2.23)}ca^{2m+1}yc = 0, \qquad \underbrace{cxa^{2m-1}ca^{2m}}_{(2.24)}ca^{2m+2}c = 0\,.$$

Thus

$$0 = \underbrace{cxa^{2m-4}[ca^{2m+1}ca^{2m+4}]yc}_{(\text{Definition of } c\in\mathfrak{C}_m^*)}$$

$$= \binom{2m+4}{4} cxa^{2m}\underbrace{[ca^{2m+1}c]\,a^{2m}yc}_{(2.20)} + \cdots = 2\binom{2m+4}{4} R' + \cdots\,,$$

where the dots denote all possible monomials other than R' containing internal segments $ca^{2m+1}c$, $ca^{2m}c$. As was remarked above, all such monomials are zero. Moreover, $2\binom{2m+4}{4} \neq 0$ since $p > 2m+3 \Rightarrow p > 2m+4$. We conclude that $R' = 0$, and thus $R = 0$ and $T = 0$. □

4.6. Theorem. *The descent from $\mathfrak{C}_m^*$, $7 \leqslant 2m+3 < p$, to $\mathfrak{C}_{m+1}^*$ is always possible in every Lie algebra L over a field of characteristic $p > 7$. In other words, if L has a sandwich of thickness greater than 1, then L contains a sandwich of thickness at least $p-4$.*

Proof. Since we are talking about the descent from $\mathfrak{C}_2^*$ to $\mathfrak{C}_{\frac{p-3}{2}}^*$, the assertion of the theorem is empty for $p = 7$.

Suppose that we are given an element $c \in \mathfrak{C}_m^*$, $7 \leqslant 2m+3 < p$. Assuming that the product $cu^{2m}c$ is not zero for some $u \in L$ (otherwise $c \in \mathfrak{C}_k^*$, $k \geqslant m+1$), for some $a \in L$ we get a new element $c_0 = [ca^{2m+1}c] \neq 0$, which by Lemma 4.2 has all the properties of c. However, in addition to this, we have the identity $c_0u^{2m}c_0v^{2m}c_0 = 0$ (Lemma 4.5) for c_0. If $c_0 \notin \mathfrak{C}_{m+1}^*$, then any non-zero element of the form $c_1 = [c_0b^{2m+1}c_0]$ is what is required in every case; that is, $c_1u^{2m}c_1 = 0$ for all $u \in L$. Let us check this.

For ease of reference, we shall change notation somewhat, assuming at the outset that

$$c_0 = [ca^{2m+1}c] \neq 0; \quad cu^kc = 0, \quad 0 \leqslant k \leqslant 2m-1; \quad cu^{2m}cv^{2m}c = 0\,. \tag{2.27}$$

The linearizations of the identities in (2.27) are also to be included there. We have to show that

$$c_0u^{2m}c_0 = 0\,. \tag{2.28}$$

For this, we need identities involving c_0 and c:

$$c_0u^{2m}ca^{2m}c = 0, \quad ca^{2m}cu^{2m}c_0 = 0\,, \tag{2.29}$$

whose truth is asserted in Lemma 4.4. Moreover,

$$c_0u^kc = cu^kc_0 = 0, \quad k < 2m; \quad c_0u^{2m}c = cu^{2m}c_0\,, \tag{2.30}$$

since this follows from the simple calculations

$$\underbrace{c_0}_{(2.20)}\, u^kc = 2ca^{2m+1}\underbrace{cu^kc}_{(2.27)} - (2m+1)\underbrace{ca^{2m}cau^kc}_{(2.27)} = 0\,,$$

$$c_0u^{2m}c - cu^{2m}c_0 = [c_0u^{2m}c] = -\underbrace{[uc_0u^{2m-1}c]}_{(2.30)} = 0\,.$$

We set $A_\delta = cu^{2m}ca^{2m+1+\delta}ca^{2m-1-\delta}c$ for $\delta = 0$ and 1. Since

$$\underbrace{cu^{2m}ca^\delta[ca^{2m+3}]a^{2m-1-\delta}c = 0}_{(2.27)}\,,$$

we have a new linear homogeneous system

$$\binom{2m+3}{2}A_0 - (2m+3)A_1 = 0\,,$$

$$\binom{2m+3}{3}A_0 - \binom{2m+3}{2}A_1 = 0$$

with determinant

$$\Delta = -\frac{1}{3}\binom{2m+3}{2}\binom{2m+4}{2} \neq 0\,.$$

Thus

$$A_\delta = 0\,. \tag{2.31}$$

Finally,

$$\begin{aligned}
(m+1)\,\underbrace{c_0u^{2m}aca^{2m}c}_{(2.29)} &= (m+1)\,\underbrace{c_0}_{(2.20)}\,au^{2m}ca^{2m}c \\
&= (m+1)\{2ca^{2m+1}cau^{2m}ca^{2m}c - (2m+1)ca^{2m}ca^2u^{2m}ca^{2m}c\} \\
&= -\underbrace{c[ca^{2m+2}]u^{2m}ca^{2m}c}_{(2.27)} = -cu^{2m}[ca^{2m+2}]ca^{2m}c \\
&= -\underbrace{A_1}_{(2.31)} + \underbrace{(2m+2)cu^{2m}a\bar{B}_1(a)}_{(2.27)} + \underbrace{\cdots}_{(2.27)} = 0\,.
\end{aligned}$$

Thus

$$c_0 u^{2m} a c a^{2m} c = 0 . \tag{2.32}$$

Everything is now ready to conclude the proof:

$$\underset{(2.20)}{\underline{c_0 u^{2m} \quad c_0}} = \underset{(2.30)}{\underline{2c_0 u^{2m} c a^{2m+1} c}} - (2m+1)\underset{(2.32)}{\underline{c_0 u^{2m} a c a^{m2} c}} = 2cu^{2m} c_0 a^{2m+1} c$$

$$= \mu_0 A_0 + \mu_1 A_1 + \mu_2 c u^{2m} \underset{(2.26)}{\underline{a c a^{2m} c a^{2m+1} c}} = 0 .$$

This establishes the truth of identity (2.28). □

4.7. Corollary. *Every Lie algebra L with E_n, $n < p$, and possessing a thick sandwich has a non-zero abelian ideal.*

The *proof* follows immediately from Theorem 4.6 and the considerations of 1.5.5. □

§ 5. Commentary

The account of the material in the first two sections is fairly transparent in the author's original paper [137], and was improved slightly in J. Wiegold's lectures [273]. Incidentally, a very short and convenient approach to the locally nilpotent radical was developed there in collaboration with N. Blackburn, B. H. Neumann, D. Held and M. F. Newman; we have followed this in § 3 of Chap. 1.

The contents of § 3 are taken from [135], which in its time stimulated work on completing the proof of Theorem 1.7.1 (in the case of a Lie alebra with E_{p-1}).

Theorem 4.6, although not all that complicated, puts a firm foundation under the theory of sandwiches. Because of it, the accomplishment of the descent from $\mathfrak{C}_1^*$ to $\mathfrak{C}_2^*$ looks like a pure technicality, perhaps more difficult than the proof of Theorem 4.6 itself (in fact, it is considerably more difficult) and requiring some restriction on the class of algebras under discussion; but completely feasible.

Chapter 3

Local Analysis on Thin Sandwiches

Let L be a Lie algebra over a field F with E_n, $n < p = \text{char } F$. Although we have at our disposal a thin sandwich c of L as given by Theorem 2.2.1, at first glance there are constraints on our means of extending the set of thin sandwiches (which is necessary for constructing thick sandwiches), despite the fact that Lemma 2.4.2 tells us that $[cx^3c] \in \mathbb{C}_1^*$ for all $x \in L$. In fact, the extreme situation where $[cx^3c] = \lambda(x)c$, $\lambda(x) \in F$, and $\lambda(a) \neq 0$ for some $a \in L$, is possible in principle. This is conceivable in another case of interest to us, namely that of a finite-dimensional simple Lie algebra; true, the nilpotency of the sandwich algebra $\mathfrak{C}(L)$ makes additional manoeuvres possible. On the other hand, if $[cx^3c] = \lambda(x)c$, it follows that $[cx^3c][cy^3c] = \lambda(x)\lambda(y)c^2 = 0$, and this identity in x, $y \in L$ provides us with a footbridge for the descent from $\mathfrak{C}_1^*$ to $\mathfrak{c}_2^*$.

We have considered only the model question up to this point. However, the aim of local analysis in the neighbourhood of a thin sandwich or of finitely many such is to get guidance towards what could be a larger number of footbridges of this type. This chapter is technical; however, the study of sandwich algebras in the concluding section enables us to state and prove an exhaustive result for $p \gg n$.

Throughout the whole chapter, it is assumed that $L \subset A(L)$.

§ 1. A first Footbridge Between Thin and Thick Sandwiches

1.1. Proposition. *Let L be any Lie algebra over a field of characteristic $p > 5$. If L contains a thin sandwich c such that*

$$cu_1^2cu_2^2c \dots cu_m^2c = 0; \qquad m \geqslant 2 ,$$

for all $u_i \in L$, then L has a thick sandwich.

Two auxiliary assertions precede the proof of this proposition. The first of them is of independent significance.

1.2. Lemma. *Let L be any Lie algebra over a field F of characteristic $p > 3$. Let c, c_0 be elements of L such that*

$$c_0^2 = c^2 = cc_0 = c_0c = 0 .$$

Then $c_1^2 = 0$, *where* $c_1 = [c_0a^2c]$, $a \in L$. *If moreover*

$$[c_0u^2cv^2c] = 0$$

for arbitrary $u, v \in L$, *then*

$$T = c_1u^2c_1v^2c_1 = 0 .$$

Proof. We note that

$$cc_0 = 0 \Rightarrow [ucc_0] = 0 \Rightarrow u\underline{cc_0} + \underline{c_0c}u - cuc_0 - c_0uc = 0 ,$$

whence it follows that

$$cuc_0 + c_0uc = 0; \qquad [c_0a^2c] = -[ca^2c_0]$$

$$([c_0a^2c] = -[a\underline{c_0ac}] = [acac_0] = -[ca^2c_0]) . \tag{3.1}$$

When $c_0 = c$, we get the known property $cuc = 0$ of the sandwich c, but this case is of no interest since then $c_1 = [ca^2c] = -[a\underline{cac}] = 0$. Further,

$$3cac_0\underline{a^2c_0ac} = \underset{(3.1)}{\underline{cac_0}}[c_0a^3]c + 3ca\underline{c_0ac_0}a^2c + cac_0a^3\underline{c_0c} - ca\underline{c_0^2}a^3c$$

$$= -c_0ac\underline{[c_0a^3]c} = 0 .$$

The first of the relations we need follows immediately from this:

$$c_1^2 = [[c_0a^2]c]^2 = -c[c_0a^2]^2c$$

$$= -\underline{cc_0}a^2c_0a^2c + 2\underline{cc_0}a^3c_0ac + 2ca\underline{c_0ac_0}a^2c - \underline{cc_0}a^4c_0c - ca^2\underline{c_0^2}a^2c$$

$$+ 2cac_0a^3\underline{c_0c} + 2ca^2\underline{c_0ac_0}ac - ca^2c_0a^2\underline{c_0c} - 4\underline{cac_0a^2c_0ac} = 0$$

(we have simply established that the last term is zero, while the first eight are zero for trivial reasons; it would not be worthwhile to write them out).

Suppose now that $[c_0u^2cv^2c] = 0$ is an identity in $u, v \in L$. Linearizing with respect to u or v yields the relations

$$[wc_0ucv^2c] = -[c_0wucv^2c] = 0 ,$$

$$[w[c_0v^2c]uc] = -[c_0v^2cwuc] = 0 .$$

Since w is arbitrary, we get the identity

$$c_0ucv^2c = 0 , \tag{3.2}$$

from the first relation, and $[c_0v^2c]uc = 0$ from the second; we rewrite it as

$$cv^2cuc_0 = 0 . \tag{3.3}$$

In fact,

$$0 = -\underline{c[c_0v^2]}uc + [c_0v^2]\underline{cuc}$$

$$= -\underset{(3.1)}{\underline{cv^2c_0uc}} + 2\underset{(3.1)}{\underline{cvc_0vuc}} - \underline{cc_0}v^2uc = cv^2cuc_0 - 2\underset{(3.2)}{\underline{c_0vcvuc}} .$$

By the definition of c_1, we have

$$c_1c - cc_1 = [c_1c] = [c_0a^2\underline{c^2}] = 0\,,$$

$$[wc_1c] = -[c_1wc] = -[c_0a^2\underline{cwc}] = 0 \Rightarrow c_1c = cc_1 = 0\,,$$

$$[wc_1c] = 0 \Rightarrow c_1wc + cwc_1 = 0 \quad \text{(analogue of identity (3.1))}\,.$$

However, by assumption we have in fact that $[c_1v^2c] = 0$, which means

$$[wc_1uc] = -[c_1wuc] = 0 \Rightarrow c_1uc = 0\,.$$

Resumé: we have

$$c_1c = 0, \qquad cc_1 = 0, \qquad cuc_1 = 0, \qquad c_1uc = 0\,. \tag{3.4}$$

Now

$$0 = [c_1u^2c] = [c_1u^2]c - c[c_1u^2] = c_1u^2c - \underbrace{2uc_1uc}_{(3.4)}$$

$$+ \underbrace{u^2c_1c}_{(3.4)} - \underbrace{cc_1u^2}_{(3.4)} + \underbrace{2cuc_1u}_{(3.4)} - cu^2c_1\,,$$

so that

$$c_1u^2c = cu^2c_1\,. \tag{3.5}$$

Reverting to the pair c_0, c_1, we note first that $[c_1c_0] = [c_0a^2\underline{cc_0}] = 0$. Further,

$$[wc_1c_0] = -[c_1wc_0] = -[c_0a^2\underbrace{cwc_0}_{(3.1)}] = [\underline{c_0a^2c_0}wc] = 0\,,$$

that is, $c_1c_0 = 0$, so that $c_0c_1 = 0$ as well. However, at the same time, $[wc_1c_0] = 0 \Rightarrow c_1wc_0 + c_0wc_1 = 0$. Thus

$$c_1c_0 = 0, \qquad c_0c_1 = 0, \qquad c_0uc_1 + c_1uc_0 = 0\,. \tag{3.6}$$

Having noted the almost obvious identities

$$c_0u^2c_0u^2c_0 = 0, \qquad c_0ucu^2c_0 = c_0u^2cuc_0\,, \tag{3.7}$$

that follow from the chain of relations

$$0 = \underline{c_0[c_0u^4]c_0} = \underline{c_0^2}u^4c_0 - 4\,\underline{c_0uc_0}u^3c_0 + 6\,c_0u^2c_0u^2c_0$$

$$- 4c_0u^3\underline{c_0uc_0} + c_0u^4\underline{c_0^2}\,,$$

$$0 = \underline{c_0[cu^3]c_0} = \underline{c_0c}u^3c_0 - 3\,c_0ucu^2c_0 + 3\,c_0u^2cuc_0 - c_0u^3\underline{cc_0}\,,$$

we add to them some more significant connections between c, c_0 and c_1:

$$c_0ac_1v^2c = 0, \qquad cv^2c_1ac_0 = 0\,. \tag{3.8}$$

In fact,

$$c_0ac_1v^2c = c_0a\underline{[c_0a^2]}cv^2c - c_0ac\underline{[c_0a^2]}v^2c$$

$$= -2\,c_0a^2\underset{(3.2)}{\underline{c_0acv^2}}c + \underline{c_0ac_0}a^2cv^2c + c_0a^3\underline{c_0c}v^2c$$

$$- c_0a\underline{cc_0}a^2v^2c + 2\,c_0\underset{(3.1)}{\underline{acac_0}}av^2c - \underset{(3.7)}{\underline{c_0aca^2c_0}}v^2c$$

$$= -2\,\underline{c_0ac_0}acav^2c - c_0a^2\underset{(3.1)}{\underline{cac_0}}v^2c = c_0a^2\underset{(3.2)}{\underline{c_0acv^2c}} = 0\,.$$

Similarly, the second identity in (3.8) is proved using (3.3).

After all these preliminary remarks, we return immediately to the proof that $T = 0$. Clearly,

$$T = c_1u^2[c_0a^2]cv^2c_1 - c_1u^2c[c_0a^2]v^2c_1 = T_1 - T_2\,.$$

The summands on the right-hand side are symmetrical. They are both zero. We prove this for the example of T_1, writing it in the form

$$T_1 = c_1u^2[c_0a^2]\underset{(3.5)}{\underline{cv^2c_1}} = c_1u^2[c_0a^2]c_1v^2c$$

$$= c_1u^2c_0a^2\underline{c_1}v^2c - 2\,c_1u^2\underset{(3.8)}{\underline{ac_0ac_1v^2c}} + c_1u^2a^2\underset{(3.6)}{\underline{c_0c_1}}v^2c$$

$$= c_1u^2c_0a^2\underline{[c_0a^2]}cv^2c - c_1u^2c_0a^2c\underline{[c_0a^2]}v^2c$$

$$= c_1u^2c_0a^2c_0a^2cv^2c - 2\,c_1u^2c_0a^3\underset{(3.2)}{\underline{c_0acv^2c}} + c_1u^2c_0a^4\underline{c_0c}v^2c$$

$$- c_1u^2c_0a^2\underline{cc_0}a^2v^2c + 2\,c_1u^2c_0a^2cac_0av^2c - c_1u^2c_0a^2ca^2c_0v^2c$$

$$= T_1' + 2T_1'' - T_1'''\,,$$

where

$$T_1' = c_1u^2c_0a^2c_0a^2cv^2c\,,$$

$$T_1'' = c_1u^2c_0a^2cac_0av^2c\,,$$

$$T_1''' = c_1u^2c_0a^2ca^2c_0v^2c\,.$$

The dénouement is upon us. Using (3.2), we get that

$$6\,T_1' = 6\,T_1' - 4\,c_1u^2c_0a^3\underset{(3.2)}{\underline{c_0acv^2c}} = \underset{(3.2)}{\underline{c_1u^2c_0[c_0a^4]cv^2c}} = 0\,.$$

Further,

$$T_1'' = c_1u^2\underset{(3.7)}{\underline{c_0a^2cac_0}}av^2c = c_1u^2\underset{(3.1)}{\underline{c_0aca^2}}c_0av^2c = -\underset{(3.5)}{\underline{c_1u^2c}}ac_0a^2c_0av^2c$$

$$= -\underset{(3.8)}{\underline{cu^2c_1ac_0}}a^2c_0av^2c = 0\,.$$

We rewrite, using (3.8) and (3.2):

$$\begin{aligned}6T_1''' &= 6T_1''' + \underbrace{4cu^2c_1ac_0}_{(3.5)}a^3c_0v^2c \\ &= 6T_1''' + 4c_1u^2\underbrace{cac_0}_{(3.1)}a^3c_0v^2c + 4c_1u^2c_0a^3\underbrace{c_0ac}_{(3.1)}v^2c \\ &= 6T_1''' - 4c_1u^2c_0aca^3c_0v^2c - 4c_1u^2\underline{c_0a^3cac_0}v^2c \\ &= c_1u^2\underline{c_0[ca^4]}c_0v^2c = 0\,. \quad \square\end{aligned}$$

1.3. *Proposition* 1.1 *holds for* $m = 2$.

Proof. Suppose that

$$c \neq 0, \qquad c^2 = cuc = cu^2cv^2c = 0\,. \tag{3.9}$$

The analogous situation for thick sandwiches was considered in 2.4.6 (see identity 2.27), but here we find ourselves in more restricted circumstances. As before (see 2.4.3), we use the notation $\bar{X} = x_kx_{k-1} \ldots x_2x_1$ for $X = x_1x_2 \ldots x_{k-1}x_k$. We have to establish the existence of a sandwich c_1 such that $c_1u^2c_1 = 0$, or, as is the same thing, $[c_1u^3c_1] = 0$ for all $u \in L$.

By (3.9), we have

$$\begin{aligned}&c[ca^5]c = 0\,, \\ &cu^2ca^\delta[ca^5]a^{1-\delta}c = 0, \qquad \delta = 0, 1\,, \\ &ca^{1-\delta}[ca^5]a^\delta cu^2c = 0, \qquad \delta = 0, 1\,.\end{aligned}$$

Expanding $[ca^5]$ according to (1.5), and again taking (3.9) into account, as well as the condition $p > 5$, we get five relations:

$$\begin{aligned}&ca^2ca^3c - ca^3ca^2c = 0\,, \\ &cu^2ca^3ca^3c = cu^2ca^4ca^2c = 0\,, \\ &ca^3ca^3cu^2c = ca^2ca^4cu^2c = 0\,.\end{aligned} \tag{3.10}$$

We divide the rest of the discussion into four parts.

I. We write

$$\begin{aligned}&A_1 = ca^2cuvaca^2c, \qquad A_2 = caucva^2ca^2c\,, \\ &B_1 = cau^2ca^3ca^2c, \qquad B_2 = ca^2uca^2uca^2c, \qquad B_3 = cauca^3uca^2c\,, \\ &B_4 = ca^3cau^2ca^2c, \qquad B_5 = ca^2ca^2u^2ca^2c, \qquad B_6 = ca^3cu^2ca^3c\,.\end{aligned}$$

Sometimes the letters A_1, A_2 denote values of the elements written out above for special values of u or v. Elements a, u, v standing in a row in A_i are permutable. For example,

$$ca^2cuvaca^2c = ca^2cvuaca^2c + \underbrace{ca^2c[uv]aca^2c}_{(3.9)} = ca^2cvuaca^2c\,.$$

Because of this, we find from the relations

$$0 = c[ca^3u]vca^2c = 3A_1 + 3A_2, \qquad 0 = caucv[ca^4]c = 6A_2$$

that

$$A_1 = 0, \qquad A_2 = 0 \, .$$

Similarly, $\bar{A}_1 = 0$, $\bar{A}_2 = 0$. Thus, the elements a and u in the monomials B_i and $\bar{B}_i$ may be assumed to commute (all additional commutators lead to the elements $A_k = \bar{A}_k = 0$). We find relations between B_i and $\bar{B}_i$ (based on (3.9), (3.10)), by expanding all the relevant commutators according to (1.5) (or without it) in the associative algebra $A(L)$:

$$\underline{c[ca^4u]uca^2c} = 0 \Rightarrow 3B_2 - 2B_3 + 2B_4 - 3B_5 = 0 \, ,$$

$$\underline{cu[ca^4u]ca^2c} = 0 \Rightarrow 2B_1 + 3B_2 - 2B_3 = 0 \, ,$$

$$\underline{c[ca^3u^2]aca^2c} = 0 \Rightarrow -3B_1 - 6B_2 + 6B_3 - B_4 + 3B_5 = 0 \, ,$$

$$\underline{[ca^3cu^2c]}a^3c = 0 \Rightarrow 3\bar{B}_4 - 3B_1 - 2B_6 = 0$$

(in the last case $[ca^3cu^2c] = -\underline{[aca^2cu^2c]} = 0$). For example, $[ca^4c] = \underline{ca^4u} - \underline{4aca^3u} + 6a^2ca^2u - 4a^3cau + \underline{a^4cu} - uca^4 + 4uaca^3 - 6ua^2ca^2 + \underline{4ua^3ca} - \underline{ua^4c}$. On multiplying $[ca^4u]$ on the left by c and on the right by uca^2c, the underlined summands give monomials that are zero either because initial segments are $c^2 = 0$, $cvc = 0$, or because final segments are $cvuca^2c = 0$ (here $v = a$ or u).

To the four relations of the form

$$\sum \alpha_i B_i + \sum \beta_i \bar{B}_i = 0$$

as written out above we can add the four that are symmetrical to them:

$$\sum \alpha_i \bar{B}_i + \sum \beta_i B_i = 0 \, ,$$

if we write out explicitly what is obtained from the relations $\underline{ca^2cu[ca^4u]c} = 0$, $\ldots$, $ca^3\underline{[ca^3cu^2c]} = 0$ when the commutators are expanded. As a result we get (since $\bar{B}_5 = B_5$, $\bar{B}_6 = B_6$) 10 independent linear relations that can be written as

$$\begin{aligned} &2B_3 = 2B_1 + 3B_2, \qquad B_4 = -B_1 - 3B_2 \, , \\ &3B_5 = -4B_1 - 6B_2, \quad 2B_6 = -6B_1 - 9B_2 \, , \\ &\bar{B}_i = B_i, \qquad 1 \leqslant i \leqslant 6 \, . \end{aligned} \tag{3.11}$$

Applying relations (2.20), (3.9) and $A_1 = 0$, $A_2 = 0$, we get

$$[ca^3c]u^2[ca^3c] = -6B_4 - 6\bar{B}_4 + 9B_5 + 4B_6 \, ,$$

which gives in conjunction with (3.11) that

$$[ca^3c]u^2[ca^3c] = -12B_1 \, . \tag{3.12}$$

In the immediate future, a is fixed, so it is convenient to write

$$B_i = B_i(u), \qquad \bar{B}_i = \bar{B}_i(u) \, .$$

II. Suppose that $c_1 = [ca^3c] \neq 0$. Then c_1 is a sandwich (Lemma 2.4.2). Assuming that condition (3.9) is satisfied, we prove that c_1 satisfies the identities

$$c_1u^2c_1v^2c_1 = 0, \qquad c_1u^2c_1v^3c_1w^2c_1 = 0 .$$

In fact, using (3.12), we find that

$$\underline{c_1u^2c_1}v^2c_1 = -12B_1(u)v^2c_1 = -12c \ldots ca^3\underline{ca^2cv^2[ca^3c]}$$

$$= -24c \ldots ca^3\underbrace{ca^2cv^2ca^3}_{(3.9)}c + 36c \ldots ca^3\underline{ca^2cv^2aca^2c}$$

$$= 36c \ldots ca^3\underline{A_1} = 0 .$$

Further,

$$\underbrace{c_1u^2c_1}_{(3.12)}\underbrace{v^3c_1w^2c_1}_{(3.12)} = \alpha B_1(u)v^3B_1(w) = \alpha B_1(u)v^3\bar{B}_1(w)$$

$$= \alpha c \ldots c \ldots \underline{ca^2cv^3ca^2c} \ldots c \ldots c = 0 ,$$

since

$$0 = -\underbrace{[vcv^2ca^2c]}_{(3.9)} = [cv^3ca^2c] = [cv^3c]a^2c - ca^2[cv^3c]$$

and thus

$$4ca^2cv^3ca^2c = 2\,\underline{ca^2[cv^3c]a^2c} + 6\,ca^2\underbrace{vcv^2ca^2c}_{(3.9)} + 6ca^2\underbrace{cv^2cva^2c}_{(3.9)}$$

$$= 2ca^2ca^2[cv^3c]$$

$$= 4\underbrace{ca^2ca^2cv^3c}_{(3.9)} - 6\underline{ca^2ca^2vcv^3c} \equiv -\underbrace{c[ca^4]vcv^2c}_{(3.9)} = 0 .$$

III. For ease of reference we shall assume now that the identities proved in II for c_1 are valid for the original sandwich c:

$$cu^2cv^2c = 0, \qquad cu^2cv^3cw^2c = 0 . \tag{3.13}$$

We set $c_1 = [ca^3c]$ and show that $c_1u^2c_1v^3c_1 = 0$ (a, u, v and w are arbitrary elements of L). In fact we consider the monomials:

$$F_1 = ca^2ca^3cv^3ca^3c, \qquad F_2 = ca^2ca^3cv^3aca^2c ,$$

$$F_3 = ca^2ca^3cv^2acva^2c, \qquad F_4 = ca^2ca^3cv^2a^2cvac ,$$

$$F_5 = ca^2ca^3cva^2cv^2ac, \qquad F_6 = ca^2ca^3cva^3cv^2c .$$

Since $ca^2ca^3c = ca^3ca^2c$ (see (3.10)), it follows from (3.13) firstly that all permutations of v and a are permissible in the segments cv^3ac, cv^2a^2c, cva^3c, cva^2c, cv^2ac of the F_i. Further,

$$\underline{ca^\delta[ca^5]a^{1-\delta}cv^3cw^2c} = 0, \qquad \delta = 0, 1 \Rightarrow ca^2ca^4cv^3w^2c = 0 ,$$

$$ca^2ca^\delta[ca^5]a^{1-\delta}cv^3c = 0, \qquad \delta = 0, 1 \Rightarrow ca^2ca^3ca^3cv^3c = 0 .$$

Using this remark, and bearing in mind (3.10), (3.13) throughout, we get the following relations:

$$0 = \underline{ca^2ca^3c[cv^3a^2]ac} = -F_1 + 2F_2 - 6F_3 + 3F_4 - 3F_5\,,$$

$$0 = \underline{ca^2ca^3ca[cv^3a^2]c} = -F_2 + 3F_3 - 6F_4 + 6F_5 - 3F_6\,,$$

$$0 = \underline{ca^2ca^3c[ca^3v^2]vc} = -3F_3 + 3F_4 - 6F_5 + 2F_6\,,$$

$$0 = \underline{ca^2c[ca^5]v^2cvac} = -10F_4\,,$$

$$0 = \underline{ca^2c[ca^5]vcv^2ac} = -10F_5\,,$$

$$0 = ca^2c[ca^5]avcv^2c = -10F_6\,.$$

We have $p > 5$, so that $F_i = 0$, $1 \leqslant i \leqslant 6$. Thus, using (3.12) and (3.10) again, we come to the required relation

$$\begin{aligned} c_1u^2c_1v^3c_1 &= -12B_1(u)v^3c_1 = -12cau^2ca^3ca^2cv^3\underline{[ca^3c]} \\ &= 36cau^2ca^3\underset{(3.13)}{\underline{ca^2cv^3ca^2}}ca - 24cau^2\underset{(3.10)}{\underline{ca^3ca^2cv}}{}^3ca^3c \\ &\quad + 36cau^2\underline{ca^3ca^2}cv^3aca^2c = -24cau^2\underline{ca^2ca^3cv^3ca^3c} \\ &\quad + 36cau^2\underline{ca^2ca^3cv^3aca^2c} = -12cau^2(2F_1 - 3F_2) = 0\,. \end{aligned}$$

IV. A double operation of type $c \to [ca^3c]$ has led us, with recurrent change of notation, to a sandwich c satisfying the identities

$$cu^2cv^2c = 0, \qquad cu^2cv^3c = 0\,. \tag{3.14}$$

Now either $cu^2c = 0$, in which case there is nothing to do, or else $c_0 = [ca^3c] \neq 0$ for some $a \in L$. In that case we have

$$\underset{(3.12)}{\underline{c_0u^2c_0}} = -12cau^2\underset{(3.10)}{\underline{ca^3ca^2c}} = -12cau^2\underset{(3.14)}{\underline{ca^2ca^3c}} = 0\,,$$

that is, c_0 is the required thick sandwich. □

1.4. Proof of Proposition 1.1. Suppose that the index $m \geqslant 2$ in the requirements of the proposition is chosen to be the smallest possible and that

$$c_1 = [ca_1^3ca_2^2c \ldots ca_{m-1}^2c] \neq 0$$

for some $a_1, a_2, \ldots, a_{m-1} \in L$.

We prove first that c_1 is a sandwich, that is, that $c_1^2 = 0$. For $m = 2$, in which case $c_1 = [ca_1^3c]$, this has already been done (Lemma 2.4.2). We proceed by induction on m (we should choose a different index for the induction, but this is unimportant). Suppose that it has already been established that $c_0^2 = 0$, where

$$c_0 = [ca_1^3ca_2^2c \ldots ca_{m-2}^2c]\,.$$

Then we find ourselves in the conditions of Lemma 1.2:

$$c_1 = [c_0 a_{m-1}^2 c], \quad \text{and clearly} \quad c_0 c = c c_0 = 0 .$$

Therefore

$$c_1^2 = 0 .$$

For $m > 2$, we get from Lemma 1.2 again that

$$c_1 u^2 c_1 v^2 c_1 = 0 ,$$

since

$$c a_1^2 c \ldots c a_m^2 c = 0 \Rightarrow [c_0 u^2 c v^2 c] = 0 \qquad (a_{m-1} = u,\, a_m = v) .$$

Thus, we have once more hit the conditions of Proposition 1.1, but with $m = 2$ this time. All we need do now is refer to 1.3. □

§ 2. A Second Footbridge Between Thin and Thick Sandwiches

This is the title that we give to two assertions that are nowhere used in their entirety, but are very useful in all of their three parts. The proofs are of the same sort.

Recall that L is assumed to be embedded in $A(L)$.

2.1. *Let* c_1 and c_2 *be thin sandwiches of the Lie algebra L. We consider an element* c_0 *of one of the following types:*

a) $c_0 = [c_2 c_1]$;

b) $c_0 = [c_1 a^3 c_1]$;

c) $c_0 = [a c_1 c_2]$, *if* $[c_1 c_2] = 0$.

Then the following relations hold:

$$c_0^2 = 0, \qquad c_0 c_1 = c_1 c_0 = 0, \qquad c_0 u c_1 + c_1 u c_0 = 0, \qquad u \in L . \tag{3.15}$$

For, in case a) we have $c_0^2 = 0$ almost by definition (see Proposition 1.5.4). Moreover, $c_0 c_1 = [c_2 c_1^2] = 0$ and $c_0 c_1 = c_2 c_1^2 - \underline{c_1 c_2 c_1} = 0$. In case b), it is enough to refer to Lemma 2.4.2 and to note that $c_0 c_1 = -\underline{c_1 [c_1 a^3] c_1} = 0$, $c_1 c_0 = \underline{c_1 [c_1 a^3] c_1} = 0$. Case c) is somewhat less trivial:

$$c_0^2 = -c_1 [a c_2]^2 c_1 = c_1 c_2 a^2 \underline{c_2 c_1} = c_1 c_2 a^2 c_1 c_2$$

$$= \underline{c_1 [c_2 a^2] c_1} c_2 + 2 c_1 a c_2 a \underline{c_1 c_2} - c_1 a^2 \underline{c_2 c_1 c_2} = 2 c_1 a c_2 a \underline{c_2 c_1} = 0 .$$

Furthermore, $c_0 c_1 = -\underline{c_1 [a c_2] c_1} = 0$, $c_1 c_0 = c_1 [a c_2] c_1 = 0$. The relation $0 = [u c_0 c_1] = \underline{u c_0 c_1} + \underline{c_1 c_0} u - c_0 u c_1 - c_1 u c_0$ concludes the verification of (3.15) in all three cases.

Lemma. *Let L be an arbitrary Lie algebra over a field F of characteristic* $p > 5$ *containing thin sandwiches. Then L has a thick sandwich when any of the following* *conditions holds*:

a) $c_1^2 = c_2^2 = 0$, $c_0 = [c_2 c_1] \neq 0$, $[c_0 u^2 c_1] = 0$ *for all* $u \in L$;

b) $c_1^2 = 0$, $c_0 = [c_1 a^3 c_1] \neq 0$ *for some* $a \in L$, $[c_1 w^3 c_1 u^2 c_1] = 0$ *for all* $w, u \in L$;

c) $c_1^2 = c_2^2 = 0$, $[c_1 c_2] = 0$, $c_0 = [a c_1 c_2]$ *for some* $a \in L$, $c_2 c_1 u^2 c_1 = 0$ *for all* $u \in L$.

In all three cases, $c_0^2 = 0$.

Proof. In case b), linearization of the equation $[c_1 w^3 c_1 u^2 c_1] = 0$ with respect to w leads immediately to the identity $c_1 v^2 c_1 u^2 c_1 = 0$, which by Proposition 1.1 guarantees the existence of a thick sandwich.

In cases a), c) we have

$$c_0 u c_1 = 0, \qquad c_1 u c_0 = 0, \qquad c_0 u^2 c_1 = c_1 u^2 c_0 . \tag{3.16}$$

For, $[c_0 u^2 c_1] = 0 \Rightarrow [w c_0 u c_1] = -[c_0 w u c_1] = 0 \Rightarrow c_0 u c_1 = 0$. Combining this with (3.15) and going over to associative form for $[c_0 u^2 c_1]$ (the way of thinking in the proof of Lemma 1.2), we get (3.16). The argument is similar in case c):

$$c_2 c_1 u^2 c_1 = 0 \Rightarrow [c_0 u^2 c_1] = [a \underline{c_2 c_1 u^2 c_1}\,] = 0 \Rightarrow c_0 u c_1 = 0 \quad \textit{etc.}$$

Case a) quickly leads to the existence of a thick sandwich. Since

$$c_0 c_2 = [c_2 c_1] c_2 = \underline{c_2 c_1 c_2} - c_1 \underline{c_2^2} = 0, \qquad c_2 c_0 = 0 ,$$

we have

$$\begin{aligned} c_0 u^2 c_0 v^2 c_0 &= c_0 u^2 \underset{(3.16)}{\underline{c_2 c_1 v^2 c_0}} - \underset{(3.16)}{\underline{c_0 u^2 c_1}} c_2 v^2 c_0 \\ &= c_0 u^2 \underline{c_2 c_0}\, v^2 c_1 - c_1 u^2 \underline{c_0 c_2}\, v^2 c_0 = 0 , \end{aligned}$$

so that Proposition 1.1 is applicable with c replaced by c_0.

In case c), the identity

$$c_2 c_1 u^2 c_1 = 0 , \tag{3.17}$$

does not give us a thick sandwich straightaway, although the final result will be that. We argue along the following lines. Since $c_1 c_2 = c_2 c_1$, we have

$$0 = \underset{(3.17)}{\underline{c_2 c_1 w u c_1}} \Rightarrow [w[u c_2 c_1] u c_1] = -[u \underline{c_2 c_1 w u c_1}] = 0 \Rightarrow 0 [u c_2 c_1] u c_1$$

$$= u c_2 \underline{c_1 u c_1} + \underset{(3.17)}{\underline{c_1 c_2 u^2 c_1}} - c_2 u \underline{c_1 u c_1} - c_1 u c_2 u c_1 ,$$

whence

$$c_1 u c_2 u c_1 = 0. \tag{3.18}$$

We now have

$$0 = c_1 [c_2 u^2] c_1 = \underset{(3.17)}{\underline{c_1 c_2 u^2 c_1}} - 2 \underset{(3.18)}{\underline{c_1 u c_2 u c_1}} + c_1 u^2 c_2 c_1 ,$$

that is,

$$c_1u^2c_1c_2 = 0\,. \tag{3.19}$$

Further, we get for $c_0 = [ac_2c_1]$ (compare with $c_0c_1 = 0$):

$$c_0c_2 = \underline{ac_2c_1c_2} - c_1\underline{ac_2^2} - c_2\underline{ac_1c_2} + \underline{c_2c_1}\,ac_2 = -\underline{c_2ac_2}c_1 + c_1\underline{c_2ac_2} = 0\,.$$

Similarly, $c_2c_0 = 0$, and finally $\underbrace{c_0u^2c_1c_2}_{(3.16)} = c_1u^2\underline{c_0c_2} = 0\,.$

We give the relations thus obtained a number:

$$c_0c_2 = 0, \qquad c_2c_0 = 0, \qquad c_0u^2c_1c_2 = 0. \tag{3.20}$$

Since

$$\begin{aligned}\underbrace{2c_0u^2c_1ac_2}_{(3.16)} = 2c_1u^2\underline{c_0}\,ac_2 &= 2\underbrace{c_1u^2c_1c_2}_{(3.19)}a^2c_2 + 2c_1u^2ac_1\underline{c_2ac_2}\\ &\quad - 2c_1u^2c_1\underline{ac_2ac_2} - 2c_1u^2c_2ac_1ac_2\\ &= c_1u^2\underline{c_2[c_1a^2]c_2} - \underline{c_1u^2c_2c_1}\,a^2c_2 - c_1u^2c_2a^2c_1c_2\,,\end{aligned}$$

that is, $2c_0u^2c_1ac_2 = -c_1u^2c_2a^2c_2c_1$, we have

$$\begin{aligned}2c_0u^2c_0 &= 2c_0u^2ac_2c_1 - \underline{2c_0u^2c_1ac_2} - 2c_0u^2c_2ac_1 + \underbrace{2c_0u^2c_1c_2}_{(3.20)}a\\ &= 2c_0u^2ac_2c_1 + c_1u^2c_2a^2c_2c_1 - 2c_0u^2c_2ac_1\,.\end{aligned}$$

Thus,

$$\begin{aligned}2c_0u^2c_0v^2c_0 &= 2c_0u^2ac_2\underbrace{c_1v^2c_0}_{(3.16)} + c_1u^2c_2a^2c_2\underbrace{c_1v^2c_0}_{(3.16)} - 2c_0u^2c_2a\underbrace{c_1v^2c_0}_{(3.16)}\\ &= 2c_0u^2a\underbrace{c_2c_0}_{(3.20)}v^2c_1 + c_1u^2c_2a^2\underbrace{c_2c_0}_{(3.20)}v^2c_1 - 2c_0u^2c_2a\underline{c_0}v^2c_1\\ &= -2c_0u^2c_2a^2\underbrace{c_2c_1v^2c_1}_{(3.17)} - 2c_0u^2\underline{c_2ac_2}c_1av^2c_1\\ &\quad + 2c_0u^2\underline{c_2ac_2}ac_1v^2c_1 + 2c_0u^2c_2\underline{ac_1ac_2}v^2c_1\\ &= -c_0u^2\underline{c_2[c_1a^2]c_2}v^2c_1 + \underbrace{c_0u^2c_2c_1}_{(3.20)}a^2c_2v^2c_1\\ &\quad + c_0u^2c_2a^2\underbrace{c_1c_2v^2c_1}_{(3.17)} = 0\,.\end{aligned}$$

We have arrived once again at a situation where the conditions of Proposition 1.1 apply.

2.2. Proposition. *Let L be a Lie algebra over a field of characteristic $p > 5$ containing thin sandwiches c_1 and c_2, and either satisfying E_n, $n < p$, or else of finite dimension over F.*

Then L has thick sandwiches whenever one of the following three conditions holds:

a) $c_0 = [c_2 c_1] \neq 0$, $[c_0 u^2 c_1 uvc_1] = 0$ for all $u, v \in L$;

b) $c_0 = [c_1 a^3 c_1] \neq 0$, $c_1 w^2 c_1 u^2 c_1 uvc_1 = 0$ for all $u, v, w \in L$;

c) $c_0 = [ac_1 c_2] \neq 0$, $[c_1 c_2] = 0, c_2 c_1 u^2 c_1 uvc_1 = 0$ for all $u, v \in L$.

Proof. By Lemma 2.1, we are justified in assuming that $[c_0 a_1^2 c_1] \neq 0$, where $c_0 = [hc_1]$ by assumption: $h = c_2$, $[c_1 a^3]$ or $[ac_2]$ (if convenient, the element a defining c_0 can be assumed fixed).

By the first assertion in Lemma 1.2, relations (3.15) guarantee that $[c_0 a_1^2 c_1] \in \mathfrak{C}_1^*$. More generally, all elements of the form

$$e_m = [c_0 u_1^2 c_1 u_2^2 c_1 \ldots c_1 u_m^2 c_1] \tag{3.21}$$

satisfy the relation $e_m^2 = 0$, since by induction $e_{m-1}^2 = 0$, $e_m = [e_{m-1} u_m^2 c_1]$, $e_m c_1 = c_1 e_m = 0$. To use the first assertion in Lemma 1.2, c must be replaced by c_1 and c_0 by e_{m-1}.

We shall assume for a while that $e_m = 0$ for all choices of the elements $u_k \in L$, $k = 1, 2, \ldots, m$. We take m to be the smallest possible, so that $m \geqslant 2$ (since $[c_0 a_1^2 c_1] \neq 0$), and thus $c_1' = [ca_{m-1}^2 c_1] \neq 0$, with $c = [c_0 a_1^2 c_1 a_2^2 c_1 \ldots c_1 a_{m-2}^2 c_1]$ for some $a_k \in L$. Then

$$c^2 = cc_1 = c_1 c = 0, \qquad [cu^2 c_1 v^2 c_1] = 0,$$

and by Lemma 1.2 we have the identity

$$c_1' u^2 c_1' v^2 c_1' = 0,$$

which, by Proposition 1.1, guarantees the existence of a thick sandwich.

Thus, we just have to prove that the following relation is an identity in $u_1, u_2, \ldots, u_m \in L$:

$$[c_0 u_1^2 c_1 u_2^2 c_1 \ldots c_1 u_m^2 c_1] = 0.$$

It is natural that the restrictions imposed on L in the assumptions of the proposition must swim to the surface at exactly this point. We must begin to work with full power on the identity

$$[c_0 u^2 c_1 uvc_1] = 0,$$

given by assumption, and which has remained in shadow up to now. Partial linearization of it with respect to u—the isolation of the coefficient of λ in the decomposition

$$[c_0 (u + \lambda v)^2 c_1 (u + \lambda v) vc_1] = T_0 + \lambda T_1 + \lambda^2 T_2$$

—leads to the identity

$$T_1 = [c_0 (uv + vu) c_1 uvc_1] + [c_0 u^2 c_1 v^2 c_1] = 0.$$

Since $c_0 = [hc_1]$ and $c_1 (uv + vu) c_1 = 2c_1 uvc_1$,

$$2[hc_1 uvc_1 uvc_1] + [hc_1 u^2 c_1 v^2 c_1] = 0.$$

Interchanging u and v here, and again noting that $c_1vuc_1 = c_1uvc_1$, we find that

$$2[hc_1uvc_1uvc_1] + [hc_1v^2c_1u^2c_1] = 0\,.$$

Comparing these two identities, we conclude that

$$[hc_1v^2c_1u^2c_1] = [hc_1u^2c_1v^2c_1]\,.$$

Arguing by induction, we assume that the element

$$[hc_1u_1^2c_1u_2^2c_1 \ldots c_1u_k^2c_1]\,, \qquad k \geqslant 2\,,$$

is symmetric in $u_1, u_2, \ldots, u_k$, that is,

$$[hc_1u_{\pi 1}^2c_1u_{\pi 2}^2c_1 \ldots c_1u_{\pi k}^2c_1] = [hc_1u_1^2c_1u_2^2c_1 \ldots c_1u_k^2c_1], \qquad \pi \in S_k\,,$$

and then prove it for the element with index $k+1$. Setting

$$f = [hc_1u_1^2c_1 \ldots c_1u_{k-2}^2]\,, \qquad u_{k-1} = x, \qquad u_k = y, \qquad u_{k+1} = z$$

for convenience, and using the relation

$$[fc_1x^2c_1y^2c_1] = [fc_1y^2c_1x^2c_1]\,,$$

which is given by the induction hypothesis, we will of course have at our disposal relations of the form

$$[fc_1abc_1ghc_1] = [fc_1ghc_1abc_1], \tag{3.22}$$

since

$$c_1abc_1 = \tfrac{1}{2}\{c_1(a+b)^2c_1 - c_1a^2c_1 - c_1b^2c_1\}\,,$$

$$c_1ghc_1 = \tfrac{1}{2}\{c_1(g+h)^2c_1 - c_1g^2c_1 - c_1h^2c_1\}\,.$$

Further,

$$c_1[c_1y^2z]zc_1 = c_1y^2c_1z^2c_1 + 2c_1yzc_1yzc_1\,,$$

$$c_1[c_1z^2y]yc_1 = c_1z^2c_1y^2c_1 + 2c_1yzc_1yzc_1\,,$$

whence it follows that

$$c_1y^2c_1z^2c_1 - c_1z^2c_1y^2c_1 = c_1[c_1y^2z]zc_1 - c_1[c_1z^2y]yc_1\,. \tag{3.23}$$

We arrive at a chain of relations:

$$[fc_1x^2\underbrace{c_1y^2c_1z^2c_1}_{(3.23)}] - [fc_1x^2\underbrace{c_1z^2c_1y^2c_1}_{(3.23)}]$$

$$= \underbrace{[fc_1x^2c_1[c_1y^2z]zc_1]}_{(3.22)} - \underbrace{[fc_1x^2c_1[c_1z^2y]yc_1]}_{(3.22)}$$

$$= [f\underbrace{c_1[c_1y^2z]zc_1}_{(3.23)}x^2c_1] - [f\underbrace{c_1[c_1z^2y]yc_1}_{(3.23)}x^2c_1]$$

$$= [fc_1y^2c_1z^2c_1x^2c_1] - [\underbrace{fc_1z^2c_1y^2c_1}_{(3.22)}x^2c_1] = 0\,,$$

from which it follows that the z^2 and y^2 in the product

$$[fc_1x^2c_1y^2c_1z^2c_1]$$

are allowed to change places. Together with the induction hypothesis, this means that all the monomials u_i^2 in the element e_m of the form (3.21) may be permuted at will; this for every finite m.

We have reached the logical conclusion. Two cases are to be considered:

1) L is a Lie algebra with E_n, $n < p$. Since

$$e_m = [hc_1u_1^2c_1 \dots c_1u_m^2c_1] = [h[c_1u_1^2][c_1u_2^2] \dots [c_1u_m^2]c_1],$$

it follows that

$$\begin{aligned} n!e_n &= \sum_{\pi\in S_n} [hc_1u_{\pi 1}^2c_1u_{\pi 2}^2c_1 \dots c_1u_{\pi n}^2c_1] \\ &= \sum_{\pi\in S_n} [h[c_1u_{\pi 1}^2][c_1u_{\pi 2}^2] \dots [c_1u_{\pi n}^2]c_1] \\ &= \Bigg[h\underbrace{\left\{\begin{matrix}[c_1u_1^2][c_1u_2^2] & \dots & [c_1u_n^2] \\ 1 \quad\quad 1 & \dots & 1\end{matrix}\right\}c_1}_{(1.7)}\Bigg] = 0\,. \end{aligned}$$

Thus, there exists a smallest index $m \leqslant n$ such that $e_m = 0$.

2) $\dim_F L < \infty$. Let $(a_1, \dots, a_r)$ be a basis of the linear space L over F, so that every segment $c_1u_i^2c_1$ in the expression (3.21) is a linear combination of monomials of the form $c_1a_k^2c_1$, $c_1a_ka_lc_1$. If we take into account the fact that

$$c_1a_ka_lc_1 = \tfrac{1}{2}\{c_1(a_k+a_l)^2c_1 - c_1a_k^2c_1 - c_1a_l^2c_1\},$$

we can conclude that the existence of an element $e_m \neq 0$ with arbitrarily large m implies that of a non-zero element of this type with $u_i = a_k$ or $a_k + a_l$, $1 \leqslant i \leqslant m$. However, when $m > r + \binom{r}{2}$, it must happen that $u_i = u = u_j$ for some different numerals i and j. Since the monomials u_k are permutable, on combining u_i^2 and u_j^2 we get the expression

$$e_m = [hc_1u_1^2c_1 \dots \underline{c_1u^2c_1u^2c_1} \dots c_1u_m^2c_1]\,.$$

Since

$$\begin{aligned} 0 = \underline{c_1[c_1u^4]c_1} &= \underline{c_1^2}u^4c_1 - 4\underline{c_1uc_1}u^3c_1 + 6c_1u^2c_1u^2c_1 - 4c_1u^3\underline{c_1uc_1} \\ &\quad + c_1u^4\underline{c_1^2}\,, \end{aligned}$$

we have

$$c_1u^2c_1u^2c_1 = 0$$

(see also (3.7)), and therefore $e_m = 0$ for $m \leqslant \binom{r+1}{2} + 1$. $\square$

2.3. Corollary. *Let L be a Lie algebra over a field F of characteristic $p > 5$, which either satisfies E_n for $n < p$ or is finite-dimensional over F. Suppose further that c is a thin sandwich in L such that*

$$[cu^3c][cv^3c] = 0 \tag{3.24}$$

identically in $u, v \in L$. Then L has a thick sandwich.

Proof. Since $c \in \mathfrak{C}_1^*(L)$, we have $c_0 = [ca^3c] \neq 0$ for some $a \in L$. At the same time, it follows from (3.24) that

$$[cxw^2c][cu^3c] = \sum_i [cu_i^3c][cu^3c] = 0$$

for all x, w, $u \in L$, so that

$$\begin{aligned} 3[xcw^2\underline{cvucu^2}\,c] &= -[xcw^2cv\underline{[cu^3]c}] = -[xcw^2cv[cu^3c]] \\ &= [cxw^2cv[cu^3c]] = -[v[cxw^2c][cu^3c]] = 0\,. \end{aligned}$$

This means that

$$cw^2cvucu^2c = 0\,.$$

If we take into account the fact that

$$cvucu^2c = -\tfrac{1}{3}cv[cu^3]c = -\tfrac{1}{3}c[cu^3]vc = -cu^2cuvc\,,$$

we will get the identity $cw^2cu^2cuvc = 0$, corresponding to condition b) of Proposition 2.2. □

It was for the sake of the memorable condition (3.24), essentially, that the version of it figuring in Proposition 2.2b) was analyzed. We also had in mind Corollary 2.3 (which we shall need later) in the preamble to the Chapter, when local analysis on thin sandwiches and a footbridge for the descent to $\mathfrak{C}_2^*$ were first mentioned.

§ 3. Two Necessary Lemmas

Identities involving one or two thin sandwiches played the dominant rôle in the two preceding sections. Whole families of thin sandwiches now come into the action, tied together by "multiplicative" conditions.

3.1. Lemma. *Let L be any Lie algebra over a field F of characteristic $p > 5$. Further, let $c_1, c_2, \ldots, c_m$ $(m \geqslant 3)$ be any family of thin sandwiches in L, possibly with repeats. If the element*

$$c = [c_1c_2 \ldots c_{m-1}c_m] \neq 0$$

is such that $[cc_i] = 0$ for $1 \leqslant i \leqslant m$ and $cc_k = cc_l = 0$, where k and l are two fixed indices with $1 \leqslant k < l \leqslant m-1$, then L has a thick sandwich.

Proof. Setting

$$c_0 = [c_1 c_2 \ldots c_{m-1}]$$

and noting that

$$c = [c_0 c_m]; \qquad c c_0 = 0, \qquad c_0 u c + c u c_0 = 0$$

(see the beginning of the proof of Lemma 1.2), we claim that

$$c_0 u c = 0, \qquad c u c_0 = 0. \tag{3.25}$$

For this we write c_0 as a linear combination of associative monomials of the form

$$A = c_{i_1} \ldots c_{i_s} c_k c_{j_1} \ldots c_{j_t} c_l \ldots, \qquad B = \ldots c_l \ldots c_k \ldots$$

in $A(L)$, and use the relations given to us by assumption:

$$c c_k = 0, \qquad c_k c = 0, \qquad c_k u c + c u c_k = 0\,,$$

$$c c_l = 0, \qquad c_l c = 0, \qquad c_l u c + c u c_l = 0\,.$$

Since $cc_i = c_i c$, $1 \leqslant i \leqslant m$, by using the stylized form of formula (1.4), that is,

$$u c_{i_1} \ldots c_{i_s} = \Sigma A'_\nu [u A''_\nu]\,,$$

$$A'_\nu = \alpha_\nu c'_{i_1} \ldots c'_{i_\nu}, \qquad A''_\nu = c'_{i_{\nu+1}} \ldots c'_{i_s}; \{i'_1, \ldots i'_s\} = \{i_1, \ldots, i_s\}\,,$$

we find easily that

$$cuA = c\left(\sum A'_\nu [u A''_\nu]\right) c_k c_{j_1} \ldots c_{j_t} c_l \ldots$$

$$= \sum A'_\nu \underline{c [u A''_\nu] c_k} c_{j_1} \ldots c_{j_t} c_l \ldots$$

$$= -\sum A'_\nu c_k [u A''_\nu] \underline{c c_{j_1} \ldots c_{j_t}} c_l \ldots$$

$$= -\sum A'_\nu c_k [u A''_\nu] c_{j_1} \ldots c_{j_t} \underline{c c_l} \ldots = 0\,.$$

Similarly, $cuB = 0$, and therefore $cuc_0 = 0$, that is, (3.25) is proved. It follows immediately from (3.25) that

$$[c_m c_0 u^2 c_0] = -[c u^2 c_0] = \underset{(3.25)}{\underline{[u c u c_0]}} = 0.$$

All the conditions of Lemma 2.1a) hold with c_2 replaced by c_m and c_1 by c_0, so that the existence of a thick sandwich in L is guaranteed. □

3.2. Lemma. *Let L be any Lie algebra over a field F of characteristic $p > 3$, let $c_1, c_2, c_3, c_4 \in \mathfrak{C}^*_1(L)$ be thin sandwiches and $c = [a c_1 c_2 c_3 c_4] \neq 0$ an element of L that is unchanged under permutation of the c_i:*

$$c = [a c_{\pi 1} c_{\pi 2} c_{\pi 3} c_{\pi 4}], \qquad \pi \in S_4\,.$$

Then c is a thick sandwich of L.

Proof. In what follows, $\{i, j, k, l\} = \{1, 2, 3, 4\}$. It follows from the conditions of the lemma that $[cc_i] = [ac_jc_kc_lc_i^2] = 0$. Further, $cc_i = [ac_jc_kc_l]\underline{c_i^2} - \underline{c_i[ac_jc_kc_l]c_i} = 0 \Rightarrow 0 = [ucc_i] = -cuc_i - c_iuc$, that is,

$$cc_i = c_ic = 0; \qquad c_iuc + cuc_i = 0, \qquad 1 \leqslant i \leqslant 4. \tag{3.26}$$

In particular,

$$\begin{aligned} c^2 &= \underset{(3.26)}{\underline{c[ac_kc_l]c_i}} - \underline{cc_i}[ac_kc_l]c_j - \underline{cc_j}[ac_kc_l]c_i + \underline{cc_j}c_i[ac_kc_l] \\ &= -c_i[ac_kc_l]\underline{cc_j} = 0\,, \end{aligned}$$

that is, c is a sandwich and thus $cuc = 0$.

The required identical relation $cu^2c = 0$ will be established if we write one of the occurrences of c as a linear combination of associative monomials and then check that

$$\begin{aligned} &\text{a)} \quad cuvc_ic_jc_k = 0\,; \\ &\text{b)} \quad cu^2c_ic_jac_kc_l = 0\,; \\ &\text{c)} \quad cu^2c_ivc_jc_kc_l = 0\,; \\ &\text{d)} \quad cu^2ac_ic_jc_kc_l = 0\,. \end{aligned} \tag{3.27}$$

We have successively:

$$\begin{aligned} \text{a)} \quad cu^2c_ic_jc_k &= \underset{(3.26)}{\underline{c[c_iu^2]c_jc_k}} + 2\,\underset{(3.26)}{\underline{cuc_i}}\,uc_jc_k - \underset{(3.26)}{\underline{cc_i}}\,u^2c_jc_k \\ &= -c_j[c_iu^2]\underline{cc_k} - 2c_iu\underline{cuc_j}c_k = 2c_iuc_ju\underline{cc_k} = 0\,. \end{aligned}$$

Clearly, $cu^2c_ic_jc_k = 0 \Rightarrow cuv\,c_ic_jc_k = 0$.

$$\begin{aligned} \text{b)} \quad cu^2c_ic_jac_kc_l &= \underset{(3.26)}{\underline{c[c_iu^2]c_jac_kc_l}} + 2\,\underset{(3.26)}{\underline{cuc_i}}\,uc_jac_kc_l - \underline{cc_i}u^2c_jac_kc_l \\ &= -c_j[c_iu^2]\underline{cac_k}c_l - 2c_iu\underline{cuc_j}\,ac_kc_l = c_j[c_iu^2]c_k\underline{acc_l} \\ &\quad + 2c_iuc_ju\underline{cac_k}c_l = -2c_iuc_juc_ka\underline{cc_l} = 0\,. \end{aligned}$$

$$\begin{aligned} \text{c)} \quad cu^2c_ivc_jc_kc_l &= \underset{a)\,(3.27)}{\underline{c[c_iu^2]vc_jc_kc_l}} + 2\,\underset{(3.26)}{\underline{cuc_i}}\,uvc_jc_kc_l - \underset{(3.26)}{\underline{cc_i}}\,u^2vc_ic_kc_l \\ &= -2c_i\underset{a)\,(3.27)}{\underline{ucuvc_jc_kc_l}} = 0\,. \end{aligned}$$

$$\begin{aligned} \text{d)} \quad cx^3c_ic_jc_kc_l &= \underset{a)\,(3.27)}{\underline{cx[c_ix^2]c_jc_kc_l}} + \underset{e)\,(3.27)}{\underline{2cx^2c_ixc_jc_kc_l}} - \underset{(3.26)}{\underline{cxc_i}\,x^2c_jc_kc_l} \\ &= c_i\underset{a)\,(3.27)}{\underline{xcx^2c_jc_kc_l}} = 0\,. \end{aligned}$$

Linearization with respect to x, which is possible for $p > 3$ (see 3.27a), shows that

$$cx^3 c_i c_j c_k c_l = 0 \Rightarrow cu^2 ac_i c_j c_k c_l = 0 . \quad \square$$

Remark. The principle of the proof, much used and typical, can be interpreted thus: if $c_0 = [ac_1 c_2] = [ac_2 c_1] \neq 0$ and $c_1^2 = c_2^2 = 0$, then c_0 is a sandwich and $c_0 uc_1 c_2 = 0$.

In fact, $c_0 c_i = 0$, $c_0 uc_i + c_i uc_0 = 0$ for $i = 1, 2$, so that $c_0^2 = c_0 [ac_1 c_2]$

$= \underline{c_0 ac_1} c_2 = -c_1 a\underline{c_0 c_2} = 0$. Further, $\underline{c_0 uc_1} c_2 = -c_1 u\underline{c_0 c_2} = 0$.

If now $p > 5$ and $c = [c_0 c_3 c_4]$ is the element from the statement of the lemma, then clearly $[cc_i] = 0$ for $i = 0, 3, 4$, and $cc_2 = 0$. Moreover, $cc_0 = c_0 c$ $= c_0 [ac_3 c_4 c_1 c_2] = \underline{c_0 [ac_3 c_4] c_1} c_2 = -c_1 [ac_3 c_4] \underline{c_0 c_2} = 0$. All the conditions of Lemma 3.1 are satisfied, and the existence of a thick sandwich is guaranteed. This weak form of Lemma 3.2 is completely sufficient for our purposes.

§ 4. Sandwich Algebras

We are interested in finitely generated sandwich algebras (see Definition 1.5.4), these being, apparently, an absolutely indispensible tool in the study of Engel Lie algebras. We formulate the main results of this section as follows.

4.1. Theorem. *Every finitely generated sandwich Lie algebra with* E_n *over a field* F, $n < p = \operatorname{char} F$, *is nilpotent.*

Corollary. *Every Lie algebra with* E_n *over a field* F *of characteristic* $p > n + [n/2]$ *is locally nilpotent.*

The proof follows at once from Theorem 4.1 and Corollary 2.3.2. $\square$

It is natural to use induction on the cardinal $r + 1$ of the set $\{x_0, x_1, \ldots, x_r\}$ of sandwiches generating L in proving the theorem. It is easy to prove without using E_n that $L^{r+2} = 0$ for $r = 0, 1, 2$, and $L^6 = 0$ for $r = 3$. We argue by contradiction, that is, we assume that the theorem is false; it is useful to introduce the following definition.

Definition. An arbitrary (that is, not necessarily Engel) Lie algebra L generated by a set of sandwiches $\{x_0, x_1, \ldots, x_r\}$ will be said to be *extremal* if L is not nilpotent but every subalgebra with fewer generators (sandwiches) is nilpotent.

Proposition. *Every extremal Lie algebra* L *over a field* F *of characteristic* $p > 5$ *contains a thick sandwich.*

Every Lie algebra with a non-zero abelian ideal contains sandwiches of arbitrary thickness. Therefore, in the course of the proof of the proposition, we assume that L is a non-nilpotent (extremal) sandwich Lie algebra with $Z(L) = 0$, embedded in the associative algebra $A(L)$ and not containing abelian ideals.

The theorem is deduced from the proposition following the usual plan: going over to the factor-algebra $\bar{L} = L/R(L)$ by the locally nilpotent radical $R(L)$ and using the fact that images of sandwiches of thickness l are sandwiches of thickness $\geqslant l$ (or are zero), we may assume at the outset that L is an extremal algebra with $R(L) = 0$; by the proposition, L has thick sandwiches, and their thicknesses can rise to p-4, by Theorem 2.4.6; by 1.5.5 and E_n, this is enough for the construction of a non-trivial abelian ideal; and the contradiction thus achieved gives the desired result.

Thus, our aim comes down to proving the proposition, and we shall be occupied with this until the end of the section.

Remark. It is useful to have thick sandwiches for the following reason also.

The linear space $\mathfrak{C}_t(L)$ spanned by the thick sandwiches is an ideal, in every sandwich Lie algebra L.

This is because

$$c^2 = 0, \qquad cu^2c = 0, \qquad c_0^2 = 0 \Rightarrow [cc_0]u^2[cc_0]$$

$$= cc_0u^2cc_0 + c_0cu^2c_0c - c_0\underline{cu^2c}c - cc_0u^2c_0c$$

$$= (\underline{cc_0[cu^2]c_0} + \underline{2cc_0ucuc_0} - \underline{cc_0c}u^2c_0)$$

$$+ (\underline{c_0[cu^2]c_0c} + 2c_0u\underline{cuc_0c} - c_0u^2\underline{cc_0c})$$

$$- (\underline{c[c_0u^2]c_0c} + 2cu\underline{c_0uc_0}c - cu^2\underline{c_0^2}c) = 0\,.$$

4.2. Lemma. *Let L be an extremal Lie algebra. Then L has a thick sandwich, or else it has sandwiches c, c_0, c_1, c_2 satisfying the following conditions:*

(i) $[cc_0c_1c_2] = [cc_0c_2c_1] \neq 0$:

(ii) $[c_0c_1] = 0, \qquad [c_0c_2] = 0$

(*we do not assert that* $[c_1c_2] = 0$.)

Proof. By definition, L is generated by a set $\{x_0, x_1, \ldots, x_r\}$ where $x_i^2 = 0$ in each case, such that $S = \{x_1, \ldots, x_r\}$ generates a nilpotent subalgebra M. This subalgebra contains a central sandwich, namely some sufficiently long commutator c_0 in $x_1, \ldots, x_r$. Since $[c_0x_i] = 0$, $1 \leqslant i \leqslant r$ and $Z(L) = 0$, we have $[x_0c_0] \neq 0$. We consider now any S-continuation of $[x_0c_0]$, in the sense of Definition 1.5.8. Let us suppose it is

$$[\widetilde{x_0c_0}] = [x_0c_0x_{i_1} \ldots x_{i_s}]\,.$$

The process of "subtwisting" the continuation $[\widetilde{x_0c_0}]$ described in 1.5.8 leads to an element

$$h = [x_0c_0f_1 \ldots f_m] = [x_0c_0f_{\pi 1} \ldots f_{\pi m}] \neq 0, \qquad \pi \in S_m\,,$$

where $f_j = f_j(x_1, \dots, x_r)$ is a sandwich and $[hx_i] = 0$, $1 \leqslant i \leqslant r$. Again we have $Z(L) = 0 \Rightarrow [c_0 x_0 f_1 \dots f_m x_0] = -[hx_0] \neq 0$. Clearly, this is possible only for $m \geqslant 2$.

If $m > 2$, because of the permutability relations $[c_0 f_j] = 0$ (corollary of $[c_0 x_i] = 0$, $1 \leqslant i \leqslant r$), h is capable of playing the rôle of c in Lemma 3.2. Thus, the absence of a thick sandwich in L implies that $m = 2$. Setting $x_0 = c, f_1 = c_1, f_2 = c_2$ and preserving c_0, we thereby satisfy (i) and (ii). □

4.3. Lemma. *An extremal algebra L without thick sandwiches contains an element of the form*

$$e = [cc_0c_1c_2cc^{(0)}c^{(1)}c_{i_1}c^{(2)}c_{i_2} \dots c^{(m-1)}c_{i_{m-1}}c^{(m)}c_{i_m}] \neq 0$$

for arbitrarily large m, where $c^{(0)} = [ac_0]$, $a^2 = 0$; $i_1 \in \{1, 2\}$, $i_k \in \{0, 1, 2\}$, $k > 1$; $c^{(j)} = [a_j c]$, $a_j^2 = 0$.

Proof. To start off with, we choose a sandwich $c' = [cc_0c_1c_2] \neq 0$ as in Lemma 4.2. It follows from properties (i) and (ii) that $c'c_i = 0 = c_ic'$, while if $[cc_0c_1c_2c] = 0$, Lemma 3.1 leads to a thick sandwich and thus to a contradiction. There remains the possibility that $[cc_0c_1c_2c] \neq 0$, and then by Lemma 3.1 again we have $[cc_0c_1c_2c]c_0 \neq 0$, that is, $[cc_0c_1c_2cac_0] \neq 0$, while we may assume that $a^2 = 0$ (L is a sandwich algebra so that every element $a \in L$ has the form $a = \sum a_i$, where $a_i^2 = 0$, and $[xa] \neq 0 \Rightarrow [xa_i] \neq 0$ for some i). This means that $[cc_0c_1c_2cc^{(0)}] \neq 0$:

$$c^{(0)} = [ac_0], \qquad c_0c^{(0)} = c^{(0)}c_0 = 0, \qquad c^{(0)}uc_0 + c_0uc^{(0)} = 0\,. \tag{3.28}$$

At this point, as well as the obvious relation $[cc_0c_1c_2c\underline{c^{(0)}c}] = 0$, we have

$$[\underline{cc_0}c_1c_2cc^{(0)}c_i] = [cc_kc_lc_icc^{(0)}c_i] = [\underline{cc_kc_lc_i}c^{(0)}cc_i] = [cc_ic_2\underline{c_0c^{(0)}}cc_i] = 0$$

(here $\{i, k, l\} = \{0, 1, 2\}$). By Lemma 3.1,

$$[cc_0c_1c_2cc^{(0)}]c \neq 0\,,$$

that is,

$$[cc_0c_1c_2cc^{(0)}c^{(1)}] \neq 0, \qquad c^{(1)} = [a_1c], \qquad a_1^2 = 0\,.$$

We begin a new round:

$$[\underline{cc_0c_1}c_2cc^{(0)}c^{(1)}c_0] = [cc_1c_2\underset{(3.28)}{\underline{c_0cc^{(0)}}}c^{(1)}c_0] = -[cc_1c_2c^{(0)}\underline{cc_0c^{(1)}c_0}] = 0\,.$$

Furthermore,

$$c^{(1)}c = cc^{(1)} = 0, \qquad c^{(1)}uc + cuc^{(1)} = 0\,,$$

so that $[cc_0c_1c_2cc^{(0)}\underline{c^{(1)}c}] = 0$ and $[cc_0c_1c_2cc^{(0)}\underline{c^{(1)}uc}] = -[cc_0c_1c_2\underline{cc^{(0)}c}uc^{(1)}] = 0$. By Lemma 3.1, the absence of a thick sandwich in L implies that

$$[cc_0c_1c_2cc^{(0)}c^{(1)}c_{i_1}] \neq 0, \qquad i_1 \in \{1, 2\}$$

(bearing in mind that $i_1=1$ or 2). We have thus already obtained our element e with index $m=1$.

Eschewing superfluous notation, we bring in the induction hypothesis that an element e with index m has been found, and construct one with index $m+1$. We note first that

$$e=[cc_0c_1c_2c[c^{(0)}c^{(1)}c_{i_1}\ldots c_{i_{m-1}}c^{(m)}c_{i_m}]]\,, \tag{3.29}$$

because, in expanding the inner commutator in $A(L)$, we cannot allow ourselves to replace the dots in $[cc_0c_1c_2c\ldots]$ either by $c^{(k)}$, $k>0$ (because $cc^{(k)}=0$), or by c_{i_k} (since $[cc_0c_1c_2cc_{i_k}]=[c\ldots\underline{c_{i_k}cc_{i_k}}]=0$), $1\leqslant k\leqslant m$. It follows from (3.29) that $[ec]=0$. Moreover,

$$[ec_j]=[cc_0c_1c_2c[c^{(0)}c^{(1)}c_{i_1}\ldots c_{i_{m-1}}c^{(m)}]c_{i_m}c_j]=0\,.$$

For, when $i_m=j$ this is clear, while for $i_m=i\neq j$ we have

$$[ec_j]=[cc_kc_ic_jc[c^{(0)}\ldots c^{(m)}]c_ic_j]=\tfrac{1}{2}[cc_kc_ic_j(c+[c^{(0)}\ldots c^{(m)}])^2c_ic_j]\,.$$

But

$$[cc_kc_ic_ju^2c_ic_j]=[cc_k\underline{c_ic_j[c_iu^2]}c_j]+2[\underset{\text{(Lemma 4.2)}}{\underline{cc_kc_ic_j}}\;uc_iuc_j]-[cc_k\underline{c_ic_jc_i}u^2c_j]$$

$$=2[cc_kc_j\underline{c_iuc_i}uc_j]=0\,.$$

Bringing all this together, we write:

$$[ec_j]=0,\qquad j=0,1,2;\qquad [ec]=0. \tag{3.30}$$

By Lemma 3.1 as applied to the element (3.29), we have to accept that $ec\neq0$, that is $[ec^{(m+1)}]\neq0$, where $c^{(m+1)}=[a_{m+1}c]$ for some sandwich a_{m+1}. Since

$$[ec^{(m+1)}uc]=-[ecuc^{(m+1)}]=0\,,$$

$$[ec^{(m+1)}[c^{(0)}\ldots c^{(m)}c_{i_m}]]$$

$$=[cc_0c_1c_2\underline{c[c^{(0)}\ldots c_{i_m}]c^{(m+1)}[c^{(0)}\ldots c_{i_m}]}]=0\,,$$

($[c^{(0)}\ldots c^{(m)}c_{i_m}]$ is a sandwich, since $c^{(0)},\ldots,c^{(m)},c_{i_m}$ are sandwiches), we get from Lemma 3.1 again that $[ec^{(m+1)}c_{i_{m+1}}]\neq0$ for some $i_{m+1}\in\{0,1,2\}$. The induction is complete: we have constructed an element of the form indicated in the statement of the lemma with index $m+1$. □

4.4. Lemma. *The expression for e in Lemma* 4.3 *does not contain intervals of the form*

$$c_jc^{(s+1)}c_ic^{(s+2)}c_jc^{(s+3)}c_ic^{(s+4)}c_jc^{(s+5)}c_ic^{(s+6)}c_j,$$

where $\{i,j\}\subset\{0,1,2\}$.

Proof. Decreasing m if necessary and taking it equal to $s+6$, we assume that

$$e=[e'c^{(s)}c_jc^{(s+1)}c_ic^{(s+2)})\ldots c_jc^{(s+5)}c_ic^{(s+6)}c_j]\,,$$

where e' is an element of the same type as e, but with index $s-1$. The same applies to the dashed elements e'', e''' below. We write e in any way as

$$e=[e''c^{(\alpha)}c_\nu c^{(\beta)}c_\mu c^{(\gamma)}c_\nu\ldots]\,,\qquad \{\nu,\mu\}=\{i,j\},$$

and then turn attention to the relations

$$[e''c^{(\alpha)}c_\nu c_\mu]=\underset{(3.30)}{\underline{[e'''c_\mu]}}=0\,, \tag{3.31}$$

$$[e''c^{(\alpha)}c_\nu c^{(\gamma)}c^{(\beta)}]=-[e''c^{(\alpha)}c_\nu ca_\gamma a_\beta c]=-\underset{(3.30)}{\underline{[e'''c\,a_\gamma a_\beta c]}}=0\,. \tag{3.32}$$

We have

$$\begin{aligned}0&=\underline{[e''c^{(\alpha)}c_\nu[c_\mu c^{(\beta)}c^{(\gamma)}]}c_\nu\ldots]\\&=[e''c^{(\alpha)}c_\nu c_\mu c^{(\beta)}c^{(\gamma)}c_\nu\ldots]+[e''c^{(\alpha)}c_\nu c^{(\gamma)}c^{(\beta)}c_\mu c_\nu\ldots]\\&\quad-[e''c^{(\alpha)}c_\nu c^{(\gamma)}c_\mu c^{(\beta)}c_\nu\ldots]-[e''c^{(\alpha)}c_\nu c^{(\beta)}c_\mu c^{(\gamma)}c_\nu\ldots]\,.\end{aligned}$$

Since the first two summands are zero, we get a skew-symmetric element e with respect to the $c^{(k)}$, $s+1\leqslant k\leqslant s+6$:

$$e=[e''c^{(\alpha)}c_\nu c^{(\beta)}c_\mu c^{(\gamma)}c_\nu\ldots]=-[e''c^{(\alpha)}c_\nu c^{(\gamma)}c_\mu c^{(\beta)}c_\nu\ldots]\,.$$

In particular,

$$\begin{aligned}e&=[e'c^{(s)}\ldots c_jc^{(\alpha)}c_ic^{(\beta)}c_jc^{(\gamma)}c_ic^{(\delta)}c_j\ldots]\\&=[e'c^{(s)}\ldots c_jc^{(\gamma)}c_ic^{(\delta)}c_jc^{(\alpha)}c_ic^{(\beta)}c_j\ldots]\,,\end{aligned} \tag{3.33}$$

since $\begin{pmatrix}\alpha&\beta&\gamma&\delta\\ \gamma&\delta&\alpha&\beta\end{pmatrix}$ is an even permutation.

Finally, we observe that

$$\begin{aligned}e&=[e'c^{(s)}\ldots c_jc^{(\alpha)}c_ic^{(\beta)}c_j\ldots]=[e'c^{(s)}\ldots c_j[c^{(\alpha)}c_i[c^{(\beta)}c_j]]\ldots]\\&=[e'c^{(s)}\ldots[c^{(\alpha)}c_i[c^{(\beta)}c_j]]c_j\ldots]\,.\end{aligned}$$

Thus, setting

$$f_k=[c^{(s+2k-1)}c_i[c^{(s+2k)}c_j]],\qquad k=1,2,3\,,$$

we arrive at the expression

$$e=\underset{(3.33)}{[e'c^{(s)}c_j\underline{f_1f_2f_3}]}=[e'c^{(s)}c_jf_{\pi1}f_{\pi2}f_{\pi3}];\qquad \pi\in S_3\,.$$

Since in addition $[f_k,c_j]=0$, we have an element

$$e=[af_1f_2f_3f_4]=[af_{\pi1}f_{\pi2}f_{\pi3}f_{\pi4}]\neq0\,,\qquad \pi\in S_4\,,$$

with $a=[e'c^{(s)}]$, $f_4=c_j$, $f_k^2=0$, $1\leqslant k\leqslant4$, that satisfies the conditions of Lemma 3.2 and is therefore a thick sandwich. This contradiction concludes the proof. □

While the construction of elements of type e went purely formally, the bugaboo threat to Lemma 3.1 from a mythical thick sandwich worked without a hitch. Lemma 4.4 alters the situation somewhat. A scarcely more difficult analysis, which we now perform, gives Lemma 3.1 the green light.

4.5. Lemma. *If the extremal Lie algebra L has no thick sandwiches, there is a thin sandwich of the form*

$$d = [cc_0c_1c_2cc^{(0)}c^{(1)}c_1c^{(2)}c_0c^{(3)}c_2c^{(4)}c_0] \neq 0 .$$

The commutator d is skew-symmetric in $c^{(1)}$, $c^{(2)}$ and independently in $c^{(3)}$, $c^{(4)}$.

Proof. We consider the sandwich e from the statement of Lemma 4.3. By Lemma 4.4, the lower indices 0, 1, 2 can occur in e as often as is desired if m is sufficiently large. In particular, there exists a thin sandwich of the form

$$e_1 = [cc_0c_1c_2cc^{(0)}c^{(1)}c_{i_1}c^{(2)} \ldots c^{(k)}c_{i_k}c^{(k+1)}c_0] \neq 0, \{i_1, \ldots, i_k\} = \{1, 2\} ,$$

with the smallest possible k (by Lemma 4.4, k is not more than 6). By construction, $i_1 \in \{1, 2\}$, so that $k \geqslant 1$.

Clearly, no adjacent pair of indices i_s, i_{s+1} can be the same ($c_ic^{(s+1)}c_i = 0$), that is,

$$(i_1, \ldots, i_k) = (1, 2, 1, 2, \ldots) \quad \text{or} \quad (2, 1, 2, 1, \ldots) .$$

The device used during the proof of Lemma 4.4 (see the simple calculation after relation (3.32)) shows that e_1 is skew-symmetric with respect to $c^{(s)}$ and $c^{(s+1)}$, $2 \leqslant s \leqslant k-1$ (consider relations of the type $c_i[c_jc^{(s)}c^{(s+1)}]c_i = 0$ and use the property $[ec_i] = 0$, $[ec^{(s)}c^{(s+1)}] = 0$). Moreover,

$$[c_{i_1}c^{(1)}c^{(2)}] = -[c_{i_1}ca_1a_2c] = [bc]$$

is an element of the same type as $c^{(1)}$, $c^{(2)}$, ... The same sort of reasoning applies to $[c_{i_k}c^{(k)}c^{(k+1)}]$. Therefore, since k is minimal,

$$[cc_0c_1c_2cc^{(0)}[c_{i_1}c^{(1)}c^{(2)}]c_{i_2} \ldots c^{(k+1)}c_0] = 0 ,$$
$$[cc_0c_1c_2cc^{(0)}c^{(1)}c_{i_1}c^{(2)} \ldots c_{i_{k-1}}[c_{i_k}c^{(k)}c^{(k+1)}]c_0] = 0 . \tag{3.34}$$

Since $c^{(1)}c^{(2)} = -ca_1a_2c$, $[cc_0c_1c_2cc^{(0)}c] = 0$, $[cc_0c_1c_2cc^{(0)}c_{i_1}] = 0$ and $[e'c] = 0$, $[e'c_{i_k}] = 0$, where $e' = [cc_0c_1c_2c \ldots c_{i_{k-1}}]$ (see the proof of Lemma 4.3), it follows that when the inner commutators are expanded, the elements $c^{(s)}$ and $c^{(s+1)}$ cannot stand alongside each other. Therefore, it follows from the two relations (3.34) that e_1 is skew-symmetric with respect to $c^{(1)}$, $c^{(2)}$ and $c^{(k)}$, $c^{(k+1)}$ respectively. Thus

$$e_1 = (\operatorname{sgn} \pi)[cc_0c_1c_2cc^{(0)}c^{(\pi 1)}c_{i_1}c^{(\pi 2)}c_{i_2} \ldots c^{(\pi(k+1))}c_0], \qquad \pi \in S_{k+1} .$$

Recalling the process of constructing sandwiches of type e from Lemma 4.3 (one writes $e_1 = [cc_0c_1c_2c[c^{(0)}c^{(1)} \ldots c_{i_k}c^{(k+1)}c_0]]$ and carries out a small discussion on the subject of the applicability of Lemma 3.1), we arrive at a sandwich

$$e_2 = [e_1c^{(k+2)}c_{j_1} \ldots c^{(k+l+1)}c_{j_l}c^{(k+l+2)}c_0] \neq 0 , \qquad \{j_1, \ldots j_1\} = \{1, 2\} ,$$

with smallest possible suffix $l \geqslant 1$. Lemma 4.4 again guarantees that $l \leqslant 6$. As in the case of e_1, the property of skew-symmetry is checked thus:

$$e_2 = (\operatorname{sgn} \sigma) [e_1 c^{(\sigma(k+2))} c_{j_1} \ldots c_{j_l} c^{(\sigma(k+l+2))} c_0] ;$$

where σ is any element of the symmetric group S_{l+1} acting on $\{k+2, k+3, \ldots, k+l+2\}$.

Up to this point, everything has gone in a more or less unique fashion. From here on, two possibilities arise.

First Possibility. The relation

$$[cc_0c_1c_2c \ldots c^{(\pi k)} c_{i_k} [c_0 c^{(\pi(k+1))} c^{(\sigma(k+2))}] c_{j_1} \ldots c_{j_l} c^{(\sigma(k+l+2))} c_0] = 0$$

is satisfied for all permutations π and σ in two groups isomorphic to S_{k+1} and S_{l+1} respectively. This will hold when $i_k = j_1$, for example. In this situation, the sandwich e_2 is skew-symmetric with respect to all the elements $c^{(\nu)}$, $\nu = 1, 2, \ldots, k+l+2$, since

$$[c \ldots c_{i_k} c_0 c^{(\pi(k+1))} c^{(\sigma(k+2))} c_{j_1} \ldots] = 0 ,$$

$$[c \ldots c_{i_k} c^{(\sigma(k+2))} c^{(\pi(k+1))} c_0 c_{j_1} \ldots] = 0$$

(we used arguments like this earlier). In particular,

$$e_2 = \pm [cc_0c_1c_2cc^{(0)} \ldots c^{(s)} c_0], \qquad s = 1, 2, \ldots, k+l+2 ,$$

so that $[e_2 c^{(s)}] = 0$. Since $c_0 c^{(0)} = 0$, we have $[e_2 c^{(0)}] = 0$. The relations $[e_2 c_i] = 0$, $i = 0, 1, 2$ and $[e_2 c] = 0$ are verified in the same way as for the general commutator e in Lemma 4.3. Thus, the product e_2 commutes with all the factors occurring in it. Moreover—and here e_2 is significantly different from e_1—c_0 occurs three times in e_2, and $e_2 c_0 = 0$ for the obvious reason. This means that we find ourselves in the conditions where Lemma 3.1 applies, and this guarantees the existence of a thick sandwich in L. Recalling the statement of our lemma, we have to acknowledge that the following must hold.

Second Possibility. There exist permutations π, σ such that

$$[cc_0c_1c_2cc^{(0)} \ldots c^{(\pi k)} c_{i_k} [c_0 c^{(\pi(k+1))} c^{(\sigma(k+2))}] c_{j_1} \ldots c_0] \neq 0 .$$

With the notation

$$f^{(1)} = c^{(\pi 1)}, \ldots, f^{(k)} = c^{(\pi k)} ,$$

$$f^{(k+1)} = [c_0 c^{(\pi(k+1))} c^{(\sigma(k+2))}] = [b'c] ,$$

$$f^{(k+2)} = c^{(\sigma(k+3))}, \ldots, f^{(k+l+1)} = c^{(\sigma(k+l+2))} \tag{3.35}$$

and since $i_k \neq j_1$, we get a sandwich

$$e' = [cc_0c_1c_2cc^{(0)} f^{(1)} c_i f^{(2)} c_j \ldots c_{j'} f^{(k)} c_{i'} f^{(k+1)} c_{j'} \ldots f^{(k+l+1)} c_0] \neq 0 ,$$

where

$$(j', i', j') = \begin{cases} (j, i, j) & \text{for odd } k \\ (i, j, i) & \text{for even } k \,, \end{cases}$$

with an alternating rule for the indices 1, 2 and having all the properties of the sandwich e from Lemma 4.3. However, as we know from the proof of Lemma 4.4 (and again referring to the simple calculation after (3.32)), the alternating rule for the indices guarantees skew-symmetry with respect to $f^{(s)}$ and $f^{(s+1)}$, for $2 \leqslant s \leqslant k+l$. We make only partial use of skew-symmetry: we substitute a selected element $f^{(k+1)}$ for $f^{(2)}$ and expand the commutator (3.35) at the same place, having first extracted c_0 from it, as it were from a Trojan horse. We get

$$e' = \pm\{[c\ldots c^{(0)}f^{(1)}c_i c^{(\pi(k+1))}c_0 c^{(\sigma(k+2))}c_j\ldots] + [c\ldots c^{(0)}f^{(1)}c_i c^{(\sigma(k+2))}c_0 c^{(\pi(k+1))}c_j\ldots]\} \neq 0\,.$$

This means that at least one of the summands is non-zero, and we come to the important conclusion that the minimal index k in the sandwich e_1, and thus in e_2, is 1.

The index k has lost its functional value and we use it from now on in a new sense: after the next change of notation, we get a sandwich

$$e_3 = [cc_0c_1c_2cc^{(0)}c^{(1)}c_{i_1}c^{(2)}c_0c^{(3)}c_{i_2}\ldots c^{(k)}c_{i_{k-1}}c^{(k+1)}c_0] \neq 0$$

with alternating law for the lower indices $i_1, \ldots, i_{k-1} \in \{1, 2\}$ and with minimal acceptable $k \geqslant 3$. This last condition (the minimality) ensures the skew-symmetry of e_3 in $c^{(3)}, c^{(4)}, \ldots, c^{(k+1)}$ (a repeat of the arguments for $e_1$and e_2). The sandwich e_3 is skew-symmetric with respect to $c^{(1)}$ and $c^{(2)}$ also:

$$[cc_0c_1c_2cc^{(0)}c^{(1)}c_{i_1}c^{(2)}c_0\ldots] + [cc_0c_1c_2cc^{(0)}c^{(2)}c_{i_1}c^{(1)}c_0\ldots]$$

$$= -[cc_0c_1c_2\underline{cc^{(0)}[c_{i_1}c^{(1)}c^{(2)}]}c_0\ldots]$$

$$= [cc_0c_1c_2\underline{cc_0}[c_{i_1}c^{(1)}c^{(2)}]c^{(0)}\ldots] = [cc_0c_1c_2\underline{c_0cc_0}\ldots] = 0\,.$$

However, in general $c^{(2)}$ and $c^{(3)}$ cannot be interchanged, and therefore it is impossible to apply the ubiquitous Lemma 3.1, since this would interfere with the perspective that $[e_3c^{(1)}] \neq 0$ or $[e_3c^{(2)}] \neq 0$.

We come now to a further application of the preceding construction. By using a formula of type (3.29) and writing e_3 in the form

$$e_3 = [cc_0c_1c_2cc_0^{(0)}c^{(3)}c_{i_2}\ldots c^{(k)}c_{i_{k-1}}c^{(k+1)}c_0]\,,$$

where $c_0^{(0)} = [c^{(0)}c^{(1)}c_{i_1}c^{(2)}c_0]$ is an element of type (3.28) fulfilling the function of $c^{(0)}$, we can construct a sandwich

$$e_4 = [e_3c^{(k+2)}c_{i_k}c^{(k+3)}\ldots c^{(k+l+1)}c_{i_{k+l-1}}c^{(k+l+2)}c_0] \neq 0$$

with the smallest allowable l, $1 \leqslant l \leqslant 6$ (remember Lemma 4.4).

Two possibilities again arise, similar to those above. One of them leads immediately (via Lemma 3.1) to a thick sandwich and is therefore rejected, while the second gives rise to a sandwich

$$[cc_0c_1c_2cc_0^{(0)}c^{(3)} \ldots c^{(k)}c_{i_{k-1}}[c_0c^{(k+1)}c^{(k+2)}]c_{i_k} \ldots c^{(k+l+2)}c_0] \neq 0$$

(the permutations π and σ have been omitted in the interests of brevity) with alternating rule for the indices $i_2, \ldots, i_{k-1}, i_k, \ldots, i_{k+l-1} \in \{1, 2\}$. We use the consequent skew-symmetry with respect to $c^{(4)}, \ldots c^{(k)}, f = [c_0c^{(k+1)}c^{(k+2)}]$, to put f instead of $c^{(4)}$, and thus we get an element

$$[cc_0c_1c_2cc_0^{(0)}c^{(3)}c_{i_2}fc_{i_3} \ldots c^{(k+l+2)}c_0] \neq 0 .$$

Hence

$$[cc_0c_1c_2c[c^{(0)}c^{(1)}c_{i_1}c^{(2)}c_0]c^{(3)}c_{i_2}[c_0c^{(k+1)}c^{(k+2)}]]$$
$$= [cc_0c_1c_2cc_0^{(0)}c^{(3)}c_{i_2}f] \neq 0 .$$

We expand the inner commutators at this stage, remembering which terms are automatically zero, setting $c^{(4)} = c^{(k+1)}$ or $c^{(k+2)}$, $c^{(5)} = c^{(k+2)}$ or $c^{(k+1)}$ and fixing one of the two summands:

$$[cc_0c_1c_2cc^{(0)}c^{(1)}c_{i_1}c^{(2)}c_0c^{(3)}c_{i_2}c^{(4)}c_0c^{(5)}] \neq 0 .$$

Because of the alternating rule for the sequence of indices (we recall that we have proceeded from the superstructure to e_3), the initial segment $i_1 = 1$, $i_2 = 2$ or $i_1 = 2$, $i_1 = 1$ has been preserved. Since the indices 1 and 2 are equally suitable, we stick with the first possibility, for definiteness. Rejecting the factor $c^{(5)}$, we arrive at a sandwich

$$d = [cc_0c_1c_2cc^{(0)}c^{(1)}c_1c^{(2)}c_0c^{(3)}c_2c^{(4)}c_0] \neq 0 ,$$

as mentioned in the statement of the lemma. The fact that it is skew-symmetric with respect to $c^{(1)}$, $c^{(2)}$ and with respect to $c^{(3)}$, $c^{(4)}$ follows from the construction (and is fairly obvious). □

4.6. Proof of the Proposition in 4.1. As a starting-point we choose a sandwich as in the statement of Lemma 4.5. Assuming that the extremal Lie algebra L is devoid of thick sandwiches, we consider two possible cases.

Case 1. $[dc^{(1)}] = 0$, $[dc^{(2)}] = 0$. Here, skew-symmetry of d with respect to $c^{(3)}$, $c^{(4)}$ gives that $[dc^{(i)}] = 0$, $1 \leqslant i \leqslant 4$. Moreover, $c_0c^{(0)} = 0 \Rightarrow [dc^{(0)}] = 0$. The relations $[dc] = 0$, $[dc_j] = 0$, $0 \leqslant j \leqslant 2$, have already been checked for all sandwiches like those in Lemma 4.3, and d is one such. Finally, $dc_0 = 0$, and since the degree of d in c_0 is three, application of Lemma 3.1 gives the desired conclusion.

Case 2. $[dc^{(1)}] \neq 0$ or $[dc^{(2)}] \neq 0$. Since d is skew-symmetric in $c^{(1)}$ and $c^{(2)}$, we can restrict attention to the element

$$d_1 = [dc^{(1)}] = [cc_0c_1c_2cc^{(0)}c^{(1)}c_1c^{(2)}c_0c^{(3)}c_2c^{(4)}c_0c^{(1)}] \neq 0 , \tag{3.36}$$

which can be written (see (3.29)) in the form

$$d_1 = [cc_0c_1c_2ch_1]\,, \tag{3.37}$$

where

$$h_1 = [c^{(0)}c^{(1)}c_1c^{(2)}c_0c^{(3)}c_2c^{(4)}c_0c^{(1)}]\,.$$

In the form (3.37), the element d_1 is a pretender to the rôle of the sandwich in Lemma 3.1 with parameters $m = 6$, $k = 1$, $l = 5$. However, the fact that the conditions of the lemma are satisfied cannot be established without leaning heavily on the expression for d_1 as a commutator (3.36) of length 15. This is clear since we want to extract relations from the special form of d_1 that blockade the general construction in Lemma 4.3.

Thus, it is sufficient to convince ourselves that

$$d_1c = cd_1 = 0, \qquad [d_1c_i] = 0, \qquad i = 0, 1, 2\,. \tag{3.38}$$

For simplicity, and for the formal deduction of the required relation, we use the notation

$$e_+ = [cc_0c_1c_2cc^{(0)}] = [cc_ic_jc_kcc^{(0)}]\,,$$

$$e = e\begin{pmatrix} s_1 & s_2 & \dots & s_m \\ i_1 & i_2 & \dots & i_m \end{pmatrix} = [e_+c^{(s_1)}c_{i_1} \dots c^{(s_m)}c_{i_m}]\,.$$

In our case $\{s_1, \dots, s_m\} \subset \{0, 1, 2, 3, 4\}$, $\{i_1, \dots, i_m\} \subset \{0, 1, 2\}$. Moreover,

$$c^{(0)} = [ac_0], \qquad c^{(s)} = [a_sc], \qquad s > 0\,,$$

so that, in view of the properties of the element e deduced above, we have

$$c_0c^{(0)} = c^{(0)}c_0 = 0, \qquad c_0uc^{(0)} + c^{(0)}uc_0 = 0\,; \tag{3.39}$$

$$cc^{(s)} = c^{(s)}c = 0, \qquad cuc^{(s)} + c^{(s)}uc = 0\,; \tag{3.40}$$

$$[e_+c_i] = 0, \qquad i \in \{0, 1, 2\}, \qquad [e_+c] = 0\,; \tag{3.41}$$

$$[ec_i] = 0, \qquad i \in \{0, 1, 2\}, \qquad [ec] = 0\,. \tag{3.42}$$

From this, there follow the useful derivative relations

$$[e_+c^{(s)}c^{(t)}] = 0, \qquad [ec^{(s)}c^{(t)}] = 0\,; \tag{3.43}$$

$$e\begin{pmatrix} s & \dots \\ 0 & \dots \end{pmatrix} = 0\,; \tag{3.44}$$

$$e\begin{pmatrix} \dots & r & s & t & \dots \\ \dots & i & j & i & \dots \end{pmatrix} = -e\begin{pmatrix} \dots & r & t & s & \dots \\ \dots & i & j & i & \dots \end{pmatrix}; \tag{3.45}$$

$$e\begin{pmatrix} \dots & s & t & s & \dots \\ \dots & i & j & k & \dots \end{pmatrix} = -e\begin{pmatrix} \dots & s & t & s & \dots \\ \dots & j & i & k & \dots \end{pmatrix}. \tag{3.46}$$

In fact,

$$[e_+ c^{(s)} c^{(t)}] = -[\underbrace{e_+ c\, a_s a_t c}_{(3.41)}] = 0\,,$$

$$[e c^{(s)} c^{(t)}] = -[\underbrace{ec\ \ a_s a_t c}_{(3.42)}] = 0\,,$$

$$e\begin{pmatrix} s & \dots \\ 0 & \dots \end{pmatrix} = [e_+ c^{(s)} c_0 \dots] = [c c_1 c_2 c_0 \underbrace{c c^{(0)} c^{(s)} c_0}_{(3.39)} \dots]$$

$$= -[c c_1 c_2 \underline{c_0 c c_0}\, c^{(s)} c^{(0)} \dots] = 0\,,$$

$$e\begin{pmatrix} \dots r\ s\ t \dots \\ \dots i\ j\ i \dots \end{pmatrix} + e\begin{pmatrix} \dots r\ t\ s \dots \\ \dots i\ j\ i \dots \end{pmatrix}$$

$$= [\dots c^{(r)} \underline{c_i c^{(s)} c_j c^{(t)} c_i} \dots] + [\dots c^{(r)} \underline{c_i c^{(t)} c_j c^{(s)} c_i} \dots]$$

$$= -[\dots c^{(r)} \underline{c_i [c_j c^{(s)} c^{(t)}] c_i} \dots] + \Bigg[\underbrace{\Bigg[e\begin{pmatrix} \dots r \\ \dots i \end{pmatrix}, c_j\Bigg]}_{(3.42)} c^{(s)} c^{(t)} c_i \dots\Bigg]$$

$$+ \Bigg[\underbrace{\Bigg[e\begin{pmatrix} \dots r \\ \dots i \end{pmatrix} c^{(s)} c^{(t)}\Bigg]}_{(3.43)} c_j c_i \dots\Bigg] = 0\,,$$

$$e\begin{pmatrix} \dots s\ t\ s \dots \\ \dots i\ j\ k \dots \end{pmatrix} + e\begin{pmatrix} \dots s\ t\ s \dots \\ \dots j\ i\ k \dots \end{pmatrix}$$

$$= [\dots \underline{c^{(s)} c_i c^{(t)} c_j c^{(s)}}\, c_k \dots] + [\dots \underline{c^{(s)} c_j c^{(t)} c_i c^{(s)}}\, c_k \dots]$$

$$= -[\dots \underline{c^{(s)} [c^{(t)} c_i c_j] c^{(s)}}\, c_k \dots] + [\underbrace{\dots c^{(s)} c^{(t)}}_{(3.43)} c_i c_j c^{(s)} c_k \dots]$$

$$+ \Bigg[\underbrace{\Bigg[e\begin{pmatrix} \dots s \\ \dots j \end{pmatrix}, c_i\Bigg]}_{(3.42)} c^{(t)} c^{(s)} c_k \dots\Bigg] = 0$$

(in the penultimate commutator, the first three dots denote $e\begin{pmatrix} \dots r \\ \dots l \end{pmatrix}$ or e_+).

We introduce another auxiliary relation, which we shall need soon:

$$\Bigg[e\begin{pmatrix} 1 \\ 2 \end{pmatrix} [c_1 c^{(2)} c_0 c^{(3)}] c^{(4)} c_0 c^{(1)} c_1\Bigg] = 0\,. \tag{3.47}$$

To prove this, we must expand the inner commutator as a linear combination of eight associative monomials in $A(L)$, make a substitution and carry out a small number of transformations:

$$\left[e\begin{pmatrix}1\\2\end{pmatrix}[c_1c^{(2)}c_0c^{(3)}]c^{(4)}c_0c^{(1)}c_1\right]$$

$$=\Big[\underbrace{e\begin{pmatrix}1\\2\end{pmatrix}c_1}_{(3.42)}c^{(2)}c_0c^{(3)}c^{(4)}\dots\Big]-\Big[\underbrace{e\begin{pmatrix}1&2\\2&1\end{pmatrix}c_0}_{(3.42)}c^{(3)}c^{(4)}\dots\Big]$$

$$-\Big[\underbrace{e\begin{pmatrix}1\\2\end{pmatrix}c_0}_{(3.42)}c_1c^{(2)}c^{(3)}c^{(4)}\dots\Big]+\Big[\underbrace{e\begin{pmatrix}1\\2\end{pmatrix}c_0}_{(3.42)}c^{(2)}c_1c^{(3)}c^{(4)}\dots\Big]$$

$$+\Big[\underbrace{e\begin{pmatrix}1\\2\end{pmatrix}c^{(3)}c^{(2)}c_1}_{(3.43)}c_0c^{(4)}\dots\Big]+\Big[\underbrace{e\begin{pmatrix}1&3\\2&0\end{pmatrix}c_1}_{(3.42)}c^{(2)}c^{(4)}\dots\Big]$$

$$-\Big[e\begin{pmatrix}1&3\\2&1\end{pmatrix}c^{(2)}\underline{c_0c^{(4)}}c_0c^{(1)}c_1\Big]-e\underbrace{\begin{pmatrix}1&3&2&4&1\\2&0&1&0&1\end{pmatrix}}_{(3.45)}$$

$$=e\underbrace{\begin{pmatrix}1&3&2&1&4\\2&0&1&0&1\end{pmatrix}}_{(3.45)}=-e\underbrace{\begin{pmatrix}1&3&1&2&4\\2&0&1&0&1\end{pmatrix}}_{(3.46)}=e\underbrace{\begin{pmatrix}1&3&1&2&4\\0&2&1&0&1\end{pmatrix}}_{(3.44)}=0\,.$$

We come now to our deduction of relations (3.38):

a) $[ud_1c]=-[d_1uc]=-[\underbrace{dc^{(1)}uc}_{(3.40)}]=[d\underline{c}uc^{(1)}]=0\,.$

Since u is an arbitrary element of L, we have $d_1c=0$, and the equation

$$d_1c-cd_1=[d_1c]=[\underbrace{dc^{(1)}c}_{(3.40)}]=0$$

implies that $cd_1=0$ as well;

b) $[\underbrace{d_1}_{(3.36)},c_0]=[c\dots c^{(4)}\underline{c_0c^{(1)}c_0}]=0\,;$

c) $$[d_1,c_2]=e\underbrace{\begin{pmatrix}1&2&3&4&1\\1&0&2&0&2\end{pmatrix}}_{(3.45)}=-e\begin{pmatrix}1&2&3&1&4\\1&0&2&0&2\end{pmatrix}$$

$$=e\underbrace{\begin{pmatrix}1&2&1&3&4\\1&0&2&0&2\end{pmatrix}}_{(3.46)}=-e\underbrace{\begin{pmatrix}1&2&1&3&4\\0&1&2&0&2\end{pmatrix}}_{(3.44)}=0\,;$$

$$\text{d)}\quad [d_1, c_1] = e\begin{pmatrix}1&2&3&4&1\\1&0&2&0&1\end{pmatrix} = [e_+ c^{(1)} c_1 c^{(2)} c_0 c^{(3)} c_2 c^{(4)} \dots]$$

$$= [e_+ c^{(1)} [c_1 c^{(2)}] c_0 c^{(3)} c_2 c^{(4)} \dots] + [\underbrace{e_+ c^{(1)} c^{(2)} c_1}_{(3.43)} c_0 c^{(3)} c_2 c^{(4)} \dots]$$

$$= [e_+ c^{(1)} [c_1 c^{(2)} c_0] c^{(3)} c_2 c^{(4)} \dots] + \left[\underbrace{e\begin{pmatrix}1\\0\end{pmatrix}}_{(3.44)} [c_1 c^{(2)}] c^{(3)} c_2 c^{(4)} \dots\right]$$

$$= [e_+ c^{(1)} [c_1 c^{(2)} c_0 c^{(3)}] c_2 c^{(4)} \dots] + [\underbrace{e_+ c^{(1)} c^{(3)}}_{(3.43)} [c_1 c^{(2)} c_0] c_2 c^{(4)} \dots]$$

$$= [e_+ c^{(1)} [c_1 c^{(2)} c_0 c^{(3)} c_2] c^{(4)} c_0 c^{(1)} c_1] + \underbrace{\left[e\begin{pmatrix}1\\2\end{pmatrix} [c_1 c^{(2)} c_0 c^{(3)}] c^{(4)} c_0 c^{(1)} c_1\right]}_{(3.47)}$$

$$= [e_+ c^{(1)} [c_1 c^{(2)} c_0 c^{(3)} c_2 c^{(4)}] c_0 c^{(1)} c_1] + [\underbrace{e_+ c^{(1)} c^{(4)}}_{(3.43)} [\dots] c_0 c^{(1)} c_1]$$

$$= [e_+ \underline{c^{(1)} [c_1 c^{(2)} c_0 c^{(3)} c_2 c^{(4)} c_0] c^{(1)}} c_1] + \left[\underbrace{e\begin{pmatrix}1\\0\end{pmatrix}}_{(3.44)} [c_1 c^{(2)} c_0 c^{(3)} c_2 c^{(4)}] c^{(1)} c_1\right]$$

$$= 0.$$

This completes the proof of (3.38). □

§ 5. Commentary

5.1. The arguments contained in 4.6 could be simplified greatly by considering an S-continuation $\tilde{d}$ of the sandwich d. Here

$$S = \{c, c_0, c_1, c_2, c^{(0)}, c^{(1)}, c^{(2)}, c^{(3)}, c^{(4)}\}.$$

If $\tilde{d} = [\dots c^{(k)}]$, $0 \leqslant k \leqslant 4$, a difficulty arises at first sight, since we do not know how often $c^{(k)}$ occurs in $\tilde{d}$. However,

$$[\dots c^{(0)} u c_0] = -[\dots c_0 u c^{(0)}] \text{ and } [\dots c^{(k)} u c] = -[\dots c u c^{(k)}] \text{ for } k \geqslant 1,$$

so that either $\tilde{d}c_0 = 0$ (or $\tilde{d}c = 0$), and Lemma 3.1 is applicable, or else $\tilde{d} = [\dots c_0]$ (or $\tilde{d} = [\dots c]$), and everything is again in order.

Unfortunately, the cardinality $|S| = 9$ is too big, and if $r \leqslant 8$ (the r is that from Definition 4.1), we cannot *a priori* guarantee the existence of $\tilde{d}$ by following Definition 1.5.8. True, sandwiches in S are very special and connected by many relations; but the analysis of these is perhaps equivalent to the very argument which we would like to be rid of.

5.2. It is striking that the existence proof for a sandwich like d, which is of very prosaic form—despite being the key to Proposition 4.1—should go along such a roundabout route. The whole of the combinatorial treatment in 4.2–4.5 smacks a little of witchdoctery. However, as we have convinced ourselves, incantations have their effect, confirming a maxim of G. K. Chesterton: "What is astonishing about miracles is that they sometimes happen".

At the same time, the author has been fully aware right from the start that Proposition 4.1, whose formulation does not involve the condition E_n, is too strong for the achievement of our more restricted aim—the proof of Theorem 4.1. A not completely successful attempt to give up general extremal algebras in favour of Engel Lie algebras was introduced at the end of § 2 in [137] (it involved the above "witchdoctor-like" arguments!). As a prophylactic against possible errors in the future, it is not without interest, and we shall dwell on this theme a little now, which merits the title:

5.3. False Sandwiches. Let us experiment with a more active use of the arguments used in Lemma 4.2. In an extremal Lie algebra L generated by thin sandwiches $x_0, x_1, \ldots, x_r$ and devoid of thick sandwiches, the lemma guarantees (in different notation) the existence of a sandwich of the form

$$c_1' = [a_0 x_0 a_1 a_1'] = [a_0 x_0 a_1' a_1] \neq 0 ,$$

where a_0, a_1, a_1' are commutators in the nilpotent subalgebra L_0 generated by $x_1, \ldots x_r$ such that $[a_0, L_0] = 0$ and $[c_1', L_0] = 0$. By assumption, $Z(L) = 0$, so that

$$c_1 = [a_0 x_0 a_1 a_1' x_0] \neq 0.$$

Arguing by induction, we assume that we have available a sandwich

$$c_k = [c_k' x_0], \qquad c_k' = [c_{k-1}' x_0 a_k a_k'] = [c_{k-1}' x_0 a_k' a_k] \neq 0 ,$$

where

$$a_k, a_k' \in L_0, \qquad [c_{k-1}', L_0] = 0, \qquad [c_k', L_0] = 0 \qquad (c_0' = a_0) .$$

We go over to an $\{x_1, \ldots x_r\}$-continuation $\tilde{c}_k = [c_k x_{j_1} \ldots x_{j_s}]$ *of* c_k, and then to a subtwist

$$c_{k+1}' = [c_k b_1 b_2 \ldots b_l] \neq 0, \qquad b_1, \ldots b_l \in L_0 ,$$

By definition of continuation and subtwisting (see 1.5.8), we have

$$c_{k+1}' = [c_k b_{\pi 1} b_{\pi 2} \ldots b_{\pi l}], \qquad \pi \in S_l ,$$

where $[c_{k+1}', L_0] = 0$.

Since $Z(L) = 0$,

$$c_{k+1} = [c_{k+1}' x_0] = [c_k' x_0 b_1 \ldots b_l x_0] \neq 0 ,$$

that is, $l \geqslant 2$. On the other hand, if $l > 2$ we have from Lemma 3.2 that c_{k+1}' is a thick sandwich (for $l \geqslant 4$ this is obvious, while for $l = 3$ we have to use the fact that $[c_k' b_j] = 0$ for $j = 1, 2, 3$ and $c_{k+1}' = -[x_0 c_k' b_1 b_2 b_3]$, from the induction hypothesis). The conditions in which we find ourselves leave the single possibility that $l=2$. Setting $a_{k+1}=b_1$, $a'_{k+1}=b_2$, we get a sandwich $c_{k+1}=[c'_{k+1}, x_0]$ with the required properties.

The induction therefore works, and we have proved:

Lemma. *Every extremal sandwich Lie algebra with no thick sandwiches has a sandwich of the form*

$$c_m=[a_0x_0a_1a'_1x_0a_2a'_2x_0 \dots x_0a_ma'_mx_0]$$

for arbitrarily large m.

The question arises: is there a real possibility of reducing the proof of the lemma to contradicting E_n? (Arguments of a length similar to those in § 4 are not counted).

Since $[ux_0a_ia'_ix_0]=[u[x_0a_ia'_i]x_0]$, we have

$$c_m=[a_0y_1y_2 \dots y_{m-1}y_mx_0] ,$$

where $y_i=[x_0a_ia'_i]$, and thus $[y_ix_0]=0$. Together with $x_0, a_1, a'_1, \dots a_m, a'_m$, the y_i are sandwiches, and the number of different elements among them is clearly not more than $\rho(\rho-1)$, where $\rho=\dim_F L_0$. It is more appropriate to write the element appearing in the statement of the lemma in the form

$$c_m=[a_0x_0y_{i_1}y_{i_2} \dots y_{i_{m-1}}y_{i_m}] ,$$

$$\{i_1, i_2, \dots i_m\} \subset \{1, 2, \dots, s\}, s \leqslant \rho(\rho-1) . \tag{3.48}$$

Remembering that we are aiming to prove Theorem 4.1, namely that a sandwich Lie algebra with n-th Engel condition is locally nilpotent, we would like to use induction on n. With L_1 standing for the subalgebra generated by the sandwiches $y_1, y_2, \dots y_s$, application of the inductive statement for $n-1$, of the collecting process from Proposition 1.5.7, and of the properties of the ideal $E_{n-1}(L_1)$ (see the beginning of § 3 of Chapter 2) leads to the expression

$$y_{i_1}y_{i_2} \dots y_{i_{m-1}}y_{i_m}=\sum W_i[u_iv_i^{n-1}] , \tag{3.49}$$

where $u_i, v_i \in L_1$, and the W_i are elements of the associative algebra A_1 on the same generating set $\{y_1, y_2, \dots y_s\}$. In all probability, this formula will not provide opportunities for using the commutativity relations $[y_i, x_0]=0$. We also have

$$c_m=\sum_j [u'_jv'^{n-1}_j] , \tag{3.50}$$

where the u'_j and v'_j are elements of the Lie algebra generated by L_1 and $c_0=[a_0x_0]$. If $v'_j \in L_1$ for every j, then using identity (1.2) and the Engel condition E_n, we find that

$$nc_m=\sum_j n[u''_jx_0v'^{n-1}_j]=\sum_j \left[u''_j \begin{Bmatrix} x_0 & v'_j \\ 1 & n-1 \end{Bmatrix}\right]=0 .$$

Unfortunately, the inclusion $v'_j \in L_1$ does not follow from anything, and the correct formulae (3.49) and (3.50) turn out to be useless. On page 17 of [137], however, the square brackets on the right-hand side of (3.49) are missing, and apart from notation it can be written as

$$y_{i_1} y_{i_2} \dots y_{i_m} = \sum W_i v_i^{n-1}, \qquad W_i \in A_1, \quad v_i \in L_1,$$

which is totally illegal: there is a mix-up between the rôles of $A(L_1)$ and A_1, and also an implicit identification of sandwiches in L_1 with sandwiches in L. The reason, roughly speaking, is that L_1 can have a principal ideal J generated by a "thick" sandwich e defined by the identities

$$[eu^s e] = 0; \qquad s = 1, 2, \dots,$$

whereas not even the relations

$$e^2 = 0, \qquad eue = 0$$

in A_1 characterizing "real" thin sandwiches follow from them.

In their time, all these considerations were the subject of discussions with Professors V. V. Morozov (in 1960) and K. Iwasawa (in 1970). The author remains forever grateful to them for their constructive interest in the proof of RBP. The correct answer to the naturally-arising question in connection with the proof of Theorem 3 in [137] has inevitably led in some measure to the arguments repeated in § 4. In fact, the casket can be opened rather more easily; but see Chap. 5 for that.

5.4. Lemmas 3.1 and 3.2, being an essential element in the whole construction, have been dubbed "necessary" by analogy with the "not avoidable" I. G. Petrovskii glade in the neighbourhood of Abramtsev near Moscow (all typical tourist itineraries take in this glade)[1]. Just as the footbridges in the first two sections led from $\mathfrak{C}_1^*$ to $\mathfrak{C}_2^*$, both "necessary" lemmas are extracted from the original paper [137], where they were forgotten in the solid mass of text. The proofs of assertions in § 4 differ from the auxiliary text [143] only in the terminology, which makes for easy verification of manipulations.

[1] Translator's remark: This is virtually untranslatable! It is a witticism based on the fact that "necessary" is "neobkhodimy" and "not avoidable" is "ne obkhodimy".

Chapter 4

Proof of the Main Theorem

We have in mind Theorem 1.7.1; all the other results formulated towards the end of Chap. 1 flow from it. No new ideas are needed for the proof of the main theorem. Contrary to a principle of Bourbaki, we have only to prepare ourselves for new calculations and a long chain of deduction.

We recall that we have a Lie algebra L with zero centre over a field of characteristic $p > 5$ and belonging to one of the two following types:

1) L satisfies E_n, $n < p$;

or

2) $\dim_F L < \infty$.

Further, L is embedded in the associative algebra $A(L)$ and contains at least one thin sandwich. We have to find an element $c \neq 0$ in L such that $cu^k c = 0$ for all $u \in L$, $k = 0, 1, 2$. The rest is by courtesy of Theorem 2.4.6, proved earlier.

§ 1. Pairs of Thin Sandwiches

1.1. Proposition. *Assume that the Lie algebra L (of finite dimension or satisfying E_n) contains a thin sandwich but is devoid of thick sandwiches. Then L has a pair of elements c_1, c_2 such that*

$$c_1 c_2 \neq 0, \quad c_1^2 = c_2^2 = [c_1 c_2] = 0\,; \tag{4.1}$$

$$c_1 u^2 c_1 c_2 = 0, \qquad c_2 u^2 c_1 c_2 = 0 \tag{4.2}$$

for all $u \in L$.

Conditions (4.1) are easy to satisfy by referring to Corollary 3.2.3. Under our assumptions, for every thin sandwich c there are elements a_1, $a_2 \in L$ such that $[ca_1^3c][ca_2^3c] \neq 0$. Simply put $c_1 = [ca_1^3c]$, $c_2 = [ca_2^3c]$ and note that $[c_1 c_2] = [ca_1^3c[ca_2^3c]] = -3[ca_1^3\underline{ca_2c}\,a_2^2c] = 0$.

The situation is very much more complicated with identities (4.2), and it will be necessary for us to go repeatedly from one pair of thin sandwiches satisfying (4.1) to another. We split the subsequent argument into a number of lemmas.

1.2. Lemma. *Let c_0, c_1, c_2 be any thin sandwiches such that c_0 and $c=[c_0c_1c_2]$ satisfy conditions* (4.1) (*in fact, we only need that $c_0c \neq 0$*), *but not* (4.2). *Then*

$$[c_0c_1c_2[c_0a^3c_0]] \neq 0$$

for some $a \in L$.

Proof. Assuming the contrary, and recalling that $[c_0c_1c_2]=c$, we get the identity $[c_0wuvc_0c]=0$, whence

$$c_0uvc_0c = 0\,. \tag{4.3}$$

But in that case, we have for $i=1, 2$,

$$c_0c_iu^2c_0c = c_0[c_iu]uc_0c + c_0u[c_iu]c_0c + c_0u^2\underline{c_ic_0c} = 0\,.$$

We have used here the fact that

$$c_1c_0c = c_1cc_0 = c_1[c_0c_2c_1]c_0 = 0$$

(it is easy to see that $[c_0c_1c_2]c_0 = [c_0c_2c_1]c_0$), and

$$c_2c_0c = c_2cc_0 = c_2[c_0c_1c_2]c_0 = 0\,.$$

Therefore,

$$c_0c_iuvc_0c = 0, \quad i=1, 2\,. \tag{4.4}$$

Moreover,

$$c_0c_1c_2u^2c_0c = \underbrace{c_0c_1[c_2u]uc_0c}_{(4.4)} + \underbrace{c_0c_1u[c_2u]c_0c}_{(4.4)} + c_0c_1u^2\underline{c_2c_0c} = 0\,. \tag{4.5}$$

Thus,

$$cu^2c_0c = \underbrace{c_0c_1c_2u^2c_0c}_{(4.5)} - \underbrace{c_1c_0c_2u^2c_0c}_{(4.4)} - \underbrace{c_2c_0c_1u^2c_0c}_{(4.4)} + \underbrace{c_2c_1c_0u^2c_0c}_{(4.3)} = 0\,,$$

and this is a contradiction, since by assumption the identities $c_0u^2c_0c=0$, $cu^2c_0c=0$ cannot both be satisfied. □

1.3. Lemma. *If Proposition* 1.1 *is false, then there exist three thin sandwiches c_0, c_1, c_2 such that*

$$[c_0c_1c_2] \neq 0, \qquad [c_1c_2] = 0\,.$$

Moreover, we may assume that

$$c_0c_1c_2c_0 \neq 0\,.$$

Proof. We shall begin with a pair c_1, c_2 having properties (4.1), and consider two cases.

a) L has no elements with properties (4.1) that satisfy at least one of the identities (4.2).

In particular, $[c_2a^3c_2c_1] \neq 0$ for some $a \in L$, and thence by Proposition 3.2.2a) (in which c_2 has to be replaced by $c_2' = [c_2a^3c_2]$), we get that

$$[c_2a^3c_2c_1f[c_1b^3c_1]] = -3[c_2a^3c_2c_1fbc_1b^2c_1] \neq 0$$

for some $a, b, f \in L$. In order to achieve full consonance with Proposition 3.2.2a), we recall relations of the type

$$c_1 f b c_1 b^2 c_1 = -c_1 b^2 c_1 b f c_1$$

(see the proof of Corollary 3.2.2a). Since

$$[c_2 a^3 c_2 c_1 f[c_1 b^3 c_1]] = [f c_1 [c_2 a^3 c_2][c_1 b^3 c_1]] \neq 0\,,$$

we have

$$c_1 [c_2 a^3 c_2][c_1 b^3 c_1] \neq 0\,. \tag{4.6}$$

Setting $c_2'' = [[c_2 a^3 c_2][c_1 a^3 c_1]]$, we get $c_1 c_2'' \neq 0$ and $[c_1 c_2''] = 0$. In the case a) under discussion, this means that L is free of the identity $c_1 u^2 c_1 c_2'' = 0$, and therefore

$$[c_1 g^3 c_1 c_2''] \neq 0$$

for some $g \in L$.

Thus,

$$[e_0 e_1 e_2] \neq 0, \qquad [e_1 e_2] = 0\,,$$

where

$$e_0 = [c_2 a^3 c_2], \qquad e_1 = [c_1 b^3 c_1], \qquad e_2 = [c_1 g^3 c_1]\,,$$

which gives the first assertion of the lemma.

b) Suppose that

$$[c_1 c_2] = c_1^2 = c_2^2 = c_1 u^2 c_1 c_2 = 0, \qquad [c_2 a^3 c_2 c_1] \neq 0\,, \tag{4.7}$$

that is, one of the identities (4.2) is satisfied by the pair c_1, c_2. Again as in case a), the sandwich $[c_2 a^3 c_2 c_1]$ leads us *via* Proposition 3.2.2a) to inequality (4.6), which we write in the form

$$c_1 c_2'' = -c_1 c_2 [c_2 a^3] c_1' \neq 0\,, \tag{4.8}$$

where now

$$c_1' = [c_1 b^3 c_1], \qquad c_2' = [c_2 a^3 c_2], \qquad c_2'' = [c_2' c_1']\,,$$

It is not hard to check that properties (4.8) hold, having once observed that

$$[c_1' c_2] = [c_1 b^3 c_1 c_2] = -[\underbrace{b c_1 b^2 c_1 c_2}_{(4.7)}] = 0\,.$$

Thus

$$\begin{aligned} 0 \neq c_1 c_2'' = c_1 c_2' c_1' &= -c_1 c_2 [c_2 a^3] c_1' + c_1 [c_2 a^3] \underline{c_2 c_1'} \\ &= -c_1 c_2 [c_2 a^3] c_1' + c_1 [c_2 a^3] \underline{c_1' c_2} \\ &= -c_1 c_2 [c_2 a^3] c_1' + \underbrace{c_1 [c_2 a^3][c_1 b^3] c_1 c_2}_{(4.7)} - \underline{c_1 [c_2 a^3] c_1 [c_1 b^3]} c_2 \\ &= -c_2 c_1 [c_2 a^3] c_1'\,. \end{aligned}$$

We continue to argue in this way. Since we have two commuting sandwiches c_1, c_2'' such that $c_1c_2'' \neq 0$, we see also that

$$\underbrace{c_1u^2c_1c_2''}_{(4.8)} = -\underbrace{c_1u^2c_1c_2}_{(4.7)}[c_2a^3]c_1' = 0\,.$$

However, since we have assumed that both the identities (4.2) cannot be satisfied together, we must relinquish the identity $c_2''u^2c_1c_2'' = 0$. In this case,

$$[c_1'g^3c_1c_2''] \neq 0 \tag{4.9}$$

for some $g \in L$. In fact, if $[c_1'u^3c_1c_2''] = 0$ for all $u \in L$, then $[wc_1'uvc_1c_2''] = -[c_1'wuvc_1c_2''] = 0$ for all w, u, $v \in L$, that is, $c_1'uvc_1c_2'' = 0$ is an identity. Using it, we find that

$$c_2''u^2c_1c_2'' = \underline{c_2'c_1'u^2c_1c_2''} - c_1'c_2'u^2c_1c_2'' = -\underline{c_1'[c_2'u]uc_1c_2''} - \underline{c_1'u[c_2'u]c_1c_2''}$$

$$-\,c_1'u^2c_2'c_1\underline{c_2''} = -c_1'u^2\underline{c_2'c_1c_2'}c_1' = 0$$

despite the conclusion obtained earlier.

All we need do now is use the obvious relations

$$c_1c_1' = c_1'c_1 = 0, \qquad c_1uc_1' + c_1'uc_1 = 0$$

and rewrite (4.9) as

$$0 \neq [c_1'g^3c_1c_2''] = [c_1'g^3c_1\underline{c_2'c_1'}] = -[c_1'g^3\underline{c_1'c_2'c_1}] = [c_2'c_1''c_1]\,,$$

where $c_1'' = [c_1'g^3c_1']$ is a new sandwich, and finally $[c_1''c_1] = 0$. Apart from the notation, we again have the first assertion of the lemma, which is now fully established.

As for the second assertion, it suffices to observe the following. If $c = [c_0c_1c_2] \neq 0$, $[c_1c_2] = 0$ and $cc_0 = c_0c_1c_2c_0 = 0$, then Lemma 3.3.1 applies to c with parameters $m = 3$, $k = 1$, $l = 2$, since

$$cc_1 = [c_0c_2c_1]c_1 = [c_0c_2]\underline{c_1^2} - \underline{c_1[c_0c_2]c_1} = 0\,.$$

This contradiction (L has no thick sandwiches by assumption) completes the proof. □

The following is the chief test of the reader's endurance in regard of the proof of Proposition 1.1.

1.4. Lemma. *If Proposition* 1.1 *is false, then there are thin sandwiches* c_0, c_1, c_2, c_3 *in L such that*

$$[c_1c_2] = 0, \qquad [c_1c_3] = 0, \qquad [c_0c_1c_2c_3] \neq 0\,. \tag{4.10}$$

Moreover, we may assume that

$$[c_0c_1c_2c_3c_0] \neq 0\,. \tag{4.11}$$

The *proof* is based on arguments close to those used in § 4 of Chapter 3. However, whereas the condition $[ac] \neq 0$ on c was satisfied there without loss of generality for $a^2 = 0$ (so that the commutator $[ac]$ is a sandwich), under the present conditions we have to restrict attention to the sandwich $[ca^3c]$. Accordingly, let us agree to write $c_i^{(j)}$ for a sandwich of the form $[c_i a_j^3 c_i]$, where the a_j are elements that are fixed throughout the discussion. If we have $c_i^{(j+1)}$, then we will possibly pass over $c_i^{(j)}$, but in return we will have $c_k^{(j)}$ for $k < i$. It is clear that

$$c_i^{(j)} = [c_i a_j^3 c_i] \Rightarrow c_i c_i^{(j)} = 0, \qquad c_i^{(j)} u c_i + c_i u c_i^{(j)} = 0 \,. \tag{4.12}$$

Step 1. To begin with, we choose a sandwich $[c_0 c_1 c_2] \neq 0$ (where $[c_1 c_2] = 0$, $[c_0 c_1 c_2] c_0 \neq 0$) as in Lemma 1.3. Applying Lemma 1.2 to it, we get a sandwich

$$e_1 = [c_0 c_1 c_2 c_0^{(1)}] \neq 0$$

(in accordance with our convention, this means that $c_0^{(1)} = [c_0 a_1^3 c_0]$ for a suitable element $a_1 \in L$). Since

$$e_1 c_0 = \underbrace{[c_0 c_1 c_2] c_0^{(1)} c_0}_{(4.12)} - c_0^{(1)} [c_0 c_1 c_2] c_0 = \underbrace{- c_0^{(1)} c_0}_{(4.12)} c_1 c_2 c_0$$

$$+ c_0^{(1)} c_1 \underline{c_0 c_2 c_0} + c_0^{(1)} c_2 \underline{c_0 c_1 c_0} - c_0^{(1)} c_2 c_1 \underline{c_0^2} = 0 \,,$$

we must have $e_1 c_1 \neq 0$, otherwise Lemma 3.3.1 provides a thick sandwich. But in that case, on writing e_1 as

$$e_1 = -[c_1 c_0 [c_2 c_0^{(1)}]]$$

and applying Lemma 1.2 again, we deduce that $[e_1 [c_1 a_2^3 c_1]] \neq 0$. Thus, we get a sandwich

$$e_2 = [c_0 c_1 c_2 c_0^{(1)} c_1^{(2)}] \,.$$

Since

$$[u e_2 c_1] = -[c_0 c_1 c_2 c_0^{(1)} \underbrace{c_1^{(2)} u c_1}_{(4.12)}] = [c_0 c_2 \underline{c_1 c_0^{(1)} c_1} \, u c_1^{(2)}] = 0 \,,$$

it follows that $e_2 c_1 = 0$, while since

$$[e_2 c_i] = [e_2 c_0^{(1)}] = [e_2 c_1^{(2)}] = 0 \,,$$

Lemma 3.3.1 applies and we have to conclude that $e_2 c_2 \neq 0$. Noting that

$$e_2 = -[c_2 [c_0 c_1] [c_0^{(1)} c_1^{(2)}]] \,,$$

we use Lemma 1.2 and get a sandwich

$$e_3 = -[c_2 [c_0 c_1] [c_0^{(1)} c_1^{(2)}] c_2^{(3)}] = [c_0 c_1 c_2 c_0^{(1)} c_1^{(2)} c_2^{(3)}] \,,$$

which can also be written in the form

$$e_3 = [c_0 c_1 c_2 [c_0^{(1)} c_1^{(2)} c_2^{(3)}]], \qquad [c_2 c_1] = [c_2 [c_0^{(1)} c_1^{(2)} c_2^{(3)}]] = 0 \tag{4.13}$$

or

$$e_3 = -[c_0^{(1)}c_1c_2[c_0c_1^{(2)}c_2^{(3)}]], \qquad [c_2c_1] = [c_2[c_0c_1^{(2)}c_2^{(3)}]] = 0\,. \tag{4.14}$$

Both (4.13) and (4.14) are equivalent to conditions (4.10), apart from the notation.

Step 2. We argue by contradiction and assume that the transition from (4.10) to (4.11) is impossible. Then, it follows from (4.13) and (4.14) that

$$[e_3c_0] = 0, \qquad [e_3c_0^{(1)}] = 0\,. \tag{4.15}$$

Indeed, in the case of (4.13), we set $c_3 = [c_0^{(1)}c_1^{(2)}c_2^{(3)}]$. Then $[c_2c_1] = 0$, $[c_2c_3] = 0$ and $[c_0c_2c_1c_3] \neq 0$, that is, conditions (4.10) are satisfied. By our assumption, property (4.11) fails, that is, $[e_3c_0] = 0$. Similarly, the second relation in (4.15) is proved using (4.14).

The usual calculations

$$[ue_3c_2] = -[e_3uc_2] = -\underbrace{[c_0c_1c_2[c_0^{(1)}c_1^{(2)}c_2^{(3)}]}_{(4.13)}uc_2]$$

$$= -[c_0c_1[\ldots]\underline{c_2uc_2}] = 0$$

show that $e_3c_2 = 0$. If also $[c_0c_1c_2c_0^{(1)}[c_1^{(2)}c_2^{(3)}]] = 0$, then $e_3 = [c_0c_1c_2c_0^{(1)}c_2^{(3)}c_1^{(2)}]$, and similar calculations enable us to deduce that $e_3c_1 = 0$. However, the relations $e_3c_1 = 0$, $e_3c_2 = 0$, together with (4.15) and Lemma 3.3.1, guarantee that we have a thick sandwich. We are forced to agree that

$$e_4 = [c_0c_1c_2c_0^{(1)}c_{1,2}] \neq 0\,,$$

where $c_{1,2} = [c_1^{(2)}c_2^{(3)}]$ is a sandwich connected with c_1 and c_2 by the following relations (see (4.12)):

$$[c_{1,2}c_i] = 0, \qquad i = 1, 2; \qquad c_1c_2c_{1,2} = 0\,. \tag{4.16}$$

Further,

$$[e_4c_0] = [c_0c_1c_2\underline{c_0^{(1)}c_{1,2}c_0}] = -[\underline{c_0c_1c_2c_0}c_{1,2}c_0^{(1)}] = 0\,,$$

and it follows from (4.16) that

$$e_4 = [c_0c_1c_2[c_0^{(1)}c_{1,2}]]\,,$$

where $e_4c_i \neq 0$, $i = 1, 2$, since otherwise Lemma 3.1 leads to a thick sandwich.

The rôles of c_1 and c_2 in (4.16) are symmetrical, and so we may assume that $e_4c_1 \neq 0$ without loss. Writing e_4 in the form

$$e_4 = -[c_1[c_0c_2][c_0^{(1)}c_{1,2}]]$$

and applying Lemma 1.2, we arrive at a sandwich

$$e_5 = [e_4[c_1a_4^3c_1]] = [c_0c_1c_2c_0^{(1)}c_{1,2}c_1^{(4)}]\,.$$

At the same time,

$$e_5 = [c_0c_1c_2c_3], \qquad [c_1c_2] = [c_1c_3] = 0\,, \tag{4.17}$$

$$e_5 = -[c_0^{(1)}c_1c_2c_3'], \qquad [c_1c_2] = [c_1c_3'] = 0\,, \tag{4.18}$$

where

$$c_3 = [c_0^{(1)} c_{1,2} c_1^{(4)}], \qquad c_3' = [c_0 c_{1,2} c_1^{(4)}] .$$

Each of (4.17) and (4.18) is equivalent to conditions (4.10), apart from notation. We have laid a taboo on (4.11), so that

$$[e_5 c_0] = 0, \qquad [e_5 c_0^{(1)}] = 0 \tag{4.19}$$

which is a repetition of (4.15) with e_3 replaced by e_5. As for e_3, we write

$$[ue_5 c_1] = -[e_5 u c_1] = -[c_0 c_2 \underbrace{c_1 [c_0^{(1)} c_{1,2} c_1^{(4)}] u c_1}_{(4.17)}]$$

$$= -[c_0 c_2 [c_0^{(1)} c_{1,2} c_1^{(4)}] \underline{c_1 u c_1}] = 0 ,$$

whence $e_5 c_1 = 0$. If in this case we also had

$$[c_0 c_1 c_2 c_0^{(1)} [c_{1,2} c_1^{(4)}]] = 0 ,$$

we would have

$$e_5 = [c_0 c_1 c_2 c_0^{(1)} c_1^{(4)} c_{1,2}] ,$$

and $e_5 c_{1,2} = 0$. But together with (4.19), the relations $e_5 c_1 = 0$, $e_5 c_{1,2} = 0$ guarantee that the requirements of Lemma 3.3.3 are satisfied, and we are forced to deduce that

$$e_6 = [c_0 c_1 c_2 c_0^{(1)} c_{1,2}'] \neq 0 ,$$

where

$$c_{1,2}' = [c_{1,2} c_1^{(4)}]$$

is a sandwich connected with c_1 and c_2 by the relations

$$[c_{1,2}' c_2] = 0, \quad c_{1,2}' c_1 = 0, \quad c_1 c_{1,2}' = 0, \quad c_{1,2}' u c_1 + c_1 u c_{1,2}' = 0 . \tag{4.20}$$

In fact,

$$[c_{1,2}' c_1] = [c_{1,2} \underbrace{c_1^{(4)} c_1}_{(4.12)}] = 0 ,$$

$$c_{1,2}' c_1 = c_{1,2} \underbrace{c_1^{(4)} c_1}_{(4.12)} - c_1^{(4)} \underbrace{c_{1,2} c_1}_{(4.16)} = -\underbrace{c_1^{(4)} c_1}_{(4.12)} c_{1,2} = 0 ,$$

$$[c_{1,2}' c_2] = -[c_1^{(4)} \underbrace{c_{1,2} c_2}_{(4.16)}] = -[c_1 a_4^3 \underbrace{c_1 c_2 c_{1,2}}_{(4.16)}] = 0 .$$

We note also that

$$[e_6 c_0] = [c_0 c_1 c_2 \underbrace{c_0^{(1)} c_{1,2}' c_0}] = -[\underbrace{c_0 c_1 c_2 c_0} c_{1,2}' c_0^{(1)}] = 0 .$$

Step 3. Up to now the persistent "witchdoctor's incantations" (a term introduced in § 5 of Chap. 3) have not yielded any appreciable results, but the asymmetry in c_1

and c_2 apparent in (4.20) seems encouraging. Since

$$e_6 = [c_0 c_2 \underbrace{c_1 c_0^{(1)} c'_{1,2}}_{(4.20)}] = -[c_0 c_2 c'_{1,2} c_0^{(1)} c_1] ,$$

we have $e_6 c_1 = 0$. In order not to come under the sphere of influence of Lemma 3.3.1 once more, we must have $e_6 c_2 \neq 0$. But

$$e_6 = -[c_2 [c_0 c_1][c_0^{(1)} c'_{1,2}]] ,$$

and we win the right to apply Lemma 1.2, which gives

$$e_7 = [e_6 [c_2 a_5^3 c_2]] = [c_0 c_1 c_2 c_0^{(1)} c'_{1,2} c_2^{(5)}] \neq 0 .$$

The easily-verified relations

$$e_7 = [c_0 c_1 c_2 [c_0^{(1)} c'_{1,2} c_2^{(5)}]], \qquad [c_2 c_1] = [c_2 [c_0^{(1)} c'_{1,2} c_2^{(5)}]] = 0 ,$$

$$e_7 = -[c_0^{(1)} c_1 c_2 [c_0 c'_{1,2} c_2^{(5)}]], \qquad [c_2 c_1] = [c_2 [c_0 c'_{1,2} c_2^{(5)}]] = 0 ,$$

which are separately equivalent to (4.10), and the analogous pairs (4.13), (4.14) and (4.17), (4.18), lead us to the relations

$$[e_7 c_0] = 0, \qquad [e_7 c_0^{(1)}] = 0 ,$$

because (4.11) is prohibited. Further, $e_7 c_2 = 0$ (*cf* $e_3 c_2 = 0$, $e_5 c_1 = 0$), so that we have $e_7 c_1 \neq 0$, by Lemma 3.3.1. Writing

$$e_7 = -[c_1 [c_0 c_2][c_0^{(1)} c'_{1,2} c_2^{(5)}]]$$

and applying Lemma 1.2, we get a sandwich

$$e_8 = [e_7 [c_1 a_6^3 c_1]] = [c_0 c_1 c_2 c_0^{(1)} c'_{1,2} c_2^{(5)} c_1^{(6)}] ,$$

which can also take the form

$$e_8 = [c_0 c_1 c_2 c_0^{(1)} c'_{1,2} c^0_{1,2}]$$

with $c^0_{1,2} = [c_2^{(5)} c_1^{(6)}]$, since

$$[c_0 \underline{c_1 c_2}\, c_0^{(1)} c'_{1,2} \underline{c_1^{(6)}}\, c_2^{(5)}] = [c_0 c_2 c_1 c_0^{(1)} c'_{1,2} [c_1 a_6^3 c_1] c_2^{(5)}]$$

$$= -3 [c_0 c_2 c_1 c_0^{(1)} \underbrace{c'_{1,2} a_6 c_1}_{(4.20)} a_6^2 c_1 c_2^{(5)}]$$

$$= 3 [c_0 c_2 \underline{c_1 c_0^{(1)} c_1}\, a_6 c'_{1,2} a_6^2 c_1 c_2^{(5)}] = 0 .$$

Clearly (see (4.16)),

$$[c^0_{1,2} c_i] = 0, \qquad i = 1, 2; \qquad c_1 c_2 c^0_{1,2} = 0 . \tag{4.21}$$

Setting $c''_3 = [c_0 c'_{1,2} c_2^{(5)} c_1^{(6)}]$, $c'''_3 = [c_0^{(1)} c'_{1,2} c_2^{(5)} c_1^{(6)}]$ for short, we get the following relations by analogy with the preceding cases:

$$e_8 = [c_0 c_1 c_2 c'''_3], \qquad [c_1 c_2] = [c_1 c'''_3] = 0 ,$$

$$e_8 = -[c_0^{(1)} c_1 c_2 c''_3], \qquad [c_1 c_2] = [c_1 c''_3] = 0 ,$$

which are equivalent to (4.10). Rejecting (4.11), we get

$$[e_8 c_0] = 0, \qquad [e_8 c_0'] = 0\,.$$

Another standard verification, namely

$$[ue_8c_1] = -[e_8uc_1] = -[\underset{(4.12)}{\underline{e_7c_1^{(6)}uc_1}}] = [e_7c_1uc_1^{(6)}]$$

$$= -[\underline{c_1[c_0c_2][c_0^{(1)}c'_{1,2}c_2^{(5)}]c_1}\,uc_1^{(6)}] = 0$$

gives the relation $e_8c_1 = 0$. By Lemma 3.3.1 we have $e_8c_2 \neq 0$, and since

$$e_8 = -[c_2[c_0c_1][c_0^{(1)}c'_{1,2}c^0_{1,2}]]\,,$$

(see (4.20), (4.21)), Lemma 1.2 gives us a sandwich

$$e_9 = [e_8[c_2a_7^3c_2]] = [c_0c_1c_2c_0^{(1)}c'_{1,2}c^0_{1,2}c_2^{(7)}]\,.$$

The relations

$$e_9 = [c_0c_1c_2[c_0^{(1)}c'_{1,2}c^0_{1,2}c_2^{(7)}]] = -[c_0^{(1)}c_1c_2[c_0c'_{1,2}c^0_{1,2}c_2^{(7)}]]\,,$$

$$[c_2c_1] = [c_2[c_0^{(1)}c'_{1,2}c^0_{1,2}c_2^{(7)}]] = [c_2[c_0c'_{1,2}c^0_{1,2}c_2^{(7)}]] = 0$$

put us twice in situation (4.10), and the assumed embargo on transition from (4.10) to (4.11) leads to relations

$$[e_9c_0] = 0, \qquad [e_9c_0^{(1)}] = 0\,. \tag{4.22}$$

We note also that either

$$f = [c_0^{(1)}c_1c_2[c_0c'_{1,2}c^0_{1,2}c_2^{(7)}c'_{1,2}]] = 0\,,$$

or else f is a sandwich, and since

$$[c_1c_2] = 0, \quad [c_1[c_0c'_{1,2}c^0_{1,2}c_2^{(7)}c'_{1,2}]] = -[c_0c'_{1,2}c^0_{1,2}c_2^{(7)}\underset{(4.20)}{\underline{c'_{1,2}c_1}}] = 0\,,$$

f puts us in situation (4.10), and we are forced to assume that

$$[fc_0^{(1)}] = 0\,. \tag{4.23}$$

In all cases,

$$[e_9[c_0^{(1)}c'_{1,2}]] = 0\,. \tag{4.24}$$

For

$$[e_9[c_0^{(1)}c'_{1,2}]] = -[e_9c'_{1,2}c_0^{(1)}] + \underset{(4.22)}{\underline{[e_9c_0^{(1)}c'_{1,2}]}}$$

$$= -\underset{(4.20)}{\underline{[c_0c_1c_2[c_0^{(1)}c'_{1,2}c^0_{1,2}c_2^{(7)}]c'_{1,2}c_0^{(1)}}}$$

$$= -[c_0c_1c_2[c_0^{(1)}c'_{1,2}c^0_{1,2}c_2^{(7)}c'_{1,2}]c_0^{(1)}] = -\underset{(4.23)}{\underline{[fc_0^{(1)}]}} = 0\,.$$

Step 4. Deviating from the beaten path, we give the sandwich e_9 in the form

$$e_9 = [c_0 c_1 c_2 [c_0^{(1)} c'_{1,2}] c_{1,2}^0 c_2^{(7)}]$$

and observe that it commutes with all its components. In fact, by (4.22), (4.24) and the two obvious relations

$$[e_9 c_{1,2}^0] = 0, \qquad [e_9 c_2] = [\dots \underbrace{c_2^{(7)} c_2}_{(4.12)}] = 0\,,$$

we have only to carry out a small verification:

$$[e_9 c_1] = [c_0 c_2 \underline{c_1 [c_0^{(1)} c'_{1,2} c_{1,2}^0 c_2^{(7)}] c_1}] = 0\,.$$

Moreover,

$$[u e_9 c_2] = -[e_9 u c_2] = [c_0 c_1 c_2 [c_0^{(1)} c'_{1,2} c_{1,2}^0] \underbrace{c_2^{(7)} u c_2}_{(4.12)}]$$

$$= -[c_0 \underline{c_1 c_2 [c_0^{(1)} c'_{1,2} c_{1,2}^0] c_2} u c_2^{(7)}] = 0\,,$$

so that $e_9 c_2 = 0$. If also $e_9 c_{1,2}^0 = 0$, Lemma 3.3.1 with parameters $m = 6$, $k = 3$, $l = 5$ lead us to a thick sandwich. Therefore, $e_9 c_{1,2}^0 \neq 0$. However, if

$$[c_0 c_1 c_2 [c_0^{(1)} c'_{1,2}] c_{1,2}^0 c_2^{(7)}] = [c_0 c_1 c_2 [c_0^{(1)} c'_{1,2}] c_2^{(7)} c_{1,2}^0]\,,$$

the relation $e_9 c_{1,2}^0 = 0$ would also hold. Thus,

$$[c_0 c_1 c_2 [c_0^{(1)} c'_{1,2}]\,[c_{1,2}^0 c_2^{(7)}]] \neq 0\,,$$

which, by (4.20), can be written as

$$e_{10} = [c_0 c_1 c_2 c_0^{(1)} c'_{1,2} c''_{1,2}] \neq 0\,,$$

where $c''_{1,2} = [c_{1,2}^0 c_2^{(7)}]$. We have

$$[c''_{1,2} c_1] = 0, \qquad [c''_{1,2} c_2] = 0, \qquad c''_{1,2} c_2 = 0\,,$$
$$c''_{1,2} u c_2 + c_2 u c''_{1,2} = 0\,, \tag{4.25}$$

since

$$[c''_{1,2} c_1] = -[\underbrace{c_2^{(7)} c_{1,2}^0 c_1}_{(4.21)}] = -[c_2 a_7^3 \underbrace{c_2 c_1 c_{1,2}^0}_{(4.21)}] = 0\,,$$

$$[c''_{1,2} c_2] = [c_{1,2}^0 \underbrace{c_2^{(7)} c_2}_{(4.12)}] = 0\,,$$

$$[u c''_{1,2} c_2] = -[c_{1,2}^0 \underbrace{c_2^{(7)} u c_2}_{(4.12)}] = [\underbrace{c_{1,2}^0 c_2}_{(4.21)} u c_2^{(7)}] = 0\,.$$

It would seem that we have found an ordinary sandwich with the reasonable numeral 10. However, successive verifications show:

a) $[e_{10} c_0] = 0$, $\quad [e_{10} c_0^{(1)}] = 0$, since

$$e_{10} = [c_0 c_1 c_2 [c_0^{(1)} c'_{1,2} c''_{1,2}]] = -[c_0^{(1)} c_1 c_2 [c_0 c'_{1,2} c''_{1,2}]]$$

by (4.25) and (4.20), and

$$[c_2c_1] = [c_2[c_0^{(1)}c'_{1,2}c''_{1,2}]] = [c_2[c_0c'_{1,2}c''_{1,2}]] = 0\,,$$

that is, we hit conditions (4.10), and our assumption that the implication (4.10) ⇒ (4.11) is lacking leads inevitably to the two relations indicated:

$$\text{b)}\ [e_{10}c_i] = [c_0c_1c_2\underbrace{c_0^{(1)}c'_{1,2}c''_{1,2}c_i}_{(4.20),(4.25)}] = [c_0\underline{c_1c_2c_0^{(1)}c_i}c'_{1,2}c''_{1,2}] = 0,\ i = 1, 2;$$

$$\text{c)}\ [e_{10}c'_{1,2}] = [c_0c_1c_2c_0^{(1)}\underline{c'_{1,2}c''_{1,2}c'_{1,2}}] = 0\,.$$

This means that e_{10} commutes with all its components. And finally, the long-awaited prize:

$$e_{10} = [\underbrace{c_0c_2c_1c_0^{(1)}}_{(4.20)}c'_{1,2}c''_{1,2}] = -[c_0\underbrace{c_2c'_{1,2}}_{(4.20)}\underbrace{c_0^{(1)}c_1c''_{1,2}}_{(4.25)}]$$

$$= -[c_0c'_{1,2}\underbrace{c_2c_0^{(1)}c''_{1,2}c_1}_{(4.25)}] = [c_0c'_{1,2}c''_{1,2}c_0^{(1)}c_2c_1]\,,$$

that is,

$$e_{10} = [\ldots c_2c_1] = [\ldots c_1c_2]\,.$$

Therefore

$$e_{10}c_1 = 0, \qquad e_{10}c_2 = 0\,,$$

and Lemma 3.3.1 is—by now unreservedly—applicable; it gives a thick sandwich. The contradiction leaves one possibility: at some stage, the implication (4.10) ⇒ (4.11) is correct. □

1.5. Proof of Proposition 1.1. As before, we assume that the Lie algebra L has no thick sandwiches and no pair of thin sandwiches satisfying conditions (4.2). Moreover, reference to (4.2) is used implicitly at the relevant step, when we appeal to Lemmas 1.2–1.4. As a starting point we choose an element

$$[cc_0c_1c_2c] \neq 0, \qquad [c_0c_1] = [c_0c_2] = 0\,, \tag{4.26}$$

whose existence (with slightly different notation) was established in the key lemma 1.4. For future use, we note a standard relation involving the thin sandwich c:

$$[cu_1u_2u_3c] = [cu_{\pi 1}u_{\pi 2}u_{\pi 3}c], \qquad \pi \in S_3\,. \tag{4.27}$$

With Lemma 3.3.1 in mind, we may assert that

$$[cc_0c_1c_2c]c_0 \neq 0\,.$$

But

$$[cc_0c_1c_2c] = [c_0[cc_1][cc_2]]\,,$$

and thus we get from Lemma 1.2 that

$$d_0 = [cc_0c_1c_2cc^{(0)}] \neq 0\,,$$

$$c^{(0)} = [c_0b^3c_0], \quad c_0c^{(0)} = c^{(0)}c_0 = 0, \quad c_0uc^{(0)} + c^{(0)}uc_0 = 0 \tag{4.28}$$

(it will help to compare with (3.28)!) A small verification shows that

$$[d_0c_1] = [c\underset{(4.27)}{\underline{c_0c_1c_2c}}c^{(0)}c_1] = [cc_2c_1\underset{(4.28)}{\underline{c_0cc^{(0)}c_1}}]$$

$$= -[\underset{(4.27)}{\underline{cc_2c_1c^{(0)}c}}\,\underset{(4.26)}{\underline{c_0c_1}}] = -[cc_2c^{(0)}\underline{c_1cc_1}\,c_0] = 0\,.$$

Similarly,

$$[d_0c_2] = 0\,.$$

Further,

$$[d_0c_0] = [\ldots \underset{(4.28)}{\underline{cc^{(0)}c_0}}] = 0\,,$$

$$[ud_0c_0] = -[d_0uc_0] = -[\underset{(4.27)}{\underline{cc_0c_1c_2c}}\,\underset{(4.28)}{\underline{c^{(0)}uc_0}}] = [cc_1c_2c_0\underline{cc_0}uc^{(0)}] = 0\,,$$

and therefore

$$d_0c_0 = 0\,.$$

By Lemma 3.3.1, we get automatically that $d_0c \neq 0$, and since

$$d_0 = -[c[cc_0c_1c_2]c^{(0)}]\,,$$

a second application of Lemma 1.2 yields an element

$$d_1 = [cc_0c_1c_2cc^{(0)}c^{(1)}] \neq 0, \qquad c^{(1)} = [cb_1^3c]\,.$$

Further elements $c^{(k)}$ emerge, and it is therefore appropriate to note without further ado that

$$c^{(k)} = [cb_k^3c], \qquad cc^{(k)} = c^{(k)}c = 0, \qquad cuc^{(k)} + c^{(k)}uc = 0\,. \tag{4.29}$$

Moreover, we intend to make systematic use of the concept of S-continuation for a whole string of elements, for a suitable finite set S of sandwiches. By Theorem 3.4.1, the required S-continuations exist for an algebra with E_n (and by the analogue of Theorem 3.4.1 for finite-dimensional algebras), because of Proposition 1.5.7 and Definition 1.5.8.

To start with, we construct a $\{c_0, c_1, c_2, c^{(1)}\}$-continuation $\tilde{d}_1$ of the sandwich d_1. By relations (4.27) and (4.28), we have

$$\tilde{d}_1 = [cc_0c_1c_2c[c^{(0)}c^{(1)}\ldots]]\,, \tag{4.30}$$

that is, $\tilde{d}_1$ is in fact a $\{c_0, c_1, c_2, c^{(1)}, c\}$-continuation, and by definition of $\tilde{d}_1$ we have

$$[\tilde{d}_1c] = 0, \qquad [\tilde{d}_1c_i] = 0, \qquad i = 0, 1, 2\,.$$

It follows from Lemma 3.3.1 as applied to $\tilde{d}_1$ in the form (4.30) that $\tilde{d}_1c \neq 0$, and thus we get from Lemma 1.2 that

$$d_2 = [\tilde{d}_1c^{(2)}] \neq 0, \qquad c^{(2)} = [cb_2^3c]\,.$$

Next we construct a $\{c_0, c_1, c_2, c^{(1)}, c^{(2)}\}$-continuation $\tilde{d}_2$ of d_2, *etc.* This process gives us arbitrarily many sandwiches

$$\begin{aligned}\tilde{d}_m &= [cc_0c_1c_2cc^{(0)}c^{(1)} \ldots c^{(2)} \ldots c^{(m)} \ldots] \\ &= [cc_0c_1c_2c[c^{(0)}c^{(1)} \ldots c^{(2)} \ldots c^{(m)} \ldots]] \neq 0 .\end{aligned} \tag{4.31}$$

The number $N = N(m)$ of all sandwiches occurring in $\tilde{d}_m$, possibly with repeats, does not lend itself to calculation; however, we need not worry about that sort of question, at least when we are carrying out a non-effective proof.

When $\dim_F L < \infty$, we may call a halt at this point, since the sandwich subalgebra $\mathfrak{C} \subset L$ has nilpotency class not more than $\dim_F L$, and for $m > \dim_F L$ we automatically get a contradiction. However, the later discussions use not condition E_n itself, but rather the corollary that continuations exist.

We note that $c^{(k)}$ and $c^{(l)}$, $1 \leqslant k, l \leqslant m$, cannot occur side-by-side in the expression for $\tilde{d}_m$, since

$$\underset{(4.29)}{\underline{c^{(k)}c^{(l)}}} = -c[cb_k^3]c^{(l)} ,$$

and in that case we would have

$$\tilde{d}_m = -[cc_0c_1c_2\underline{c[c^{(0)}c^{(1)} \ldots]c[cb_k^3]c^{(l)}} \ldots] = 0 .$$

Thus, all the $c^{(k)}$ are separated by elements c_0, c_1, c_2, and since $c_jc^{(s)}c_j = 0$ for $j = 0$, 1, 2, at least two elements c_k, c_l $(k \neq l)$ from the triple c_0, c_1, c_2 must be repeated arbitrarily often (we need at least 3 such repeats). For the same reason, we can assume that $\tilde{d}_m$ is different from

$$[cc_0c_1c_2cc^{(0)} \ldots c^{(s)}] .$$

Otherwise

$$d_{m+1} = [\tilde{d}_m c^{(m+1)}] = [c \ldots c^{(s)}c^{(m+1)}] \neq 0 ,$$

which is excluded by what was said above.

Suppose that m, taken large enough as before, is such that

$$d = \tilde{d}_m = [cc_0c_1c_2cc^{(0)} \ldots c_k]$$

(we have enough flexibility in the choice of k, so that there is no loss of generality here). Although $dc_k = 0$ and c_k occurs in d sufficiently often (at least 3 times), we cannot use Lemma 3.3.1 since it could happen that $[dc^{(0)}] \neq 0$. In this case, we consider an S-continuation $\tilde{d}$ of d, where

$$S = \{c, c_0, c_1, c_2, c^{(0)}, c^{(1)}, \ldots, c^{(m)}\} ,$$

a set containing c and $c^{(0)}$. By construction, $\tilde{d}$ is a non-zero element of the centre of the sandwich Lie algebra generated by S, and $d = \tilde{d}_m$ is a non-zero element of the centralizer of the algebra generated by $S_0 = S \backslash \{c^{(0)}\}$. The fate of the proof now hangs, in the literal sense of the word, on the tip of the sandwich $\tilde{d}$.

1) $\tilde{d} = [dc^{(0)}]$. Although $\tilde{d}c^{(0)} = 0$, the degree of $\tilde{d}$ with respect to $c^{(0)}$ can only be 2, and this hinders the action of Lemma 3.3.1. But

$$[u\tilde{d}c_0] = -[\tilde{d}uc_0] = -\underbrace{[dc^{(0)}uc_0]}_{(4.28)} = [d\underline{c_0u}c^{(0)}] = 0\,,$$

so that $\tilde{d}c_0 = 0$. We need only observe now that $c^{(0)}$ and c_0 occur as components of d.

2) $\tilde{d} = [dc^{(0)} \ldots c^{(0)}]$. All the requirements of Lemma 3.3.1 are clearly satisfied.

3) $\tilde{d} = [dc^{(0)} \ldots c]$. We note that

$$\tilde{d} = [d_0 \ldots c^{(0)} \ldots c\,]\,,$$

that c occurs twice in d_0 and $\tilde{d}c = 0$. Lemma 3.3.1 applies.

4) $\tilde{d} = [dc^{(0)} \ldots c^{(s)}]$. Here $\tilde{d}c = 0$ (and Lemma 3.3.1 applies), since otherwise

$$0 \neq [a\tilde{d}c] = -[dc^{(0)} \ldots \underbrace{c^{(s)}ac}_{(4.29)}] = [dc^{(0)} \ldots cac^{(s)}]\,,$$

and we would replace $\tilde{d}$ by $[dc^{(0)} \ldots c] \neq 0$ (see Definition 1.5.8), thus reverting to case 3).

5) $\tilde{d} = [dc^{(0)} \ldots c_0]$. Since $c^{(0)}$ also occurs in d, Lemma 3.3.1 is applicable provided that $\tilde{d}c^{(0)} = 0$. If $\tilde{d}c^{(0)} \neq 0$, then

$$0 \neq [a\tilde{d}c^{(0)}] = -[dc^{(0)} \ldots \underbrace{c_0ac^{(0)}}_{(4.28)}] = [dc^{(0)} \ldots c^{(0)}ac_0]\,,$$

and, replacing $\tilde{d}$ by $[dc^{(0)} \ldots c^{(0)}]$, we revert to case 2).

The remain the less obvious cases

$$\tilde{d} = [dc^{(0)} \ldots c_1], \qquad \tilde{d} = [dc^{(0)} \ldots c_2]\,.$$

In these discussions, the indices 1 and 2 have equal rights. For definiteness, we shall assume that

$$\tilde{d} = [dc^{(0)} \ldots c_2]\,.$$

We need only discuss the case where c_2 occurs in $d = \tilde{d}_m$ (for arbitrary m here!) just once, namely in the initial segment $[cc_0c_1c_2c \ldots]$. This means that $\{k, l\} = \{0, 1\}$. This is a rather good set for us, since $c_0c_1 = c_1c_0$ (see (4.26)). We shall use this property immediately after showing that c_0 and c_1 do not appear next to each other anywhere in $\tilde{d}_m$ after the initial segment. Otherwise,

$$\begin{aligned}
\tilde{d}_m &= [\underbrace{cc_0c_1c_2c}_{(4.27)}\underbrace{c^{(0)} \ldots c^{(s)}c_0c_1}_{(4.31)} \ldots] = [cc_2c_0c_1c[c^{(0)} \ldots c^{(s)}]c_0c_1 \ldots] \\
&= [cc_2c_0\underline{c_1[c_0[c^{(0)} \ldots c^{(s)}]c]c_1} \ldots] - [cc_2c_0\underline{c_1c_0}[c^{(0)} \ldots]cc_1 \ldots] \\
&\quad + [cc_2\underbrace{c_0c_1}_{(4.26)}[c^{(0)} \ldots c^{(s)}]c_0cc_1 \ldots] + [\underbrace{cc_2c_0c_1cc_0}_{(4.26)}[c^{(0)} \ldots]c_1 \ldots] \\
&= [cc_2\underline{c_1c_0[c^{(0)} \ldots c^{(s)}]c_0}cc_1 \ldots] + [cc_2c_1\underline{c_0cc_0}[c^{(0)} \ldots]c_1 \ldots] = 0.
\end{aligned}$$

By the above remarks, as they relate to segments $c^{(r)}c^{(q)}$, $c_j c^{(r)} c_j$ and ends of the form $[\ldots c^{(s)}]$, we deduce that $\tilde{d}_m$ can be constructed in just one way:

$$\tilde{d}_m = [cc_0c_1c_2cc^{(0)}c^{(1)}c_1c^{(2)}c_0c^{(3)}c_1c^{(4)}c_0\ldots c_i], \qquad i\in\{0,1\}\,. \tag{4.32}$$

Of course, the elements a_k and $c^{(k)} = [ca_k^3c]$ are not controllable. It might appear that we should have considered a variant $\tilde{d}_m = [\ldots cc^{(0)}c^{(1)}c_0c^{(2)}c_1\ldots]$, but it drops out immediately:

$$[\underline{cc_0c_1c_2c}\,\underline{c^{(0)}c^{(1)}c_0}\ldots] = -[cc_1c_2\underline{c_0cc_0}c^{(1)}c^{(0)}\ldots] = 0\,.$$

We claim that with $m = 4$ in (4.32), the sandwich

$$d = \tilde{d}_4 = [cc_0c_1c_2cc^{(0)}c^{(1)}c_1c^{(2)}c_0c^{(3)}c_1c^{(4)}c_0] \tag{4.33}$$

is an S-continuation $\tilde{d}$ of itself when

$$S = \{c,\, c_0,\, c_1,\, c_2,\, c^{(0)},\, c^{(1)},\, c^{(2)},\, c^{(3)},\, c^{(4)}\}\,,$$

that is, d commutes with every element of S. We have

$$[dc] = [cc_0c_1c_2c\underline{[c^{(0)}c^{(1)}c_1\ldots c_1c^{(4)}c_0]c]} = 0\,,$$

$$[dc_i] = [cc_0c_1c_2c\ldots \underline{c_0c_i}] = 0, \qquad i\in\{0,1,2\}\,.$$

It follows from the relations

$$[\underline{cc_0c_1c_2cc^{(0)}[c_1c^{(1)}c^{(2)}]c_0}\ldots] = -[cc_1c_2\underline{c_0cc_0}[c_1c^{(1)}c^{(2)}]c^{(0)}\ldots] = 0\,,$$

$$[\ldots cc^{(0)}c^{(1)}c_1c^{(2)}\underline{c_0[c_1c^{(3)}c^{(4)}]c_0]} = 0\,,$$

$$[\ldots cc^{(0)}c^{(1)}\underline{c_1[c_0c^{(2)}c^{(3)}]c_1}c^{(4)}c_0] = 0$$

that d is skew-symmetric in the $c^{(r)}$ with $1 \leqslant r \leqslant 4$, so that

$$[dc^{(r)}] = 0, \qquad 1\leqslant r\leqslant 4\,.$$

Finally,

$$[dc^{(0)}] = [\ldots \underline{c_0c^{(0)}}] \doteq 0\,,$$

a trivial relation that did not follow from anything in the general construction of $\tilde{d}_m$.

In this way, the elimination of cases 1)–5) considered above has led us to a sandwich $d = \tilde{d}$ of the form (4.33), of degree 3 in c_0 and having the obvious property that $dc_0 = 0$. Inevitably, we have hit up against the conditions of Lemma 3.3.1. The conclusion forces itself upon us that none of the logically possible forms of $\tilde{d}_m$ can in fact occur, without causing a contradiction to our hypotheses. □

1.6. Remark. The condition E_n in L (the condition $\dim_F L < \infty$ respectively) is in fact used only with reference to Theorem 3.4.11 (to its finite-dimensional analogue, respectively), and then in the deduction of the "almost obvious" properties (4.1) and the proof of the "almost obvious" Lemma 1.3. All further arguments are very general.

§ 2. Thick Pairs of Thin Sandwiches

The obvious pun in the title of this section explains the following definition:

2.1. Definition. Let L be a Lie algebra embedded in $A(L)$ and possessing thin sandwiches. We shall say that the pair (c_1, c_2) of thin sandwiches in L has *thickness* r if

$$c_1c_2 \neq 0, \qquad [c_1c_2] = 0\,; \tag{4.34}$$

$$c_\delta u_1 u_2 \ldots u_s c_1 c_2 = 0, \qquad \delta = 1, 2\,, \tag{4.35}$$

for $s = 0, 1, 2, \ldots, r$ and arbitrary $u_1, u_2, \ldots, u_s \in L$, but at least one of the identities (4.35) ceases to hold for $s = r + 1$.

Every pair of thin sandwiches with property (4.34) in a Lie algebra L without thick sandwiches (as supplied by Corollary 3.2.3, for example) also satisfies (4.35) with $s = 0$ and 1:

$$c_1(c_1c_2) = \underline{c_1^2}c_2 = 0, \qquad c_2c_1c_2 = 0\,,$$

$$\underline{c_1uc_1}\,c_2 = 0, \qquad c_2\underline{uc_1c_2} = \underline{c_2uc_2}\,c_1 = 0\,.$$

The descent to a pair of thickness 2 realised in Proposition 1.1 required real effort. However, the footbridge between pairs of thickness $r \geqslant 2$ and $r + 1$ turns out to be fairly short.

2.2. Proposition. *Every Lie algebra L over a field F of characteristic $p > 5$ that either satisfies E_n with $n < p$ or is finite-dimensional over F, is embedded in $A(L)$ and devoid of thick sandwiches but contains a pair of thickness* 2, *contains a pair (of thin sandwiches) of arbitrary thickness.*

Proof. Arguing by induction, we assume that conditions (4.34) and (4.35) are satisfied by the pair (c_1, c_2) for $r \geqslant 2$, and then establish that L has a pair of thickness $r + 1$ (in fact, of thickness $\geqslant r + 1$). For this we introduce elements

$$d_1 = [ac_1c_2], \qquad d_2 = [c_1b^3c_1]\,, \tag{4.36}$$

where $a, b \in L$ are chosen so that

$$c_2c_1b^2c_1abc_1 \neq 0\,. \tag{4.37}$$

Since L has no thick sandwiches by assumption, the existence of a and b satisfying (4.37) is guaranteed by Proposition 3.2.2c). Noting that

$$c_1[c_1b^3c_1] = -3c_1bc_1b^2c_1 = 0, \qquad \underset{(4.34)}{\underline{c_1c_2}}\,[c_1b^3c_1] = c_2\underline{c_1[c_1b^3c_1]} = 0\,,$$

$$c_2[c_1b^3c_1] = [c_1b^3c_1]c_2 - [c_1b^3c_1c_2]$$

$$= [c_1 b^3 c_1] c_2 + \underset{(4.35):\delta=1}{\underline{[bc_1 b^2 c_1 c_2]}} = [c_1 b^3 c_1] c_2 ,$$

$$c_1 abc_1 b^2 c_1 = -c_1 b^2 c_1 abc_1$$

(see the proof of Corollary 3.2.3), we find:

a) $d_1^2 = 0$ by Lemma 3.2.1c) (see the first part of the proof);
b) $d_2^2 = 0$—a general property of sandwiches (see Lemma 2.4.2);

c) $[d_1 d_2] = -[d_2 d_1] = [c_1 b^3 c_1 [a\underline{c_1 c_2}]] = [c_1 b^3 c_1 [ac_2 c_1]]$

$$= [c_1 b^3 \underline{c_1 [ac_2] c_1}] - [c_1 b^3 \underline{c_1^2} [ac_2]] = 0 ;$$

d) $\underline{d_1} d_2 = c_2 \underline{c_1 a [c_1 b^3 c_1]} - c_1 \underline{ac_2 [c_1 b^3 c_1]} - c_2 a \underline{c_1 [c_1 b^3 c_1]}$

$$+ \underline{ac_1 c_2 [c_1 b^3 c_1]} = -3 c_2 \underline{c_1 abc_1 b^2 c_1} - c_1 a [c_1 b^3 c_1] c_2$$

$$= 3 c_2 c_1 b^2 c_1 abc_1 + 3 \underset{(4.35):\delta=1}{\underline{c_1 abc_1 b^2 c_1 c_2}} = \underset{(4.37)}{\underline{3 c_2 c_1 b^2 c_1 abc_1}} \neq 0 .$$

Here we have used the expression

$$d_1 = ac_1 c_2 - c_1 ac_2 - c_2 ac_1 + c_2 c_1 a . \tag{4.38}$$

Thus (d_1, d_2) is a pair of sandwiches satisfying conditions (4.34) (with new notation).

By (4.38), it follows immediately from (4.35) that

$$d_1 u_1 u_2 \ldots u_s c_1 c_2 = 0 \quad \text{for} \quad s \leqslant r-1 .$$

For $s = r$ we also have:

$$d_1 u_1 u_2 \ldots u_r c_1 c_2 = \underline{c_2 c_1 a} u_1 u_2 \ldots u_r c_1 c_2$$

$$= \underset{(4.35):\delta=2}{\underline{c_2 [c_1 au_1] u_2 \ldots u_r c_1 c_2}} - \underset{(4.35):\delta=1}{\underline{c_2 u_1 c_1 au_2 \ldots u_r c_1 c_2}}$$

$$- \underset{(4.35):\delta=1}{\underline{c_2 ac_1 u_1 u_2 \ldots u_r c_1 c_2}} + c_2 u_1 \underset{(4.35):\delta=1}{\underline{ac_1 u_2 \ldots u_r c_1 c_2}} = 0$$

Further,

e) $c_1 c_2 u_1 u_2 \ldots u_{r+1} c_1 c_2 = \underset{(4.35):\delta=1}{\underline{c_1 [c_2 u_1 u_2] u_3 \ldots u_{r+1} c_1 c_2}}$

$$+ c_1 u_1 \underset{(4.35):\delta=2}{\underline{c_2 u_2 u_3 \ldots u_{r+1} c_1 c_2}}$$

$$+ c_1 u_2 \underset{(4.35):\delta=2}{\underline{c_2 u_1 u_3 \ldots u_{r+1} c_1 c_2}}$$

$$- c_1 u_2 u_1 \underset{(4.35):\delta=2}{\underline{c_2 u_3 \ldots u_{r+1} c_1 c_2}} = 0 ;$$

f) $c_1u_0c_2u_1u_2 \ldots u_{r+1}c_1c_2 = \underset{(4.35):\delta=1}{\underline{c_1u_0[c_2u_1u_2u_3]u_4 \ldots u_{r+1}c_1c_2}}$

$$-\sum A_i + \sum B_j - c_1u_0u_3u_2u_1\underline{c_2u_4 \ldots u_{r+1}c_1c_2} = 0\,,$$

$$A_i = c_1u_0u_i\underline{c_2u_ju_ku_4 \ldots c_1c_2} = 0,\ B_j = c_1u_0u_iu_k\underline{c_2u_ju_4 \ldots u_{r+1}c_1c_2} = 0,$$

$\{i, j, k\} = \{1, 2, 3\}$;

g) $c_2u_0c_1u_1u_2 \ldots u_{r+1}c_1c_2 = 0$—an analogue of f),

h) $c_1c_2u_0u_1u_2 \ldots u_{r+1}c_1c_2 = \underline{c_1[c_2u_0u_1u_2u_3]u_4 \ldots u_{r+1}c_1c_2} + \cdots = 0$

(the dots denote terms of the type on the left-hand side of f) and terms that are obviously zero).

Using e)–h) and expression (4.38), we find that

$$d_1u_1 \ldots u_sc_1c_2 = 0, \qquad 0 \leqslant s \leqslant r+1\,,$$

while, using expression d) for d_1d_2, we also have

$$d_1u_1 \ldots u_sd_1d_2 = 0, \qquad 0 \leqslant s \leqslant r+1\,.$$

We do not get the second of the necessary identities at this stage. Moreover, $d_2u_1 \ldots u_sd_1d_2 = 0$ for $0 \leqslant s \leqslant r-1$, but it is not clear whether $d_2u_1 \ldots u_rd_1d_2 = 0$ for $r = 2$. At this point we review what has been achieved so far:

$$d_1^2 = 0,\ d_2^2 = 0, \qquad [d_1d_2] = 0, \qquad d_1d_2 \neq 0, \qquad d_1u_1 \ldots u_sd_1d_2 = 0\,,$$
$$0 \leqslant s \leqslant r+1,\ r \geqslant 2\,. \tag{4.39}$$

We introduce new elements

$$e_1 = [fd_1d_2], \qquad e_2 = [d_1g^3d_1]\,,$$

where $f, g \in L$ are chosen in accordance with Proposition 3.2.2c) so that

$$d_2d_1g^2d_1fgd_1 \neq 0\,.$$

Since properties a)–d) above were proved using (4.35) with $\delta = 1$ only, we again have

$$e_1^2 = e_2^2 = [e_1e_2] = 0, \qquad e_1e_2 = 3d_2d_1g^2d_1fgd_1 \neq 0\,. \tag{4.40}$$

We shall carry out all remaining calculations in full detail. Setting

$$h_1 = [fd_2], \qquad h_2 = [d_1g^3]\,,$$

we write

$$e_\delta = [h_\delta d_1], \qquad \delta = 1, 2\,,$$

and note that

$$h_\delta d_1d_2 = 0, \qquad \delta = 1, 2\,. \tag{4.41}$$

For,

$$h_1 d_1 d_2 = \underline{[fd_2]d_1 d_2} = \underline{[fd_2]d_2} d_1 = 0 ,$$

$$h_2 d_1 d_2 = \underline{[d_1 g^3]} d_1 d_2 = \underbrace{d_1 g^3 d_1 d_2}_{(4.39)} - 3g\underbrace{d_1 g^2 d_1 d_2}_{(4.39)} = 0$$

(it is important that the value $s = 3$ be allowed in (4.39)). Thus,

$$e_\delta u_1 u_2 \ldots u_s d_1 d_2 = \underbrace{h_\delta d_1 u_1 u_2 \ldots u_s d_1 d_2}_{(4.39)} - d_1 h_\delta u_1 u_2 \ldots u_s d_1 d_2$$
$$= - \sum_{i=1}^{s} d_1 u_1 \ldots u_{i-1} [h_\delta u_i] u_{i+1} \ldots u_s d_1 d_2$$
$$- d_1 u_1 \ldots u_s \underbrace{h_\delta d_1 d_2}_{(4.41)} = 0, \qquad 0 \leqslant s \leqslant r+1 ,$$

and in that case we also have

$$e_\delta u_1 u_2 \ldots u_s e_1 e_2 = \underline{3e_\delta u_1 u_2 \ldots u_s d_1 d_2} g^2 d_1 fgd_1 = 0, \qquad 0 \leqslant s \leqslant r+1 .$$

Combining the results so far obtained, we have

$$e_1^2 = e_2^2 = [e_1 e_2] = 0, \qquad e_1 e_2 \neq 0 ,$$
$$e_\delta u_1 u_2 \ldots u_s e_1 e_2 = 0, \qquad 0 \leqslant s \leqslant r+1 ,$$

that is, (e_1, e_2) is a pair of thin sandwiches whose thickness is 1 more than in the pair in (4.34) and (4.35). This means that the inductive step works. □

2.3. Corollary. *Every Lie algebra (finite-dimensional or satisfying* E_n*) over a field of characteristic* $p > 5$ *that has no thick sandwiches, but at least one thin sandwich, contains a pair of thin sandwiches of arbitrarily large thickness.*

The *proof* is a combination of Propositions 1.1 and 2.2.

2.4. Remark. Clearly, it is sufficient to take the thickness r of a pair of thin sandwiches in a Lie algebra with E_n to be less than n, and less than $\dim_F L$ in the finite-dimensional case: the product $u_1 u_2 \ldots u_s$ can be "condensed" to these lengths.

§ 3. Completion of the Proof of the Main Theorem

3.1. And so, let L be a Lie algebra as described in the requirements of Theorem 1.7.1 and at the beginning of this chapter. As before, we assume that L has no thick sandwiches. By Corollary 2.3, L contains an "ersatz thick sandwich", that is, a pair (c_1, c_2) of thin sandwiches with properties (4.34) and (4.35), where r is large enough (see 2.4: it is enough to take $r = n - 1$ in the case of an algebra with E_n, and $r = \dim_F L - 1$ in the finite-dimensional case; indeed, these values of r will be fixed for the rest of the account).

The extent to which thick pairs work is shown up well by the example of a simple Lie algebra L of finite dimension. Let us consider the ideal $J \neq 0$ generated by one of the thin sandwiches in our pair, c_1 shall we say. Then every element u of J is a sum of monomials of the form $[c_1 u_0 u_1 \ldots u_s]$. But L is simple, so that $J = L$ and u is a general element of L; we have

$$[uc_1c_2] = \sum [c_1 u_0 \ldots u_s c_1 c_2] = -\sum \underbrace{[u_0 c_1 u_1 \ldots u_s c_1 c_2]}_{(4.35)} = 0\,,$$

so that $c_1 c_2 = 0$, despite condition (4.34). This contradiction establishes the existence of a thick sandwich in L.

Remark. The conditions of the main theorem are such that the arguments to date allow us to exclude finite-dimensional Lie algebras over a field of characteristic $p = 7$ from further consideration.

3.2. Coming now to the general case, we introduce the ideal I_c of L generated by the element

$$c = [at^2] \neq 0, \qquad t = c_1 + c_2\,,$$

for some $a \in L$. Clearly, $c = 2[ac_1c_2]$ is a sandwich of a sort familiar to us from (4.36); however, its description in terms of t is more convenient to deal with. It follows in particular from identities (4.35) that

$$tu_1u_2 \ldots u_s t^2 = 0, \qquad cu_1u_2 \ldots u_s t^2 = 0 \tag{4.42}$$

$(0 \leqslant s \leqslant r;\ u_1, u_2, \ldots, u_s \in L)$, since

$$tu_1u_2 \ldots u_s t^2 = 2(\underline{c_1u_1u_2 \ldots u_s c_1 c_2} + \underline{c_2u_1u_2 \ldots u_s c_1 c_2})\,,$$

$$cu_1u_2 \ldots u_s t^2 = 2\underline{(at^2 - 2tat + t^2a)u_1 \ldots u_s t^2}\,.$$

A less obvious corollary, namely

$$cu_1 \ldots u_s tvtvt = 0 \tag{4.43}$$

is obtained at once if we note that

$$t^3 = 0 \Rightarrow t^2vt = tvt^2 \text{ (See (1.13))}\,,$$

$$tvt^2 = 0 \Rightarrow [tv^2t^2] = -\underline{[vtvt^2]} = 0$$

$$\Rightarrow \underline{[tv^2]t^2} - 2t[tv^2]t + t^2[tv^2] = 0$$

$$\Rightarrow -2t^2v^2t + 4tvtvt - 2t\underline{v^2t^2} + \underline{t^3}v^2 - 2t^2\underline{vtv} + t^2v^2t = 0$$

$$\Rightarrow 4tvtvt = t^2v^2t \Rightarrow 4c \ldots tvtvt = \underbrace{c \ldots t^2v^2t}_{(4.42)} = 0\,.$$

Every element $h \in I_c$ can be written in the form

$$h = \sum_s [cu_0 u_1 u_2 \ldots u_s], \qquad u_i \in L\,,$$

so that

$$[ht^2] = -\sum [\underset{(4.42)}{\underline{u_0 c u_1 \ldots u_s t^2]}} = 0\,,$$

and hence

$$ht^2 - 2tht + t^2h = 0\,.$$

But

$$ht^2 = \sum \pm u_{i_0} \ldots \underset{(4.42)}{\underline{u_{i_k} c u_{j_1} \ldots u_{j_{s-k}} t^2}} = 0\,,$$

so that

$$t^2h = 2tht \tag{4.44}$$

is an identity in $h \in I_c$. By (4.42),

$$cg^2ch^2c = 4cg^2tath^2tat\,, \tag{4.45}$$

where g, h are elements of L. On the other hand,

$$h \in I_c \Rightarrow [tah^2] \in I_c\,,$$

Using this and relation (4.42), we write

$$\begin{aligned}
0 &= \underset{(4.44)}{\underline{cg^2tat^2}}[tah^2] = 2cg^2tat[tah^2]t \\
&= 2cg^2tat[ta]h^2t - 4cg^2tath[ta]ht + 2cg^2tath^2[ta]t \\
&= 2cg^2\,\underset{(4.42)}{\underline{tat^2}}\,ah^2t - 2\underset{(4.43)}{\underline{cg^2tatath^2t}} - 4cg^2ta\,\underset{(4.44)}{\underline{tht}}\,aht \\
&\quad + 4cg^2tatha\,\underset{(4.44)}{\underline{tht}} - 2cg^2tat\underset{(4.42)}{\underline{h^2at^2}} + 2cg^2tath^2tat \\
&= -2cg^2\,\underset{(4.42)}{\underline{tat^2}}\,haht + 2cg^2tat\underset{(4.42)}{\underline{hat^2h}} + 2cg^2tath^2tat = 2cg^2tath^2tat\,.
\end{aligned}$$

Comparing this relation with expression (4.45), we conclude that

$$cg^2ch^2c = 0 \tag{4.46}$$

is an identity in L (that is, $[ucg^2ch^2c] = 0$ for all $u \in L$) with respect to g, $h \in I_c$.

3.3. We note that the identity

$$[g_0 u_1 u_2 \ldots u_s g_0] = 0, \qquad s \geqslant 0, \qquad u_i \in L$$

is satisfied for every element $g_0 \in Z(I_c)$ (the centre of I_c), since $[g_0 u_1 u_2 \ldots u_s] \in I_c$. In particular, a non-zero g_0 is a thick sandwich in L (more than that, g_0 generates a non-zero abelian ideal of L), contrary to the above proposition. We may therefore assume that $Z(I_c) = 0$, and many of the general results obtained earlier can be carried over to the Lie algebra I_c (in both the finite-dimensional case and that where E_n holds), since it allows of the embedding $h \mapsto \bar{h}$ into $A(I_c)$.

We claim that I_c has a sandwich of thickness $p - 4$. If J is a non-zero abelian ideal in I_c, then every non-zero element of it can fulfill the duties of such a sandwich. But suppose that $J = 0$. Restricting (4.46) to I_c, we get that

$$\bar{c}\bar{g}^2\bar{c}\bar{h}^2\bar{c} = 0\,, \tag{4.47}$$

which is equivalent to $[fcg^2ch^2c] = 0$, $f \in I_c$, since $Z(I_c) = 0$. Here $Z(I_c) = 0 \Rightarrow \bar{c} \neq 0$, and at the same time $\bar{c}^2 = 0$, so that $\bar{c}$ is a sandwich in I_c. If $\bar{c}\bar{h}^2\bar{c} = 0$, then $\bar{c}$ is a sandwich of thickness 2. Otherwise, the existence of a sandwich of thickness 2 in I_c is guaranteed by identity (4.47) and Proposition 3.1.1. A sandwich of $\bar{c}_0$ of I_c thickness $p - 4$ is obtained now by a formal application of Theorem 2.4.6.

A pre-image $c_0 \in I_c$ in L of the sandwich $\bar{c}_0$ is determined by the relations

$$[gc_0h_1 \ldots h_jc_0] = 0, \qquad 0 \leqslant j \leqslant p - 4\,,$$

for all $g, h \in I_c$. Since I_c contains c_0, it also contains the element $[c_0vu_1 \ldots u_i]$ for any $v, u_1, \ldots, u_i \in L$, $i \geqslant 0$, and then

$$[vc_0u_1 \ldots u_ic_0h_1 \ldots h_jc_0] = -[c_0vu_1 \ldots u_ic_0h_1 \ldots h_jc_0] = 0\,,$$

whence

$$c_0u_1 \ldots u_ic_0h_1 \ldots h_jc_0 = 0\,. \tag{4.48}$$

This identity is valid for all $u_1, \ldots, u_i \in L$, $h_1, \ldots, h_j \in I_c$. Note that by definition of c_0, the index i can be any non-negative integer when $j \leqslant p - 4$. However, in the case of a Lie algebra L with E_n, $n < p$, the index $j \geqslant 0$ can also be assumed arbitrary, since the arguments in 1.5.5 allow us to restrict the value of j in such a way that $j \leqslant n - 3 \leqslant p - 4$, without affecting the fact that $h_1, \ldots, h_j$ lie in I_c. In fact, we do not need arbitrary values for j; $j = 0, 1, 2, 3, 4, 5$ suffice. These values are also enough when $\dim_F L < \infty$; just recall the remark at the end of 3.1, which yields that $p - 4 \geqslant 11 - 4 = 7$.

Unfortunately, it cannot be discerned immediately across the barricade raised by the constructions in Proposition 3.1.1 and Theorem 2.4.6 that c_0 is a thin sandwich in L. It is simpler to carry out a smallish additional argument. With $j = 0$ and $i = 0$ or 2 in (4.48), we get

$$c_0^3 = 0, \qquad c_0u^2c_0^2 = 0\,. \tag{4.49}$$

If $c_0^2 \neq 0$ and $e = [bc_0^2] \neq 0$ say, then (see (1.14) with u replaced by b and b by c_0),

$$e^2 = [bc_0^2]^2 = c_0^2b^2c_0^2 = \underbrace{c_0(c_0b^2c_0^2)}_{(4.49)} = 0\,.$$

We consider the monomial

$$eu_1 \ldots u_ieh_1 \ldots h_je\,,$$

obtained from the left-hand side of (4.48) when c_0 is replaced by e. Substituting

$$e = bc_0^2 - 2c_0bc_0 + c_0^2b$$

in this last expression, we get 27 new monomials. We note, however, that we can view a segment of the form

$$c_0 \ldots c_0^2 ,$$

(where the dots denote any segment of elements of L) in the same way as the left-hand side of (4.48) with $j = 0$. Therefore,

$$c_0 \ldots c_0^2 = 0 ,$$

and in that case

$$eu_1 \ldots u_i eh_1 \ldots h_j e = 4eu_1 \ldots \underbrace{u_i c_0 bc_0 h_1 \ldots h_j c_0\, bc_0}_{(4.48):\, i=1} = 0 .$$

Thus, we may assume without loss of generality that the element $c_0 \in I_c$ in identity (4.48) is a thin sandwich of L.

3.4. Let e_0 be any thin sandwich in L contained in I_c such that

$$e_0 h^2 e_0 = 0$$

is an identity in L with respect to elements h of I_c. On linearizing with respect to h, we get

$$e_0 h_1 h_2 e_0 = 0 .$$

Substituting $h_1 = [e_0 u^3] \in I_c$, $h_2 = [e_0 v^3] \in I_c$ at this point for any choice of $u, v \in L$, we arrive at an identity

$$e_0 [e_0 u^3] [e_0 v^3] e_0 = 0 ,$$

which it is appropriate to write as

$$[e_0 u^3 e_0] \cdot [e_0 v^3 e_0] = 0 .$$

However, by Corollary 3.2.3, this last identity is a secure fotbridge on the road to a thick sandwich.

3.5. The contradiction just obtained shows in particular that

$$c_0 h_0^2 c_0 \neq 0$$

for some $h_0 \in I_c$, whence

$$e_0 = [c_0 b_0 h_0^2 c_0] \neq 0$$

for some $b_0 \in L$. We claim that this concrete element e_0 has the property indicated in 3.4. That will then complete the proof of the theorem.

We do not state explicitly that all the relations considered (and obtained using (4.48) only) hold in L. Since $[c_0 b_0^2] \in I_c$,

$$2\underline{c_0 h_0^2 c_0 h_0 b_0 c_0 b_0 h_0 c_0} = -\underbrace{c_0 h_0^2 c_0 h_0 [c_0 b_0^2] h_0 c_0}_{(4.48):\, i=2,\, j=3} = 0 . \tag{4.50}$$

Expanding the commutator in the expression

$$e_0^2 = -c_0 [c_0 b_0 h_0^2]^2 c_0$$

and writing out all monomials arising purely formally, with undetermined coefficients, we can check without difficulty that

$$e_0^2 = \underbrace{\alpha c_0 b_0 h_0 c_0 b_0 h_0 c_0 h_0^2 c_0}_{(4.48):\, i=2,\, j=2} + \underbrace{\beta c_0 h_0^2 c_0 b_0^2 c_0 h_0^2 c_0}_{(4.48)}$$

$$+ \underbrace{\gamma c_0 b_0 h_0 c_0 h_0^2 c_0}_{(4.48)} b_0 h_0 c_0 + \underbrace{\delta c_0 h_0^2 c_0 h_0 b_0 c_0 b_0 h_0 c_0}_{(4.50)}$$

$$+ \underbrace{\nu c_0 b_0^2 c_0 h_0^2 c_0 h_0^2 c_0}_{(4.48)} + \underbrace{\mu c_0 h_0^2 c_0 h_0^2 c_0}_{(4.48)} b_0^2 c_0 = 0$$

(the relations $c_0^2 = c_0 x c_0 = 0$, $c_0 xy c_0 = c_0 yx c_0$ are trivial). Thus e_0 is a sandwich in L.

The same considerations apply verbatim to the monomial $e_0 g^2 e_0$. For, identities of the form

$$e_0 u_1 \ldots u_i c_0 h_1 \ldots h_j c_0 = 0, \qquad u_1, \ldots, u_i \in L, \qquad h_1, \ldots, \; h_j \in I_c ,$$

are a direct consequence of relations (4.48), and this means that

$$e_0 g^2 e_0 = \sum \alpha(\ldots) e_0 g^2 \ldots c_0 \ldots b_0 \ldots c_0 \ldots ,$$

where the dots denote h^k, h^l, h^t or h^q, where $k + l + t + q = 2$ ($\alpha(\ldots)$ is an undetermined coefficient, as is the $\beta(\ldots)$ coming soon). In other words, there is no sense in considering the terms $b_0 c_0 h_0^2 c_0$ and $c_0 h_0^2 c_0 b_0$. Since g, $h \in I_c$, expansion of the remaining commutator e_0 annihilates all summands except possibly $c_0 h_0^2 c_0 b_0$, for the same reasons. Thus,

$$e_0 g^2 e_0 = \sum \beta(\ldots) c_0 h_0^2 c_0 b_0 g^2 \ldots c_0 \ldots b_0 \ldots c_0 \ldots \tag{4.51}$$

When b_0 is permuted inside each segment $c_0 \ldots c_0$, commutators of the form $[b_0 g]$, $[b_0 h_0]$ contained in I_c will arise. Except for one such commutator, there are not more than three elements of I_c in the same segment. Left of the segment under discussion, there is at least one occurrence of $c_0 h_0^2 c_0$. Therefore, the commutators $[b_0 g]$, $[b_0 h_0]$ in I_c create conditions guaranteeing that identities (4.48) apply. This means that b_0 can be moved freely within a segment $c_0 \ldots c_0$. In particular, the two appearances of b_0 can be placed in the neighbourhood of a sandwich c_0 separating them; as a result, we can rewrite formula (4.51), with the dots personifying powers of h, in the form

$$e_0 g^2 e_0 = \sum \beta(\ldots) c_0 h_0^2 c_0 g^2 h_0^k \underline{b_0 c_0 b_0} h_0^l c_0 \ldots , \tag{4.52}$$

where $k + l \leqslant 2$.

Since

$$c_0 \in I_c \Rightarrow [c_0 b_0^2] \in I_c$$

and h_0, $g \in I_c$ by assumption, the monomial

$$c_0 g^2 h_0^k [c_0 b_0^2] h_0^l c_0$$

consists entirely of elements of I_c, and $2 + k + 1 + l = 3 + k + \leqslant 5$. Thus,

$$\underbrace{c_0 h_0^2 c_0 g^2 h_0^k [c_0 b_0^2] h_0^l c_0}_{(4.48):\, i=2,\, j\leqslant 5} = 0 .$$

In this case,

$$2c_0h_0^2c_0g^2h_0^k\underline{b_0c_0b_0}\,h_0^lc_0 = -\underline{c_0h_0^2c_0g^2h_0^k[c_0b_0^2]h_0^lc_0}$$

$$+\underbrace{c_0h_0^2\,c_0g^2h_0^kb_0^2c_0h_0^lc_0}_{(4.48):\,i=4+k,\,j=l\leqslant 2}$$

$$+\underbrace{c_0h_0^2c_0g^2h_0^kc_0}_{(4.48):\,i=2,\,j=2+k\leqslant 4}\;b_0^2h_0^lc_0 = 0\,.$$

We are forced to conclude that all the monomials under the summation sign in (4.52) are zero. Therefore,

$$e_0g^2e_0 = 0$$

for all $g \in I_c$, and we are in the situation considered in 3.4. □

§ 4. Commentary

4.1. The arguments in this chapter follow exactly the main canvas of the original proof in [137]. The later attempt [143] to get rid of pairs of thickness $r > 2$ was on shaky ground (the remark on p. 12 of [143] contains incorrect calculations), and it probably arose as a reaction to the necessity for reverting to the longer old version of the proof of Theorem 3 in [137].

4.2. The concept of thick pair of thin sandwiches is introduced for the first time here, although in fact thick sandwiches and thick pairs were devised as independent objects with a time-gap of not more than two years (1956–1958). The sequence in which the remaining fragments of the proof of Theorem 1.7.1 were found is a fairly complex one. If we ignore terminological distinctions, the process of thought was as follows. Having substantiated the descent from $\mathfrak{C}_2^*$ to $\mathfrak{C}_{(p-3)/2}^*$ (§ 4 of Chapter 2) and established the identity $[uv^{n-1}]^2 = 0$ for $p > n + [n/2]$ (§ 3 of Chapter 2), the author naturally wanted to convince himself that the theorem is true at least when $p \gg n$. For this it turned out to be necessary to carry out a significant part of the local analysis on thin sandwiches. After proving Theorem 3.4.1 about sandwich algebras and simultaneously its corollary, the author observed that many of the arguments (the "witchdoctor's incantations", in the parlance of § 5 in Chap. 3), can be carried over to the general case. At the same time, the assertion of Theorem 3.4.1 was used on the basis of the concept of continuation of an arbitrary element, consisting of thin sandwiches. This was done in the key Lemma 1.4 with its 10-move $e_1 \Rightarrow e_2 \Rightarrow \cdots \Rightarrow e_{10}$. The subsequent logic is inexorably linked with the discussion of pairs of arbitrary thickness.

It is curious to note that, after all that the author had done, the main difficulties were still ahead, and the desired finale—the proof of Theorem 2.2.1—required another half-year of meditation and doubt.

Chapter 5

Evolution of the Method of Sandwiches

As usually happens, the possibilities of an apparatus devised for the solution of a sufficiently weighty problem turn out to be used in part only. For example, we have seen that the method of sandwiches, which was applied to Engel Lie algebras in the first place, has exhibited its viability in terms of Lie algebras with strong degeneracy. Constructions using sandwiches have occurred comparatively recently in the works of other authors. With time, it is almost inevitable that perfection of the original apparatus occurs, and that useful new facets emerge. A large part of the present chapter is devoted to polishing such facets of the method of sandwiches.

§ 1. Rehabilitation of False Sandwiches

1.1. We shall see now that the idea of "false sandwiches", which we rejected in 3.5.3, does in fact contain a rational kernel. Some additional arguments in the same spirit enable us to obtain two proofs of Theorem 3.4.1 (see below for the first and second versions; implicit here is that the zero-th version—not the shortest, but the most fruitful—is the one expounded in § 4 of Chap. 3). This fact is an eloquent illustration of the thought, already expressed in the pages of this book, that the method of sandwiches has a large degree of flexibility in concrete situations. We deliberately dwell on these different versions of a theorem that has already been proved, and urge the reader to search for new ways of doing it.

1.2. First Version. Our aim is to prove that *every extremal sandwich Lie algebra L with E_n over a field of characteristic $p > n$ contains a thick sandwich.*

We assume that $L = \mathrm{Lie}\langle x_0, x_1, \ldots, x_r\rangle$ is a Lie algebra generated by sandwiches $x_0, x_1, \ldots, x_r$, embedded in the associative algebra $A(L)$ and containing no non-zero abelian ideals. In fact it is convenient to consider Lie algebras satisfying a weaker condition.

Condition *). Here, L is a Lie algebra with the identical relation $[uv^n] = 0$, $n < p$, and A is the associative enveloping algebra with the same set of generators and multiplication $(a, b) \mapsto ab$, in which $u^n = 0$ for all $u \in L$. In the new situation, it is natural to define a sandwich to be an element $c \neq 0$ of L such that $c^2 = 0$ and $cuc = 0$ for all $u \in L$. A sandwich has thickness not less than m if $cu_1 \ldots u_k c = 0$ for all $k \leqslant m$ and all $u_1, \ldots, u_k \in L$.

In what follows, Loc(A) is the locally nilpotent radical (the Levitzki radical) of A and $\mathfrak{C}_2(L)$ the set of thick sandwiches of L. The following assertions hold.

1) *Let I be an ideal in A. Then the Lie algebra $\bar{L} = (L + I)/I$ with enveloping algebra $\bar{A} = A/I$ also satisfies condition *). Here $\overline{\mathfrak{C}_i(L)} \subseteq \mathfrak{C}_i(\bar{L})$, $i = 1, 2, \ldots$. If $I =$ Loc(A), we have in addition that* Loc$(\bar{A}) = 0$.

The proof is obvious. □

2) $\mathfrak{C}_2(L) \subseteq$ Loc(A).

Proof. By 1) and Corollary 4.7, which holds also for sandwiches in this new sense, we may assume without loss of generality that Loc(A) = 0. Assume that $\mathfrak{C}_2(L) \neq 0$. Corollary 2.4.7 means that L has a non-zero abelian ideal J in this case. If $u_1, \ldots, u_n \in J$, then by identity (1.7) we have

$$n! u_1 \ldots u_n = \begin{Bmatrix} u_1 & \ldots & u_n \\ 1 & & 1 \end{Bmatrix} = 0 ,$$

so that the associative power $\underbrace{JJ \ldots J}_{n}$ is 0. We consider the non-zero two-sided ideal $A^{(1)}JA^{(1)}$ of A, where $A^{(1)} = F\cdot 1 + A$ is the algebra A with identity element formally adjoined. Then

$$\underbrace{JJ \ldots J}_{n} = 0 \Rightarrow (A^{(1)}JA^{(1)})^n = 0 ,$$

since A is generated by the set L. This contradiction proves our assertion. □

3) *In every finitely generated sandwich Lie algebra L with non-nilpotent enveloping algebra A and satisfying *) there exist sandwiches $x_0, y_1, \ldots y_q$ such that*

$$[x_0 y_1] = 0, \ldots, [x_0 y_q] = 0 ,$$

but such that, for every natural number m,

$$x_0 y_{i_1} y_{i_2} \ldots y_{i_m} \neq 0$$

with a suitable choice of the indices $i_1, i_2, \ldots, i_m \in \{1, 2, \ldots q\}$.

Proof. Corollary 1.4.7 shows that L cannot be nilpotent, since A is not. Without loss of generality, we can set Loc(A) = 0 and $L =$ Lie$\langle x_0, x_1, \ldots, x_r\rangle$, where $\{x_1, \ldots, x_r\}$ generates a nilpotent subalgebra of A. In other words, L is an extremal sandwich Lie algebra with $\mathfrak{C}_2(L) = 0$ (see assertion 2)). The lemma in 3.5.3 about "false sandwiches" applies to L, and it guarantees the existence of sandwiches

$$a_1, a'_1, a_2, a'_2, \ldots, \qquad a_{2m+1}, a'_{2m+1} ,$$

such that

$$c_{2m+1} = [a_0 x_0 a_1 a'_1 x_0 a_2 a'_2 x_0 \ldots x_0 a_{2m+1} a'_{2m+1} x_0] \neq 0 .$$

Expansion of the commutator c_{2m+1} in A represents it as a sum

$$c_{2m+1} = \sum \ldots x_0 \ldots x_0 \ldots x_0 \ldots$$

of monomials, each of which contains $2m+2$ occurrences of the sandwich x_0. There are not less than two letters a_i or a_i' between two successive appearances of x_0 in a non-zero monomial. The collection of letters and intervals between occurrences of x_0 is such that only one interval can contain three letters. Thus, every non-zero monomial contains a segment

$$x_0 a_{i_1} a'_{i_1} x_0 \ldots x_0 a_{i_m} a'_{i_m} x_0 \neq 0 ,$$

where the number of different letters among the a_i and a_i' is not more than $\rho = \dim_F L_0$. Setting $y_k = [x_0 a_{i_k} a'_{i_k}]$, $1 \leqslant k \leqslant m$, we get the required assertion with $q \leqslant \rho(\rho-1)$. □

Completion of the First Version of the Proof. By 3), our original finitely generated sandwich Lie algebra $L \subset A(L)$ (which is assumed to be non-nilpotent) automatically satisfies condition *) and contains a Lie subalgebra $L_1 \subset L$ generated by elements $y_1, y_2, \ldots, y_q$, which is non-nilpotent by Proposition 1.5.7. The associative enveloping algebra on the same set of generators $\{y_1, y_2, \ldots, y_q\}$ is contained in A, so that L_1 satisfies *). We consider the ideal

$$I_1 = \{ Y \in A_1 | x_0 Y = 0 \} ,$$

(which is two-sided since $[x_0 y_i] = 0$) and the natural homomorphism $A_1 \to \bar{A}_1 = A_1/L_1$ of A_1 into the associative enveloping algebra $\bar{A}_1$ for $\bar{L}_1 = (L_1 + I)/I$. As was remarked in 1), $\bar{L}_1$ satisfies *), and since $y_{i_1} y_{i_2} \ldots y_{i_m} \notin I$ for every m, $\bar{A}_1$ is not nilpotent. Again by 3), there exist commutators $\bar{z}_1, \ldots, \bar{z}_{q_1}$ in the elements $\bar{y}_1, \ldots, \bar{y}_q$ such that, for every natural number m and some sequence $i_1, i_2, \ldots, i_m$ of suffixes,

$$\bar{y}_0 \bar{z}_{i_1} \bar{z}_{i_2} \ldots \bar{z}_{i_m} \neq 0, \qquad [\bar{y}_0 \bar{z}_k] = 0, \qquad 1 \leqslant k \leqslant q_1$$

(y_0 is one of $y_1, \ldots, y_q$). Going over to pre-images, which are sandwiches in L, we get

$$y_0 z_{i_1} z_{i_2} \ldots z_{i_m} \notin I_1 ,$$

or, as is the same thing,

$$x_0 y_0 z_{i_1} z_{i_2} \ldots z_{i_m} \neq 0 .$$

We note that $[x_0 y_0] = 0$, $[x_0 z_k] = 0$, $[y_0 z_k] \in I_1$, $1 \leqslant k \leqslant q_1$.

Next, consider the subalgebra L_2 of the original Lie algebra $L \subset A(L)$ generated by $z_1, \ldots z_{q_1}$, the associative enveloping algebra A_2 on the same generating set $\{z_1, \ldots, z_{q_1}\}$, and the ideal

$$I_2 = \{Z \in A_2 | x_0 y_0 Z = 0\} \supset I_1$$

of A_2. The Lie algebra $\bar{L}_2 = (L_2 + I_2)/I_2$ satisfies *), and its enveloping algebra $\bar{A}_2 = A_2/I_2$ is not nilpotent since $z_{i_1} z_{i_2} \ldots z_{i_m} \notin I_2$ for arbitrarily large m. Reverting to 3) again, we find commutators $\bar{t}_1, \ldots, \bar{t}_{q_2}$ in $\bar{z}_1, \ldots, \bar{z}_{q_1}$ such that there exists a sequence of indices $i_1, \ldots, i_m \in \{1, \ldots, q_2\}$ of pre-assigned length m such that

$$\bar{z}_0 \bar{t}_{i_1} \ldots \bar{t}_{i_m} \neq 0, \qquad [\bar{z}_0 \bar{t}_k] = 0, \qquad 1 \leqslant k \leqslant q_2$$

(z_0 is one of the sandwiches $z_1, \ldots, z_{q_2}$). In terms of the pre-images, we have

$$z_0 t_{i_1} \ldots t_{i_m} \notin I_2 ,$$

that is,

$$x_0 y_0 z_0 t_{i_1} \ldots t_{i_m} \neq 0 ,$$

where

$$[x_0 y_0] = [x_0 z_0] = [x_0 t_k] = 0, \quad [y_0 z_0] \in I_1, \quad [y_0 t_k] \in I_1, \quad [z_0 t_k] \in I_2 .$$

This means in particular that for $m = 1$ we have a product $x_0 y_0 z_0 t_0 \neq 0$ of four sandwiches in L whose order can be changed at will. For example, if we set

$$[z_0 y_0] = u \in I_1, \qquad [t_0 y_0] = v \in I_1, \qquad [t_0 z_0] = w \in I_2 ,$$

we find that

$$\begin{aligned} t_0 z_0 y_0 x_0 &= t_0 x_0 z_0 y_0 = t_0 x_0 (y_0 z_0 + u) = t_0 x_0 y_0 z_0 \\ &= x_0 t_0 y_0 z_0 = x_0 (y_0 t_0 + v) z_0 = x_0 y_0 t_0 z_0 \\ &= x_0 y_0 (z_0 t_0 + w) = x_0 y_0 z_0 t_0 . \end{aligned}$$

Since the condition $x_0 y_0 z_0 t_0 \neq 0$ in the Lie algebra $L \subset A(L)$ is equivalent to the existence of an element $a \in L$ such that

$$c = [a x_0 y_0 z_0 t_0] \neq 0 ,$$

we find ourselves in the conditions of Lemma 3.3.2, according to which c is a thick sandwich. This contradiction completes the proof. □

We observe that reference to E_n is confined to assertions 2) and 3).

1.3. Second Version. The premises formulated at the beginning of 1.2 are preserved; however, from the very beginning preference remains in the arguments for a pair (L, A) in which L is a Lie algebra and A its associative enveloping algebra. In particular, an element $0 \neq c \in L$ is said to be a (thin) sandwich for the pair (L, A) if $c^2 = 0$ and $cLc = 0$. There is nothing novel in this when $A = A(L)$; however, as was remarked in 3.5.3, caution must be observed when $Z(L) \neq 0$. Instead of condition *) in 1.2, we assume that the following holds:

Condition *k): (i) L is a Lie algebra, A its associative enveloping algebra; L and A are both generated by thin sandwiches of the pair (L, A);

(ii) L satisfies E_n, $n < p$;
(iii) $u^k = 0$ in A for all $u \in L$.

The exponent n is fixed, and k can take any value from 1 to n. *We intend to prove by induction on k that A is nilpotent when $k = n$, and in particular, that an extremal Lie algebra $L = \mathrm{Lie}\ \langle x_0, x_1, \ldots, x_r \rangle \subset A(L)$ with condition *n) is nilpotent* (the assertion is obvious when $k = 1$). From here on we assume that the nilpotency of an algebra A with *l) has already been established for $l < k$. Factoring A by its Levitzki radical $\mathrm{Loc}(A)$ preserves all parts (i)–(iii) of condition *k). Thus, we can assume at the outset that $\mathrm{Loc}(A) = 0$.

Assume that there is a thick sandwich in the Lie algebra L with $*k)$ which is also a thin sandwich for the pair (L, A). Then, by Corollary 2.4.7, L contains an abelian ideal $J \neq 0$. As in assertion 2) of 1.2, we conclude that $\underbrace{J \cdot J \cdot \ldots \cdot J}_{n} = 0$, and thus that $A^{(1)}JA^{(1)}$ is a nilpotent ideal of A. The contradiction thus obtained prompts the rest of the path to the proof.

Arguing by contradiction (A is assumed non-nilpotent) and using induction on r, we note that, like condition $*$) in assertion 3) of 1.2, condition $*k)$ guarantees the existence of thin sandwiches $x_0, y_1, \ldots, y_q$ (with $[x_0 y_k] = 0$, $1 \leqslant k \leqslant q$) of the pair (L, A) such that

$$x_0 y_{i_1} y_{i_2} \ldots y_{i_m} \neq 0$$

for arbitrary m and a suitable choice of the sequence $i_1, i_2, \ldots, i_m$.

Let L_1 and A_1 be the subalgebras of L and A respectively generated by $y_1, \ldots, y_q$. We consider in A_1 the ideal I generated by the set $\{u^{k-1} | u \in L_1\}$. Since x_0 lies in the centralizer of A_1, by identity (1.8) we have, for every element $\sum_i a_i u_i^{k-1} b_i$ of I (here $a_i, b_i \in A_1^{(1)} = F \cdot 1 + A_1$, $u_i \in L_1$),

$$k x_0 \sum_i a_i u_i^{k-1} b_i = \sum_i a_i \begin{Bmatrix} x_0 & u_i \\ 1 & k-1 \end{Bmatrix} b_i = 0 \,,$$

in view of $*k)$. Clearly, the pair $(\bar{L}_1, \bar{A}_1)$ with $\bar{L}_1 = (L_1 + I)/I$, $\bar{A}_1 = A_1/I$, satisfies $*(k-1)$. Thus, $\bar{A}_1$ is nilpotent by induction, that is, there is a natural number l such that $u_1 u_2 \ldots u_l \in I$ for all $u_1, u_2, \ldots, u_l \in L_1$, or equivalently $x_0 u_1 u_2 \ldots u_l = 0$. In particular, $x_0 y_{j_1} y_{j_2} \ldots y_{j_l} = 0$ for every sequence $j_1, j_2, \ldots, j_l$. However, this contradicts the earlier conclusion. □

§ 2. A Geodesic Connecting Theorems 3.4.1 and 1.7.4

Leaving Theorem 1.7.1 on one side, we endeavour now to find a direct proof of Theorem 1.7.4 on the local nilpotency of Lie algebras with E_n, $n < p$. We shall assume that $p > 5$ in what follows. The idea is to exclude the arguments of Chap. 4 by taking as auxiliary results those of § 3 in Chap. 1 (the proposition about the radical $R(L)$), §§ 1 and 2 of Chap. 2 (descent to thin sandwiches), and Theorem 3.4.1 with proof in the form of one of the versions provided in § 1 of this chapter, for example.

The following relatively simple assertion is directed at the descent to nil-elements of index 3, and it plays a key rôle:

Lemma 2.1. *A Lie algebra L with E_n, $n < p$, generated by finitely many elements $x_1, \ldots, x_d$ of nil-index* $\leqslant 3$ (that is, $(\operatorname{ad} x_i)^3 = 0$ *for* $1 \leqslant i \leqslant d$) *is nilpotent.*

Proof. Let us suppose that L is not nilpotent. By Proposition 1.3.2, we may assume without loss of generality that $R(L) = 0$. By Theorem 2.2.1, the sandwich algebra $\mathfrak{C}(L)$ is non-zero. It follows from the definition of $\mathfrak{C}(L)$ that this subalgebra is invariant under the automorphism group $\operatorname{Aut}(L)$ of L. The maps

$$\varphi_i = \exp(\lambda \operatorname{ad} x_i) \colon L \to L \,,$$

defined by the relations

$$\varphi_i(u) = u + \lambda[ux_i] + \frac{\lambda^2}{2}[ux_i^2], \qquad 1 \leqslant i \leqslant d, \qquad \lambda \in F, \qquad u \in L,$$

are easily seen to be automorphisms of L (recall that $p > 5$). Thus,

$$a \in \mathfrak{C}(L) \Rightarrow a + \lambda[ax_i] + \frac{\lambda^2}{2}[ax_i^2] \in \mathfrak{C}(L),$$

so that

$$[ax_i] + \frac{\lambda}{2}[ax_i^2] \in \mathfrak{C}(L), \qquad \lambda \neq 0.$$

Setting $\lambda = \pm 1$, we conclude that $[ax_i] \in \mathfrak{C}(L)$ for $1 \leqslant i \leqslant d$, that is, $[\mathfrak{C}(L), x_i] \subseteq \mathfrak{C}(L)$; since $x_1, \ldots, x_d$ generate L, we have

$$[\mathfrak{C}(L), L] \subseteq \mathfrak{C}(L).$$

This means that $\mathfrak{C}(L)$ is a non-zero ideal of L. By Theorem 3.4.1, this ideal is locally nilpotent, which contradicts the assumption that $R(L) = 0$. □

2.2. Proof of Theorem 1.7.4. 1) The ground field can be assumed infinite. For every extension $\tilde{F} \supset F$, the Lie algebra $L_{\tilde{F}} = L \otimes_F \tilde{F}$ also satisfies E_n, that is, $[\tilde{u}\tilde{v}^n] = 0$ for all $\tilde{u}, \tilde{v} \in L_{\tilde{F}}$. This follows from the fact that the identity $[uv^n] = 0$ in L is equivalent to the linearized identity

$$\sum_{\pi \in S_n} [uv_{\pi 1} v_{\pi 2} \ldots v_{\pi n}] = 0.$$

Since this identity is multilinear, $L_{\tilde{F}}$ also satisfies the identity

$$\sum_{\pi \in S_n} [\tilde{u}\tilde{v}_{\pi 1} \tilde{v}_{\pi 2} \ldots \tilde{v}_{\pi n}] = 0,$$

which in its turn is equivalent to $[\tilde{u}\tilde{v}^n] = 0$ (set $\tilde{v}_1 = \cdots = \tilde{v}_n = \tilde{v}$ and note that $n!$ is a non-zero element of $\tilde{F}$). Clearly, this argument fails for $n \geqslant p$.

2) So, let L be a Lie algebra over an infinite field F, and let $f(x) = f(x_1, \ldots x_m)$ be a Lie polynomial, that is, an element of a free Lie F-algebra. Further, let $f(L)$ be the set of values of f on L. Then the subspace $V_f \subset L$ spanned by $f(L)$ over F is an ideal in L.

This assertion relates to general facts from the theory of varieties of Lie algebras (see [26], Theorem 4.2.9, for example).

3) As our original Lie polynomial we take

$$f(x_0, x_1, \ldots, x_m) = [x_m[\ldots[x_2[x_1 x_0^{k_1}]^{k_2}]\ldots]^{k_m}],$$

which is not an identity in L and such that $f(L)$ consists of nil-elements of index $\leqslant 3$. The natural numbers $k_1, k_2, \ldots, k_m$ are chosen in accordance with the proof of Proposition 2.1.2, or more exactly, with the procedure in Lemma 2.1.1: there exists an index $k_1 \leqslant n-1$ such that $[ux_0^{k_1+1}] = 0$ but $[x_1 x_0^{k_1}] \neq 0$ for some $x_1 \in L$; then an index $k_2 \leqslant k_1 - 1$ is chosen such that $[u[x_1 x_0^{k_1}]^{k_2+1}] = 0$ but

$[x_2[x_1 x_0^{k_1}]^{k_2}] \neq 0$, *etc.* The process leads to the required polynomial in $m \leqslant n-3$ steps.

4) Without loss of generality, we may assume that L is a Lie algebra with E_n over an infinite field (see 1)) and such that $R(L) = 0$. By 2), the space V_f associated with the polynomial f in 3) and spanned by nil-elements of index $\leqslant 3$ is an ideal. By Lemma 2.1, this ideal is locally nilpotent, and we have a contradiction which proves the theorem. □

§ 3. The Local Nilpotency of Sandwich Algebras

Some fresh additional considerations enable us to use Proposition 3.4.1 more effectively, and to get the theorem about sandwich algebras in the following more complete form.

3.1. Theorem. *Every finitely generated sandwich Lie algebra L over a field of characteristic zero or $p > 5$ is nilpotent.*

Thus, the condition E_n turns out to be redundant in Theorem 3.4.1, and the parameter n will be used for other pu poses. Let us return to the remarks of § 1. It would be convenient for the proof to ıave available the embedding $L \subset A(L)$, but in general the existence of a centre . $(L) \neq 0$ is not excluded. Thus, we begin by working within the algebra L itself. B definition, a sandwich Lie algebra L has the form

$$L = \mathrm{Lie}\langle c_1, c_2, \ldots, c_n | (\mathrm{ad}\, c_i)^2 = 0, \qquad 1 \leqslant i \leqslant n \rangle . \tag{5.1}$$

Since

$$(\mathrm{ad}\, x)^2 = 0, \qquad (\mathrm{ad}\, y)^2 = 0 \Rightarrow (\mathrm{ad}[x, y])^2 = 0 ,$$

$$(\mathrm{ad}\, c)^2 = 0 \Rightarrow \mathrm{ad}\, c \cdot \mathrm{ad}\, u \cdot \mathrm{ad}\, c = 0, \qquad u \in L \tag{5.2}$$

(see 1.5.3), and since the nilpotency proof for L is equivalent to that for the factor-algebra

$$L/Z(L) \cong \mathrm{ad}\, L = L^* \subset A(L) ,$$

we may assume without loss of generality that

$$L \subset A_L = A(\tilde{L}) ,$$

where

$$\tilde{L}/Z(\tilde{L}) \cong L .$$

Thus, a sandwich Lie algebra L with specification (5.1) can be assumed embedded in the associative algebra A_L with the same generating set $\{c_1, \ldots, c_n\}$ and multiplication $(a, a') \mapsto aa'$. Now $[u, v] = uv - vu$ for $u, v \in L$. By (5.2) we have

$$c_i^2 = 0, \qquad c_i u c_i = 0, \qquad u \in L, \qquad 1 \leqslant i \leqslant n , \tag{5.3}$$

and it is more correct to speak of the set $\mathfrak{C}_1(L, A_L)$ of thin sandwiches of the pair (L, A_L). In particular, the generators $c_1, \ldots, c_n$ lie in this set. More generally, we follow Definition 1.5.2 and introduce the set $\mathfrak{C}_r(L, A_L)$ of thin sandwiches $c \in L$ of thickness r.

In order to avoid the necessity for appealing to the system of relations

$$cu_1 u_2 \ldots u_k c = 0, \qquad u_i \in L, \qquad k = 0, 1, \ldots, r,$$

we write

$$cL^{(r)}c = 0,$$

for short, where $L^{(r)}$ is understood to be the subspace of the algebra $A_L^{(1)} = F \cdot 1 + A_L$ with identity spanned by $1, x_1, x_1 x_2, \ldots, x_1 x_2 \ldots x_r, x_i \in L$.

In what follows we shall frequently go over from L to the sandwich algebra $L^* \subset A(L)$ with elements $x^* = \operatorname{ad} x$, $x \in L$. We split the proof of Theorem 3.1 into several stages. As a preliminary we introduce the following:

3.2. Definition. The sequence $\{\mathscr{D}^k M \mid k = 0, 1, 2, \ldots\}$ of *derived sets* of a set $M \subset L$ is defined inductively:

$$\mathscr{D}^0 M = M, \qquad \mathscr{D}^{k+1} M = \{[xy] \mid x, y \in \mathscr{D}^k M\}.$$

When $M = L$, the sequence of linear spans $L^{[k]} = \langle \mathscr{D}^k L \rangle_F$ gives the *derived series* of L:

$$L = L^{[0]} \supset L^{[1]} \supset L^{[2]} \supset \cdots$$

We shall say that a set $M \subset L$ is *soluble* if $\mathscr{D}^k M = 0$ for large enough k. The Lie algebra L is said to be *soluble* (of *solubility length* l) if $L^{[l]} \neq 0$, $L^{[l+1]} = 0$. We also speak of the *length* l *of the derived series* of a soluble algebra L.

3.3. Lemma. *The derived subalgebra* $L^{[1]} = [L, L]$ *of a finitely generated sandwich algebra* $L \subset A_L$ *is finitely generated. Moreover, if* $L = \operatorname{Lie}\langle c_1, \ldots, c_n \rangle$, *then* $L^{[1]}$ *is generated by the commutators* (*which are sandwiches*) *in the set*

$$S = \{[c_{i_0} c_{i_1} \ldots c_{i_k}] \mid 1 \leqslant i_j \leqslant n, \qquad 1 \leqslant k \leqslant n-1\}.$$

Proof. We consider an arbitrary commutator $w = [c_{i_0} c_{i_1} \ldots c_{i_m}]$ of length $m+1$ with $m \geqslant n$. Since $1 \leqslant i_j \leqslant n$, at least two of the occurrences in w are the same. Suppose that c_1 occurs twice, for example. Since multiplication in L is anticommutative, we may assume that $c_{i_0} \neq c_1$ and that

$$w = [c_{i_0} \ldots c_1 c_{i_k} \ldots c_{i_l} c_1 \ldots c_{i_m}]$$

$$= \sum_{s=k+1}^{l} [c_{i_0} \ldots c_1 c_{i_k} \ldots [c_{i_s} c_1] \ldots c_{i_l} \ldots c_{i_m}], \tag{5.4}$$

since $c_1^2 = 0$ and $c_1 u c_1 = 0$. Then, by pushing the commutators $[c_{i_s} c_1]$ to the extreme right position, we can only increase their weight. Thus, $w = \sum_j [u_j v_j]$, where u_j and v_j are commutators in $c_{i_0}, \ldots, c_{i_m}$ of length at least 2. Induction on m now proves that the u_j and v_j lie in the subalgebra $\operatorname{Lie}\langle S \rangle$ generated by S. This proves that $w \in \operatorname{Lie}\langle S \rangle$ also. □

3.4. Lemma. *If the sandwich Lie algebra* $L = \text{Lie}\langle c_1, c_2, \ldots, c_n\rangle$ *is soluble, then it is nilpotent.*

Proof. By assumption, $L^{[l+1]} = 0$ for some natural number l. By Lemma 3.3, the derived subalgebra $K = L^{[1]}$ of L is a finitely generated sandwich Lie algebra and

$$L^{[l+1]} = 0 \Rightarrow K^{[l]} = 0\,,$$

so that induction on l (with the obvious starting-point $l = 0$) enables us to state that K is nilpotent:

$$K^m = \underbrace{[K, K, \ldots, K]}_{m} = 0$$

for large enough m. We claim that $L^{(n+1)m} = 0$. For, every left-normed commutator

$$w = [\underline{c_{i_0} c_{i_1} \ldots c_{i_n}}\, \underline{c_{j_0} c_{j_1} \ldots c_{j_n}} \ldots \underline{c_{k_0} c_{k_1} \ldots c_{k_n}}]$$

of length $N = (n + 1)m$ with respect to $c_1, \ldots, c_n$ splits up into m segments of length $n + 1$, each of which contains at least two identical components (which are sandwiches):

$$w = [c_{i_0} \ldots \underline{c_i \ldots c_i} \ldots \underline{c_j \ldots c_j} \ldots \underline{c_k \ldots c_k}]\,.$$

As in the case of (5.4), we have

$$[uc_s c_{s_1} \ldots c_{s_q} c_s \ldots] = \sum_{v=2}^{q} [uc_s c_{s_1} \ldots [c_{s_v} c_s] \ldots c_{s_q} \ldots]\,,$$

that is, every segment "generates" a commutator of length at least 2 independently of the others, and w can be expressed as a linear combination of products

$$[\ldots [c_i c_{i_\alpha}] \ldots [c_j c_{j_\beta}] \ldots [c_k c_{k_\omega}] \ldots]\,,$$

containing not less than m elements of K. In other words, $w \in K^m = 0$ and $L^{(n+1)m} = 0$. □

Remark. Under the assumptions of the lemma, not only L is nilpotent, but A_L is too. This follows immediately from the proof or from Proposition 1.5.7.

We introduce the notation

$$L(x; k_1, \ldots, k_s) = xL^{(k_1)} xL^{(k_2)} \ldots xL^{(k_s)}\,,$$

$$L(x; k \circ s) = L(x; \underbrace{k, \ldots, k}_{s})\,, \tag{5.5}$$

where $x \in L$ and $k_1, \ldots, k_s$ are natural numbers. By assumption, $L(x; k \circ 0) = 1$ (more generally, $L(x; k_1, \ldots, k_s) = 1$ for $s = 0$).

The following is very useful in the technical plan:

3.5. Lemma (about deletions). *The following inclusions hold for any sandwiches* b, c *of the pair* (L, A):

(i) $bcL^{(ko)}L([bc]; k_1, \ldots, k_s)cb \subseteq L(b; k_0, k_1, \ldots, k_s)b$;

(ii) $bcL^{(ko)}L([bc]; k_1, \ldots, k_s)bc \subseteq L(b; k_0, k_1, \ldots, k_s)bc$;

(iii) $cbL^{(ko)}L([bc]; k_1, \ldots, k_s)bc \subseteq cL(b; k_0, k_1, \ldots, k_s)bc$;

(iv) $cbL^{(ko)}L([bc]; k_1, \ldots, k_s)cb \subseteq cL(b; k_0, k_1, \ldots, k_s)b$.

Thus, the deletion of c is expressed by the inclusion

$$t_1 t_2 L^{(ko)} L([bc]; k_1, \ldots, k_s) t_3 t_4 \subseteq L^{(1)} L(b; k_0, k_1, \ldots, k_s) b L^{(1)}$$

for arbitrary $t_i = b$ or c, *and in particular*

$$L([bc]; k_0, k_1, \ldots, k_s)[bc] \subseteq L^{(1)} L(b; k_0, k_1, \ldots, k_s) b L^{(1)} .$$

The notation L.3.5(t) below denotes that Lemma 3.5 is being applied with the deletion of t.

Proof. a) We first check that the following inclusions (which are of independent interest) hold for every x:

$$L^{(k)}x \subseteq xL^{(k)} + L^{(k)}, \qquad xL^{(k)} \subseteq L^{(k)}x + L^{(k)} ; \tag{F0}$$

$$L^{(k)}x \subseteq L^{(1)}xL^{(k-1)} + xL^{(k)} + L^{(k-1)} ,$$

$$xL^{(k)} \subseteq L^{(k-1)}xL^{(1)} + L^{(k)}x + L^{(k-1)} ; \tag{F1}$$

$$L^{(k)}x \subseteq L^{(2)}xL^{(k-2)} + L^{(1)}xL^{(k-1)} + xL^{(k)} + L^{(k-2)} ,$$

$$xL^{(k)} \subseteq L^{(k-2)}xL^{(2)} + L^{(k-1)}xL^{(1)} + L^{(k)}x + L^{(k-2)} . \tag{F2}$$

Inclusion (F0) is obvious. Further,

$$x_1 \ldots x_k x = x_1 \ldots x_{k-1}[x_k x] + x_1 \ldots x_{k-1} x x_k ,$$

where

$$x_1 \ldots x_{k-1}[x_k x] \equiv [x_k x] x_1 \ldots x_{k-1} (\operatorname{mod} L^{(k-1)}) .$$

Similarly,

$$x_1 \ldots x_{k-1} x x_k = x_1 \ldots x_{k-2}[x_{k-1} x] x_k + x_1 \ldots x_{k-2} x x_{k-1} x_k ,$$

so that after a finite number of steps we get a congruence

$$x_1 \ldots x_{k-1} x \equiv x_1 x x_2 \ldots x_k + \sum_{i=2}^{k} [x_i x] x_1 \ldots \hat{x}_i \ldots x_k (\operatorname{mod} L^{(k-1)}) ,$$

from which the first inclusion in (F1) follows when the commutator $[x_i x]$ is expanded. The second inclusion of (F1) is obtained by displacing x to the right. The subtler relations (F2) are based on the appearance of commutators of the form $[x_i[x_j x]]$.

b) Assertions (i)–(iv) of the lemma are all of the same sort. We prove them now, starting with the case $s = 0$.

(i) $$\underbrace{bcL^{(k_0)}cb}_{\text{(F1)}} \subseteq b\underbrace{cL^{(1)}cL^{(k_0-1)}}_{\text{(5.3)}}b + b\underbrace{c^2L^{(k_0)}}_{\text{(5.3)}}b + bcL^{(k_0-1)}b$$

$$\subseteq bL^{(k_0)}b = L(b;\, k_0)b;$$

(ii) $$\underbrace{bcL^{(k_0)}bc}_{\text{(F1)}} \subseteq bL^{(k_0)}\underline{cbc} + bL^{(k_0-1)}cL^{(1)}bc + bL^{(k_0-1)}bc$$

$$\subseteq bL^{(k_0-1)}[c,\, L^{(1)}]bc + bL^{(k_0-1)}bc$$

$$\subseteq bL^{(k_0)}bc = L(b;\, k_0)bc\,.$$

Assertion (iii) is trivial for $s = 0$, and (iv) is obtained just as simply as the first two.

We shall go into details of the induction on s only in case (i). We have

$$bcL^{(k_0)}L([bc];\, k_1, \ldots, k_s)cb \subseteq bcL^{(k_0)}L([bc];\, k_1, \ldots, k_{s-1})\underbrace{cbL^{(k_s)}cb}_{\text{(iv)}:\, s=0}$$

$$+ \underbrace{bcL^{(k_0)}L([bc];\, k_1, \ldots, k_{s-1})bc}_{\text{(ii)}:\, s-1}L^{(k_s)}cb$$

$$\subseteq \underbrace{bcL^{(k_0)}L([bc];\, k_1, \ldots, k_{s-1})cb}_{\text{(i)}:\, s-1}L^{(k_s)}b$$

$$+ bL^{(k_0)}L(b;\, k_1, \ldots, k_{s-1})\underbrace{bcL^{(k_s)}cb}_{\text{(ii)}:\, s=0}$$

$$\subseteq L(b;\, k_0, k_1, \ldots, k_{s-1})bL^{(k_s)}b + bL^{(k_0)}L(b;\, k_1, \ldots, k_{s-1})bL^{(k_s)}b$$

$$\subseteq L(b;\, k_0, k_1, \ldots, k_s)b\,. \quad \square$$

3.6. Lemma. *Suppose that the integers $s \geqslant 0$, $k \geqslant 3$, the sandwiches b, c of the pair (L, A_L) and the element $X \in A_L$ are such that*

$$L(t;\, 2\circ s)tL^{(k-1)}tL^{(k)}X = 0, \qquad t = b, c\,.$$

Then

$$L([bc];\, 2\circ s)[bc]L^{(k)}[bc]X = 0\,.$$

Proof. Note that, because of the definition of $L^{(k)}$, each relation $GL^{(k)}H = 0$ means that $GL^{(i)}H = 0$ for $i \leqslant k$. In particular, we have for instance that

$$L(t;\, 2\circ s)tL^{(2)}tL^{(k)}X = 0, \qquad t = b, c\,,$$

since $k - 1 \geqslant 2$ by assumption. Similarly,

$$L(t;\, 2\circ s)tL^{(k-1)}tX = 0\,.$$

It is appropriate to begin with the general case $s > 0$. Symmetry considerations relating to b and c mean that it suffices to consider sets of elements Y, Z of two types.

a) $Y = L([bc];\, 2\circ s)bcL^{(k)}bcX$.

Together with (F1), the relations $bL^{(1)}b = 0$, $cL^{(1)}c = 0$ give that

$$Y \subseteq \underbrace{[bc]L^{(2)}L([bc];\, 2\circ(s-1))bcL^{(1)}bL^{(k-1)}cX}_{\text{L. 3.5(c)}}$$

$$+ \underbrace{[bc]L^{(2)}L([bc];\, 2\circ(s-1))bcL^{(k-1)}cX}_{\text{L. 3.5(b)}}$$

$$\subseteq \underbrace{L^{(1)}L(b;\, 2\circ s)bL^{(2)}bL^{(k)}X}_{\text{condition on sandwich } b} + \underbrace{L^{(1)}L(b;\, 2\circ s)cL^{(k-1)}cX}_{\text{condition on } c} = 0\,.$$

b) $Z = L([bc];\, 2\circ s)bcL^{(k)}cbX$.

Using F2) with x replaced by c, we get that

$$Z \subseteq \underbrace{[bc]L^{(2)}L([bc];\, 2\circ(s-1))bcL^{(2)}cL^{(k-2)}bX}_{3.5\text{(b)}}$$

$$+ \underbrace{[bc]L^{(2)}L([bc];\, 2\circ(s-1))bcL^{(k-2)}bX}_{3.5\text{(c)}}$$

$$\subseteq \underbrace{L^{(1)}L(c;\, 2\circ s)cL^{(2)}cL^{(k-1)}X}_{\text{condition on sandwich } c} + \underbrace{L^{(1)}L(b;\, 2\circ s)bL^{(k-1)}bX}_{\text{condition on b}} = 0\,.$$

For $s = 0$, when the deletion lemma does not apply, only one of the inclusions (F1) is functioning effectively:

$$Y = \underbrace{bcL^{(k)}bcX}_{\text{(F1)}} \subseteq bc(L^{(1)}bL^{(k-1)} + L^{(k-1)})cX$$

$$\subseteq \underbrace{bL^{(2)}bL^{(k-1)}X}_{\text{condition on } b} + \underbrace{bcL^{(k-1)}cX}_{\text{condition on } c} = 0\,,$$

$$Z = \underbrace{bcL^{(k)}cbX}_{\text{(F1)}} \subseteq bc(cL^{(k)} + L^{(1)}cL^{(k-1)} + L^{(k-1)})cbX$$

$$\subseteq \underbrace{bcL^{(k-1)}cL^{(1)}X}_{\text{condition on } c} = 0\,. \quad \square$$

3.7. Proposition. *Suppose that* $s \geqslant 1$ *and that*

$$M = \{b \in L \mid bL^{(1)}b = 0, \quad L(b;\, 2\circ s)b = 0\}\,.$$

Then there exists a function $f\colon \{3, 4, \ldots\} \to Z$, $f(r) \geqslant 0$, *such that*

$$L(c;\, r\circ(s+1))c = 0$$

for arbitrary $c \in \mathcal{D}^{f(r)}M$.

Proof. We propose to construct a recurrent sequence of functions $\{f_t \mid 0 \leqslant t \leqslant s\}$ such that

$$L(c;\, 2\circ(s-t))L(c;\, r\circ(t+1))c = 0 \tag{5.6}$$

for all $c \in \mathscr{D}^f t^{(r)}$ and all $r \geqslant 3$. Once this is done, the function $f = f_s$ will obviously be the one we want, since $L(c; 2\circ 0) = 1$ and relation (5.6) with $t = s$ is just the relation occurring in the statement of the proposition.

We begin the construction of the sequence $\{f_t\}$ with $t = 0$, by setting $f_0(r) = 0$. This is possible because it follows at once from the requirements of the proposition that $L(c; 2\circ s)cL^{(r)}c = 0$ for all $c \in M = \mathscr{D}^0 M$. Assuming that $f_0, f_1, \ldots, f_t$ have been constructed for $t < s$, we then construct f_{t+1} as follows. Replace the parameter r in (5.6) by

$$r_1 = r + \frac{(r-2)(r+5)}{2} .$$

We will then have

$$L(a; 2\circ(s-t))L(a; r_1\circ(t+1))a = 0$$

for all $a \in \mathscr{D}^f t^{(r_1)} M$. In particular, the weaker relation

$$L(a; 2\circ(s-t))aL^{(r_1)}L(a; r\circ t)a = 0 \tag{5.7}$$

holds.

For arbitrary $a \in \mathscr{D}^f t^{(r_1)+q-2} M$ and $\sigma_q = (q-2)(q+5)/2$, we establish the relation

$$L(a; 2\circ(s-t-1))aL^{(q)}aL^{(r_1-\sigma_q)}L(a; r\circ t)a = 0 \tag{5.8}$$

by induction on q, $2 \leqslant q \leqslant r$. For $q = 2$, relation (5.8) turns into (5.7), while for $q = r$ it is the relation we want to establish. Our problem is thus to move up from q to $q+1$ in (5.8).

We shall use the fact that

$$L(u; 2\circ(s-t-1))uL^{(q)}uL^{(r_1-\sigma_q-1)}L([uv]; r\circ t)[uv] = 0 \tag{5.9}$$

for $u \in \mathscr{D}^f t^{(r_1)+q-2} M$, $v \in M$; this is a corollary of (5.8). Indeed, by Lemma 3.5,

$$\begin{aligned} &L^{(r_1-\sigma_q-1)}L([uv]; r\circ t)[uv] \subseteq L^{(r_1-\sigma_q-1)}L^{(1)}L(u; r\circ t)uL^{(1)} \\ &= L^{(r_1-\sigma_q)}L(u; r\circ t)uL^{(1)} . \end{aligned} \tag{5.10}$$

Denoting the left-hand side of equation (5.9) by T and using (5.10), we get that

$$T \subseteq \underset{(5.8)}{\underline{L(u; 2\circ(s-t-1))uL^{(q)}uL^{(r_1-\sigma_q)}L(u; r\circ t)uL^{(1)} = 0}} .$$

We use the definition of σ_q to rewrite (5.9) in the form

$$\begin{aligned} &L(a; 2\circ(s-t-1))aL^{(q)}aL^{(q+1)}X = 0, \qquad a \in \mathscr{D}^{f_t(r_1)+q-2} M, \\ &L(b; 2\circ(s-t-1))bL^{(q)}bL^{(q+1)}X = 0, \qquad b \in \mathscr{D}^{f_t(r_1)+q-2} M , \end{aligned} \tag{5.11}$$

for every $X \in L^{(r_1-\sigma_q+1)}L([ab]; r\circ t)[ab]$. By Lemma 3.6, (5.11) yields the relation

$$L([ab]; 2\circ(s-t-1))[ab]L^{(q+1)}[ab]X = 0 ,$$

that is,

$$L(c;\ 2\circ(s-t-1)cL^{(q+1)}cL^{(r_1-\sigma_q+1)}L(c;\ r\circ t)c = 0$$

for all $c \in \mathscr{D}^{f_1(r_1)+q-1}M$.

Thus, the induction hypothesis is justified, and (5.8) is established. Since $r_1 - \sigma_r = r$, on setting

$$f_{t+1}(r) = f_t\left(r + \frac{(r-2)(r+5)}{2}\right) + r - 2\,,$$

we see that the relation we want, namely (5.6) with t replaced by $t+1$, does in fact hold. □

3.8. Lemma. *Every Lie algebra L over a field F generated by finitely many thick sandwiches $c_1, \ldots, c_n$ of the pair (L, A_L) is nilpotent.*

Proof. Consider the subspace

$$L_t = Fc_i + [L, c_i] \subset L, \qquad 1 \leqslant t \leqslant n\,.$$

Clearly, $L = L_1 + \cdots + L_n$, and $L_tL_t = 0$ since $c_t \in \mathfrak{C}_2(L, A_L)$. The discussion now splits into two parts.

1) $A_L^{(1)} = L^{(n)}$, that is, every element of A_L is an F-linear combination of products $u_1u_2 \ldots u_k$, where the $u_i \in L$, $k \leqslant n$. In fact, we need only check that every product $u_1u_2 \ldots u_{n+1}$ of elements u_i of L lies in $L^{(n)}$, where without loss of generality we may assume that $u_i \in L_{t_i}$, $1 \leqslant i \leqslant n+1$, $1 \leqslant t_i \leqslant n$. This means that there exist different numbers $1 \leqslant i \leqslant j \leqslant n+1$ such that u_i and u_j lie in the same subspace L_t. Modulo $L^{(n)}$, the factors $u_1, u_2, \ldots, u_{n+1}$ in $u_1u_2 \ldots u_{n+1}$ commute. However, as we saw at the very beginning, $u_iu_j \in L_tL_t = 0$. Thus, $L^{(n+1)} \subseteq L^{(n)}$.

2) If the lemma is false and L is not nilpotent, neither is A_L. Factoring A_L by its locally nilpotent radical $\mathrm{Loc}(A_L)$ preserves all the requirements of the lemma (images of thick sandwiches are thick sandwiches or zero) and we arrive at the situation where A_L has no nilpotent ideals. On the other hand, by Proposition 3.7,

$$c_tL^{(2)}c_t = 0 \Rightarrow cL^{(n)}cL^{(n)}c = 0$$

for every $c \in \mathscr{D}^{f(n)}\mathfrak{C}_2(L, A_L)$. It follows from item 1) that $I_c = L^{(n)}cL^{(n)}$ is an ideal of A_L such that

$$I_c^3 \subseteq L^{(n)}cL^{(n)}cL^{(n)}cL^{(n)} = 0\,.$$

This means that $\mathscr{D}^{f(n)}\mathfrak{C}_2(L, A_L) = 0$, that is, $\mathfrak{C}_2(L, A_L)$ is a soluble set. But L is generated by $\mathfrak{C}_2(L, A_L)$, and also $[\mathfrak{C}_2(L, A_L), \mathfrak{C}_1(L, A_L)] \subset \mathfrak{C}_2(L, A_L)$ (see the remark at the end of 3.4.1), so that L is the space spanned by all thick sandwiches, and the solubility of $\mathfrak{C}_2(L, A_L)$ implies that of L. By Lemma 3.4, L must be nilpotent, contrary to our assumption. □

3.9. Lemma. *In every sandwich Lie algebra L, the space $\langle \mathfrak{C}_2(L, A_L) \rangle_F$ spanned by all thick sandwiches is a locally nilpotent ideal.*

Proof. The remark at the end of 3.4.1 referred to above shows that $\langle \mathfrak{C}_2(L, A_L)\rangle_F$ is an ideal. The fact that it is locally nilpotent is an immediate consequence of Lemma 3.8. □

3.10. Lemma. *Let L be a Lie algebra generated by sandwiches $c_i \in \mathfrak{C}_1(L, A_L)$, $1 \leqslant i \leqslant n$, and S the set of all commutators of all lengths in $c_1, \ldots, c_n$. Suppose further that S_0 is a maximal subset of S generating a locally nilpotent ideal J_0 of L, and let $\phi: L \to \bar{L} = L/J_0$ be the natural map. Then no subset of $\bar{S} = (S + J_0/J_0$ generates a non-zero locally nilpotent ideal of L.*

Proof. For a contradiction, suppose that some non-zero subset of $\bar{S}$ does generate a locally nilpotent ideal in $\bar{L}$. We denote the pre-image of this set under the map $\phi: S \to \bar{S}$ by S_1, so that $S_1 \supseteq S_0$, and show that the ideal J_1 of L generated by S_1 is locally nilpotent, thus contradicting the maximality of S_0.

We observe that $J_1 = \langle J_1 \cap S\rangle_F$, choose any elements $a_1, \ldots, a_m \in J_1 \cap S$ and set $L_1 = \mathrm{Lie}\langle a_1, \ldots, a_m\rangle$. By assumption, some term $L_1^{(r)}$ of the derived series of L_1 is contained in J_0. By Definition 3.2 and Lemma 3.3., $L_1^{[r]}$ is finitely generated. Since J_0 is locally nilpotent, L_1 is soluble and thus nilpotent, by Lemma 3.4. Since every finitely generated subalgebra of J_1 is contained in a subalgebra of type L_1, J_1 is locally nilpotent. □

3.11. Completion of the Proof of the Theorem. Starting from the assumption that the theorem is false, we may assume without loss of generality that $L = \mathrm{Lie}\langle c_1, \ldots, c_n | c_i \in \mathfrak{C}_1(L, A_L)\rangle$ is an extremal sandwich Lie algebra in the sense of Definition 3.4.1. By Lemma 3.10, we may assume that no non-zero subset of the set S of all commutators in $c_1, \ldots, c_n$ generates a locally nilpotent ideal of L. At the same time, Proposition 3.4.1 guarantees that L has a thick sandwich, and this lies in S. By Lemma 3.9, the space spanned by all thick sandwiches is a non-zero locally nilpotent ideal. This contradiction completes the proof. □

§ 4. The Sandwich Radical and its Applications

4.1. Definition. Let L be any Lie algebra. We set $\mathfrak{S}_0(L) = 0$ and

$$\mathfrak{S}_1(L) = I_L(\mathfrak{C}(L))$$

for the ideal generated by all the sandwiches in L. We define an ascending chain of ideals by transfinite induction:

$$\mathfrak{S}_\alpha(L) = \bigcup_{\beta < \alpha} \mathfrak{S}_\beta(L) \text{ if } \alpha \text{ is a limit ordinal}$$

$$\mathfrak{S}_\alpha(L)/\mathfrak{S}_{\alpha-1}(L) = \mathfrak{S}_1(L/\mathfrak{S}_{\alpha-1}(L)) \text{ otherwise.}$$

We call the ideal

$$\mathfrak{S}(L) = \bigcup_\alpha \mathfrak{S}_\alpha(L)$$

the *sandwich radical* of L.

Clearly, $\mathfrak{S}(L)$ is the smallest ideal of L such that $L/\mathfrak{S}(L)$ contains no sandwiches. It can be shown that $\mathfrak{S}(L)$ is a radical in the sense of Amitsur and Kurosh. In the case where the field F is of characteristic $p > 0$, this radical can be used to make a formal distinction between the classes of Cartan algebras and of classical finite-dimensional semisimple Lie algebras: $\mathfrak{S}(L) = 0$ for a classical Lie algebra and $\mathfrak{S}(L) = L$ for an algebra L of Cartan type (see § 6 of Chap. 1). There is an even more fundamental rôle for $\mathfrak{S}$ when $p = 0$, as is shown by:

4.2. Theorem. *For any Lie algebra L over a field F of zero characteristic, the ideal* $\mathfrak{S}_1(L)$ *is locally nilpotent.*

Proof. We consider the free Lie algebra $\mathscr{L} = \mathscr{L}(C; X)$ on the free generating set $C \cup X$: $C = \{c_1, c_2, \ldots\}$, $X = \{x_1, x_2, \ldots\}$. Let J be the ideal of $\mathscr{L}$ generated by the set

$$\{[[u, c_i], c_i] \mid u \in \mathscr{L}; i = 1, 2, \ldots\} .$$

If $L = \mathscr{L}/J$, then denoting the images of the c_i and x_j by the same letters, we see that C is a set of sandwiches. To prove the theorem, it is enough to check that $I_L(C) = \mathfrak{S}_1(L)$ is a locally nilpotent ideal. Since J is homogeneous in both types of generator, L can be represented in the form

$$L = \bigoplus_{i=1}^{\infty} L_i , \tag{5.12}$$

where L_i is the subspace spanned by the commutators of length i in the alphabet $C \cup X$. There is a natural topology on L associated with the grading (5.12), namely that where the ideals $L^k = \bigoplus_{i \geqslant k} L_i$ form a basis of neighbourhoods of zero. We consider the completion of L in this topology, that is, the algebra $\hat{L}$ of formal power series of the form $\sum_{i=1}^{\infty} u_i$, $u_i \in L_i$, with the usual operations of addition and multiplication.

A consequence of the definition (1.5.4) of the sandwich subalgebra $\mathfrak{C} = \mathfrak{C}(\hat{L})$ is that it is invariant under the action of the automorphisms of $\hat{L}$ (a property that we have already used, in 2.1). We claim that $[a, x] \in \bar{\mathfrak{C}}$ if $a \in \bar{\mathfrak{C}}$, the closure of $\mathfrak{C}$, and $x \in L$. It is enough to establish the inclusion

$$[ax] \in \mathfrak{C} (\operatorname{mod} \hat{L}^m) ,$$

where m is any natural number. We consider the parametric set $\{\Phi_\lambda \mid \lambda \in F\}$ of automorphisms

$$\Phi_\lambda = \exp \lambda \operatorname{ad} x = 1 + \lambda \operatorname{ad} x + \frac{\lambda^2}{2!} (\operatorname{ad} x)^2 + \cdots$$

of $\hat{L}$ (remember that $\operatorname{char} F = 0$). Since $a \in \bar{\mathfrak{C}}$, we have $a - a' \in \hat{L}^m$ for some $a' \in \mathfrak{C}$. Since

$$\Phi_\lambda(\mathfrak{C}) \subseteq \mathfrak{C} , \quad \Phi_\lambda(\hat{L}^m) \subseteq \hat{L}^m ,$$

we have

$$\Phi_\lambda(a') \equiv \Phi_\lambda(a) = a + \lambda[ax] + \cdots \in \mathfrak{C}(\hat{L}) (\bmod \hat{L}^m) .$$

By giving the parameter λ the values 1, 2, . . . , we find that $[ax] \in \mathfrak{C} (\bmod \hat{L}^m)$. Since $C \subset \mathfrak{C}$, all homogeneous elements whose expressions involve one of the letters $c_1, c_2, \ldots$ must lie in $\hat{\mathfrak{C}}$.

Every element $u \in I_L(C)$ is contained in the subalgebra generated by its homogeneous components, which also lie in $I_L(C)$. Thus, it is enough to establish nilpotency for a Lie algebra Lie $\langle v_1, \ldots, v_d \rangle$, where $\{v_1, \ldots, v_d\}$ is a finite set of homogeneous elements of weights $l_1, \ldots, l_d$ in $I_L(C)$. Let s be a natural number such that $s > \max(l_1, \ldots, l_d)$. Since $v_1, \ldots, v_d \in \hat{\mathfrak{C}}$, there exist elements $a_1, \ldots, a_d \in \hat{\mathfrak{C}}$ such that $w_i = v_i - a_i \in \hat{L}^s$. By Theorem 3.1, $\mathfrak{C}$ is locally nilpotent. Thus, the subalgebra Lie $\langle a_1, \ldots, a_d \rangle$ is nilpotent, of class $q - 1$ say. Let $[v_{i_1}, \ldots, v_{i_q}]$ be a left-normed commutator of length q in $\{v_1, \ldots, v_d\}$. Then

$$[v_{i_1} \ldots v_{i_q}] = [(a_{i_1} + w_{i_1}) \ldots (a_{i_q} + w_{i_q})] = [a_{i_1} \ldots a_{i_q}] + w = w ,$$

where $w \in \hat{L}^{r+1}$ and $r = l_{i_1} + \cdots + l_{i_q}$. Since $\hat{L}^{r+1} \cap L_r = 0$, we have that $[v_{i_1}, \ldots v_{i_q}] = 0$. Thus Lie $\langle v_1, \ldots, v_d \rangle$ is nilpotent of class $q - 1$. □

4.3. Lemma. *Let J be an ideal of the Lie algebra L. Then $\mathfrak{S}(J) \subseteq \mathfrak{S}(L)$.*

Proof. We may factor out $\mathfrak{S}(L)$ and thus assume that $\mathfrak{S}(L) = 0$ from the outset. It is now enough to note that whenever $a \in J$ and $[[J, a], a] = 0$, it follows that $[ba^3] \in [[J, a], a] = 0$ for some $b \in L$, so that $[ba^2]^2 = a^2b^2a^2$, by (1.14). But then $[x[ba^2]^2] = [\underline{xa^2b^2}\,a^2] \in [[J, a], a] = 0$, and we conclude that $[ba^2] \in S(L) = 0$ for every $b \in L$ whenever $a \in \mathfrak{S}(J)$; thus $a^2 = 0$ and $a \in \mathfrak{S}(L) = 0$. Thus $\mathfrak{S}(J) = 0$.

A Lie algebra L is said to *satisfy a polynomial identity*, or to be a PI-*algebra*, if $f(a_1, \ldots, a_n) = 0$ for all $a_1, \ldots, a_n \in L$, where $f(t_1, \ldots, t_n)$ is a non-zero element of the free Lie algebra. Finitely generated Engel Lie algebras are not usually nilpotent (see [54, 55]), and the following result is therefore interesting in this context.

4.4. Theorem. *Every Engel Lie* PI-*algebra over a field F of characteristic zero is locally nilpotent.*

Sketch Proof. Arguing by contradiction and using Proposition 1.3.2, we may assume without loss of generality that the Engel Lie PI-algebra L does not have locally nilpotent ideals (that is, $R(L) = 0$). We have $\mathfrak{S}_1(L) = 0$ by Theorem 4.2, and this means that $\mathfrak{S}(L) = 0$. The PI-condition and the fact that $\mathfrak{S}(L) = 0$ are used to prove that $[ab^2] = 0$ for all nil-elements $a, b \in L$ of index at most $\leqslant 3$ (we omit the details).

Clearly, the subspace $J = \langle a \in L | \operatorname{ad} a)^3 = 0 \rangle$ is invariant under Aut L, and since F is of characteristic zero and L is Engel, every element $x \in L$ has associated with it the family $\{\exp t \operatorname{ad} x | t \in F\}$ of automorphisms, whence it follows that $[J, x] \subset J$ and thus that J is an ideal (see 2.1). By Proposition 2.1.2, J is non-zero, and $\mathfrak{S}(J) = 0$ by Lemma 4.3. However, the relation $[ab^2] = 0$ mentioned above has the interpretation that $[[J, b], b] = 0$ for all b such that $(\operatorname{ad} b)^3 = 0$; and thus $\mathfrak{S}(J) = J$. The contradiction thus obtained completes the proof. □

In fact, there is a very much stronger result relating to a very wide class of Lie algebras.

Definition. L is said to be a Lie algebra with *algebraic adjoint representation* if every operator $\operatorname{ad} x$ is annihilated by some polynomial $f_x(t)$, depending in general on x.

It is clear that every finite-dimensional Lie algebra has this property. Kurosh's problem about the local finite-dimensionality of a Lie algebra with algebraic adjoint representation has a negative answer in general (even in the class of Engel algebras, as we have seen), but it turns out that the natural condition requiring a global bound on the degrees of the polynomials $f_x(t)$ alters the position radically (this was known long ago in many other varieties of algebras, for example in the variety of associative algebras).

4.5. Theorem. *Every Lie* PI-*algebra with algebraic adjoint representation where the ground field is of characteristic zero is locally finite-dimensional.*

Corollary. *Every Lie algebra over a field of characteristic zero whose adjoint representation is algebraic of bounded degree is locally finite-dimensional.*

Proof. Let n be the maximum of the degrees of the polynomials annihilating the operators $\operatorname{ad} y$, $y \in L$. Then, for all $x, y \in L$, the elements

$$x_0 = x, \qquad x_1 = [xy], \ldots, \qquad x_n = [xy^n]$$

are linearly dependent. In that case,

$$P(x, y) = \sum_{\sigma \in S_{n+1}} \operatorname{sgn} \sigma \operatorname{ad} x_{\sigma(0)} \operatorname{ad} x_{\sigma(1)} \ldots \operatorname{ad} x_{\sigma(n)} = 0 .$$

Thus L satisfies the identity

$$zP(x, y) = 0 ,$$

and is therefore a PI-algebra. □

A full exposition of Zel'manov's proof [282] of Theorem 4.5, and of Theorem 4.4, would deflect us somewhat from our main theme. We shall therefore restrict ourselves to reproducing a fragment of [282]—the proof of a result whose statement includes an approximation to finite-dimensionality. As before, we assume that F has characteristic zero.

4.6. Theorem. *If the Lie algebra L over F with algebraic adjoint representation satisfies all the identities that hold in some finite-dimensional algebra, then L is locally finite-dimensional.*

The proof splits into a number of comparatively elementary steps. The first is a modification of Engel's theorem, in a form due to Jacobson.

(I) *Let A be an associative algebra of dimension m over F and generated by elements $a_1, \ldots, a_k$. Assume that every commutator of weight not more than m^{2m-2} in $\{a_i\}$ is nilpotent. Then A is nilpotent.*

We shall construct inductively an ascending chain of nilpotent subalgebras of A. Let A_0 be the subalgebra generated by a_1. Suppose that A_i has been constructed and is generated by some set W_i of commutators of weight not more than $m^{2(i-1)}$ in $\{a_j | 1 \leqslant j \leqslant k\}$. If $A_i = A$, we set $A_{i+1} = A$. Suppose that $A_i \neq A$, so that $\dim_F A \leqslant m - 1$. The subalgebra of $\mathrm{End}_F(A)$ generated by the right and left multiplications by elements from A_i has dimension not more than $(m-1)^2$, and therefore it has nilpotency class not more than $(m-1)^2$. Since $A_i \neq A$, some a_j is outside A_i. Further, for all commutators $w_1, \ldots, w_{(m-1)^2+1} \in W_i$ we have

$$[a_j, w_1, w_2, \ldots, w_{(m-1)^2+1}] = 0 .$$

Therefore, there exist commutators $w_{j_1}, \ldots, w_{j_r} \in W_i$, $0 \leqslant r \leqslant (m-1)^2$, such that

$$w = [a_j, w_{j_1}, \ldots, w_{j_r}] \notin A_i, [w, W_i] \subseteq A_i .$$

The weight of w is not more than

$$1 + r \cdot m^{2(i-1)} \leqslant 1 + (m-1)^2 m^{2(i-1)} \leqslant m^{2i} .$$

For A_{i+1} we take the subalgebra generated by A_i and w. Since w is a nil-element by assumption, and $wA_i \subseteq A_i + A_i w$ by construction, A_{i+1} is also nilpotent. At the m-th step we get $A_m = A$.

(II) *Let A be any associative algebra over F satisfying all the identities of some m-dimensional F-algebra. Suppose further that A is generated by a set $\{a_1, \ldots, a_k\}$ of elements such that every commutator of weight $\leqslant m^{2m-2}$ in the a_i is nilpotent. Then A is nilpotent.*

Let us suppose that this assertion is false. Then, without loss of generality, we may assume that A is primary (that is, the product of every pair of non-zero ideals of A is non-zero). By the Markov-Rowen theorem (see [225], for example), the centre $Z(A)$ of a primary PI-algebra A is non-zero and the ring of fractions $Z(A)^{-1}A$ is a simple finite-dimensional algebra over K, the field of fractions of the centre. By assumption, $\dim_K Z(A)^{-1}A \leqslant m$.

But clearly, the algebra $Z(A)^{-1}A$ is generated over K by the same set $\{a_1, \ldots, a_k\}$, and all the requirements of item (I) hold. Thus $Z(A)^{-1}A$ is nilpotent, contrary to our assumptions.

(III) *Let L be a Lie algebra over F satisfying all the identities of some m-dimensional algebra. If L is generated by elements $a_1, \ldots, a_k$ such that all commutators of weight at most m^{4m^2-4} in the a_i are Engel elements, then L is nilpotent.*

We cannot use Theorem 4.4, since only commutators of bounded length are Engel.

For the proof, we introduce into the discussion the algebra $A(L)$ of multiplications—the subalgebra of $\mathrm{End}_F(L)$ generated by the adjoint operators $\operatorname{ad} x$, $x \in L$.

Then $A(L)$ satisfies all the identities of some m^2-dimensional algebra (see [26], 6.3.3) and is generated by the elements $\operatorname{ad} x_i$, $1 \leqslant i \leqslant k$; moreover, every commutator of weight $\leqslant (m^2)^{2m^2-2} = m^{4m^2-4}$ in the $\operatorname{ad} a_i$ is nilpotent. By (II), $A(L)$ is nilpotent, and this means that L is likewise nilpotent.

(IV) *Let L be a Lie algebra satisfying all the identities of some m-dimensional algebra G. Assume that L is generated by elements $a_1, \ldots, a_k$ such that the adjoint operator* $\operatorname{ad} w$ *is algebraic for every commutator w of weight not more than m^{4m^2-4} in the a_i. Then L is finite-dimensional.*

We write T for the ideal of identities of G. Let $\mathscr{L}\langle X \rangle$ be the free Lie algebra on the generating set $X = \{x_i | 1 \leqslant i \leqslant k\}$ in the variety defined by the identities in T (see [26] for these concepts). The map $\phi: x_i \mapsto a_i$, $1 \leqslant i \leqslant k$, extends to a homomorphism $\phi: \mathscr{L}\langle X \rangle \to L$. We consider the set W of all commutators of weight at most m^{4m^2-4} in the x_i. By assumption, for every commutator $w \in W$ there is a polynomial $f_w(t) = t^{n_w} + \sum_{<n_w} \alpha_i(w) t^i$ annihilating $\operatorname{ad} \phi(w)$: $f_w(\operatorname{ad}(\phi(w)) = 0$. Let J be the ideal of $\mathscr{L}(X)$ generated by $\bigcup_{w \in W} \mathscr{L}\langle X \rangle (\operatorname{ad} w)^{n_w}$. Clearly, J is homogeneous in X. The factor-algebra $\mathscr{L}\langle X \rangle / J$ is generated by elements $\bar{x}_i = x_i + J$, $1 \leqslant i \leqslant k$, where every commutator $\bar{w}$ of weight not more than m^{4m^2-4} in the $\bar{x}_i$ is Engel of index n_w. By (III), $\mathscr{L}\langle X \rangle / J$ is nilpotent. Suppose that $\mathscr{L}\langle X \rangle^s \subseteq J$.

We claim that L is generated as F-space by commutators of weight less than s in $a_1, \ldots, a_k$. In fact, let $b = [a_{i_1}, \ldots, a_{i_s}]$ be a commutator of weight s in the a_i. The corresponding element $[x_{i_1}, \ldots, x_{i_s}]$ lies in J, that is,

$$[x_{i_1}, \ldots, x_{i_s}] = \sum_{w,j} u_j(w) (\operatorname{ad} w)^{n_w} P_j(w) \,, \tag{5.13}$$

where

$$w \in W, \qquad u_j(w) \in \mathscr{L}\langle X \rangle, \qquad P_j(w) \in A(\mathscr{L}\langle X \rangle) + F \cdot \mathrm{Id} \,.$$

We may assume that the elements $u_j(w)$ and the operator $P_j(w)$ are homogeneous in X, so that $s = \deg u_j(w) + n_w(\deg w) + \deg P_j(w)$. By assumption, application of ϕ to both sides of (5.13) leads to the relation

$$[a_{i_1}, \ldots, a_{i_s}] = \sum_{w,j} u_j(\phi(w)) \left\{ - \sum_{i < n_w} \alpha_i(w) (\operatorname{ad} \phi(w))^i \right\} P_j(\varphi(w)) \,.$$

Clearly, the right-hand side is a linear combination of commutators of weight less than s in $a_1, \ldots, a_k$. □

Both Theorem 4.4 and assertion (IV), which is formally stronger than Theorem 4.6, are used during the proof of the main Theorem 4.5 formulated above.

§ 5. Commentary

5.1. The contents of § 1 grew out of a correspondence between the author and E. I. Zel'manov in March 1984. Since the arguments expounded in 3.5.3 (they were originated in [137, 143]) had been used frequently and with success up to that time, there was a temptation to return to the direct proof of Theorem 3.4.1. The first version (contained in one of the author's letters) intentionally excluded all fantasies and simply contained the idea of "false sandwiches" in a "witchdoctor's double circle"—obtuse repetition of the same arguments, with small variations. The second version (in a letter of Zel'manov this time) contains a useful additional idea, namely induction on k. The arguments are shorter as a result.

5.2. The author learned of the possibility of shortening the proof of Theorem 1.7.4 as long ago as the end of 1980, in a letter from Zel'manov. Subsequently A. N. Grishkov's notes "Engel Lie algebras and the restricted Burnside problem" were in circulation; the notes absorbed, from various sources, a number of new considerations relating to RBP, including the ideas for the geodesic expounded in § 2.

5.3. It was not noticed for some long time that reference to E_n in [143] is inessential. Proposition 3.4.1 remedies this defect with no special difficulty; for the proof of Theorem 3.4.1 it takes on the burden contained in the original paper of Zel'manov [280] on Jordan pairs. We have not wrestled with characteristic $p = 5$, since we are assuming that the restriction $p > 5$ is permissible enough here. Jordan pairs are also eliminated from the papers of Grishkov mentioned above (in 5.2), although the arguments have remained almost identical to those in [280]. The deletion lemma in [280] (our Lemma 3.5) is a clever technical invention of Zel'manov that enables many different arguments to be unified.

Although it is not strictly necessary for the solution of RBP in full generality—rather, it was born and took shape under the influence of the solution—Theorem 3.1 has gained wide recognition, and pretends to a dignified place in the general theory of Lie algebras, by virtue of its intrinsic beauty. It still has to be embellished with a constructive proof, which would give a formula or an upper bound for the nilpotency class of the free sandwich algebra on d generators. The first noticeable hindrance arises already when $d = 6$, and it seems that the general case is as far from resolution as RBP itself.

An interesting idea expressed by Zel'manov in one of his letters to the author is the possibility that "every Lie algebra of characteristic zero containing a thin sandwich also contains thick sandwiches, so that there exists a function $f(m)$ of a natural argument such that every Lie algebra of characteristic not less than $f(m)$ containing a thin sandwich contains a sandwich of thickness at least m"[1].

5.4. The sandwich (or the strongly degenerate) radical $\mathfrak{S}(L)$, also denoted sometimes by $\mathfrak{N}(L)$ and $K(L)$, was first studied by V. T. Filippov in [50]. The radical $\mathfrak{S}(L)$ has the hereditary property for subalgebras: if M is a subalgebra of a Lie algebra L contained in $\mathfrak{S}(L)$, then $M = \mathfrak{S}(M)$. In particular, we have $\mathfrak{S}(J) = J \cap \mathfrak{S}(L)$ for every ideal $J \lhd L$. The properties of $\mathfrak{S}(L)$ were established and used intensively by Zel'manov in the two articles [282, 283]. He had already stated

[1] This was confirmed in the summer of 1988 by A. D. Chanyshev in Moscow, and in a stronger form by Yu. A. Medvedev in Novosibirsk (1989).

the conjecture that $\mathfrak{S}_1(L)$ is locally nilpotent, in 1979; this lies at the heart of the construction of $\mathfrak{S}(L)$. It was first confirmed by Grishkov in [60], and his arguments are reproduced *verbatim* in 4.2. The proof of Theorem 4.5 given in [282] rests heavily on the technique of Jordan pairs, whose transfer to the river-bed of Lie algebra theory would require real effort.

In [282], Zel'manov reduced the problem of the local finite-dimensionality of Lie PI-algebras with algebraic adjoint representations to a number of problems that are tractable in terms of Lie algebras with a finite grading:

$$L = \sum_{\lambda \in \Lambda} L_\lambda, \qquad [L_\lambda, L_\mu] \subseteq L_{\lambda+\mu}, \lambda, \mu \in \Lambda .$$

Here it is assumed that Λ is a torsion-free abelian group and that $|\{\lambda \in \Lambda | L_\lambda \neq 0\}| < \infty$. Later, in [283], these results were developed and yielded a complete classification, up to isomorphism, of all simple $\mathbb{Z}$-graded Lie algebras $L = \sum_{i=-n}^{n} L_i$ over a field of charaacteristic $p \geqslant 4n+1$ (or zero). A similar theorem, perhaps with a slightly different statement, concerning the class of finite-dimensional simple Lie algebras over a field of characteristic independent of $n(p > 7$, say, *etc*) will perhaps be proved at some stage in the future using the ideas concentrated in [283, 29]. A theorem of the type of 4.4 would have special value for the aims pursued in this book for Lie algebras with E_n, $n \gg p > 0$, or even 4.5 in the case where the ground field is of characteristic $p > 0$; however, investigations in this direction are too few in number for one to be able to judge the prospects with any reliability.

5.5. *Remark.* A proof of Theorem 3.1 about sandwich algebras that is more transparent and free of restrictions on the characteristic is expounded in [287].

Chapter 6

The Problem of Global Nilpotency

Theorem 1.7.4 is optimal in a very definite sense of that word. We shall show in § 3 below that it ceases to hold if the local condition is removed from the statement. In such a case, we say that a Lie algebra L with E_{p-1} (p is the characteristic of the ground field) is not globally nilpotent. The heart of the matter is that the values of the integer-valued function $c(d, n)$—the nilpotency class of the free Lie algebra $L = \mathrm{Lie}\langle x_1, \ldots, x_d | (\mathrm{ad}\, u)^n = 0\rangle$ grows indefinitely with d when $n = p - 1 > 2$. On the other hand, we shall show that $c(d, n)$ grows exponentially in n in general, when $d \geqslant 2$ of course. Although many problems arise in this connection, attention is focussed in this chapter on the watershed between the properties of local and global nilpotency. It is remarkable that when the characteristic p is significantly bigger than $n (p \gg n)$, in particular when $p = 0$, there exists a d_0 where stability occurs: $c(d, n) = c(n) = c(d_0, n)$ when $d \geqslant d_0$, and $L^{c(n)+1} = 0$; that is, L is globally nilpotent. Theorem 1.7.3 functions again here.

The contents of the chapter are taken up by the proof of four essential results, due to different authors. Particularly non-trivial constructions are expounded in §§ 3, 4. An insoluble locally finite group of exponent p is constructed in § 3. This example is not obvious, since the correspondence between finite p-groups and Lie algebras with E_{p-1} is not exact.

§ 1. The Nilpotency Class of a Lie Algebra with E_n

We shall be concerned with finitely generated Lie algebras. What we have in mind is a lower bound for the nilpotency class, so that the number of generators can be taken as two. This is a somewhat alien theme for Chapter 6; however, it does have a bearing on the problem of global nilpotency. We select the case that is closest to the circle of Burnside problems.

1.1. Theorem. *Let c_p be the nilpotency class of the 2-generator free Lie algebra $L = L(2, p - 1)$ with E_{p-1} over a field of characteristic $p > 0$. Then*

$$c_p > 2^{(\frac{1}{2} - \varepsilon)p}, \qquad (6.1)$$

where ε is a positive number tending to zero with increasing p.

More formally: *for every* $\varepsilon > 0$ *there exists* p_0 *such that estimate* (6.1) *holds for* $p > p_0$. The proof of this theorem is preceded by an explanation of the required terminology and by some auxiliary combinatorial assertions.

1.2. Notation. Throughout this section, $\mathscr{L} = \sum_{m=1}^{\infty} \mathscr{L}_m$ is the free Lie algebra on two generators over the field $\mathbb{F}_p$, where the $\mathscr{L}_m$ are homogeneous components; their dimensions are computed using the Witt formula:

$$\psi_m = \dim \mathscr{L}_m = \frac{1}{m} \sum_{d|m} \mu(d) 2^{\frac{m}{d}}. \tag{6.2}$$

Further, $\mathscr{E} = E_{p-1}(\mathscr{L}) = \sum_{n=0}^{\infty} \mathscr{E}_{p+n}$ is the ideal of $\mathscr{L}$ generated by the elements $[uv^{p-1}]$, $u, v \in \mathscr{L}$; here $\mathscr{E}_{p+n}$ is the homogeneous component of $\mathscr{E}$ spanned by elements of the form

$$\left\{ \begin{matrix} u_0 & u_1 & u_2 & \ldots & u_{p-1} \\ 1 & 1 & 1 & & 1 \end{matrix} \right\} = \sum_{\sigma \in S_{p-1}} [u_0 u_{\sigma 1} \ldots u_{\sigma(p-1)}], \tag{6.3}$$

where the u_i are homogeneous elements of $\mathscr{L}$ and $\sum_{i=0}^{p-1} \deg u_i = p + n$. It is clear from the defining relation (1.1) in Chap. 1 that the expression $\left\{ \begin{matrix} u_0 \, u_1 \ldots u_{p-1} \\ 1 \;\; 1 \qquad 1 \end{matrix} \right\}$ is symmetrical in all the arguments u_i, $0 \leqslant i \leqslant p-1$, which may be chosen from some ordered homogeneous basis of $\mathscr{L}$. Without loss of generality, we may assume that $u_0 \leqslant u_1 \leqslant \cdots \leqslant u_{p-1}$ (equality of the various components is allowed).

If ϕ_{p+n}^* is the number of elements of the form (6.3) such that

$$u_0 \leqslant u_1 \leqslant \cdots \leqslant u_{p-1}, \sum_{l=0}^{p-1} \deg u_i = p + n \,,$$

then obviously

$$\phi_{p+n} = \dim_{\mathbb{F}_p} \mathscr{E}_{p+n} \leqslant \phi_{p+n}^* \,.$$

The computation of ϕ_{p+n}^* is a purely combinational affair. For convenience we set

$$\phi_{p+n}^* = \alpha_n^* p - \beta_n^* \,,$$

where α_n^* and β_n^* are non-negative integers. Here $\alpha_n^* \leqslant \alpha_n$, where the α_n are the numbers defined by Lyndon [163] from the generating function

$$\sum_{n=0}^{\infty} \alpha_n T^n = \prod_{s=1}^{\infty} (1 - T^s)^{-\psi_s + 1} = I + T + 3T^2 + 6T^3 + I5T^4 + \cdots \tag{6.4}$$

Together with the elements in (6.3), the numbers α_n take care of the similar elements

$$\sum_{\sigma \in S_n} [u_0 u_{\sigma 1} u_{\sigma 2} \ldots u_{\sigma m}], \qquad m > p - 1 \,,$$

which we bring into play simply in order to avoid the more difficult calculations of the quantities α_n^* and β_n^*. We introduce the table from [134] as a model:

ϕ_{p+n}^* \ p	$p \geqslant 11$	$p = 7$	
$n = 0$	$p - 1$	$p - 1$	$= 6$
$n = 1$	p	p	$= 7$
$n = 2$	$3\,p - 1$	$3\,p - 1$	$= 20$
$n = 3$	$6\,p - 4$	$6\,p - 4$	$= 38$
$n = 4$	$15\,p - 13$	$15\,p - 13$	$= 92$
$n = 5$	$30\,p - 34$	$30\,p - 34$	$= 176$
$n = 6$	$70\,p - 90$	$70\,p - 90$	$= 400$
$n = 7$	$145\,p - 214$	$145\,p - 214$	$= 801$
$n = 8$	$320\,p - 518$	$319\,p - 511$	$= 1772$
$n = 9$	$672\,p - 1188$	$669\,p - 1166$	$= 3517$
$n = 10$	$1447\,p - 2737$	$1438\,p - 2670$	$= 7396$

We supplement this table with some values of the ψ_n and α_n:

n	11	12	13	14	15	16	17	18	19
ψ_n	186	335	630	1161	2182	4080	7710	14532	27594
α_n	3038	6458	13533	28491	59551	124612	259612	540815	1123412

Further, we shall disregard the β_n^*, which are positive, and will be satisfied with a rough upper bound:

$$\phi_{p+n} < \alpha_n^* p \leqslant \alpha_n p, \qquad n > 0\,. \tag{6.5}$$

When working with formal power series of the type

$$f(T) = \sum_{i \geqslant 0} a_i T^i, \qquad a_0 = 1, a_i \in \mathbb{Z}, \qquad a_i \geqslant 0, \qquad i \geqslant 1;$$

$$g(T) = \sum_{i \geqslant 0} b_i T^i, \qquad b_0 = 1, \qquad b_i \geqslant 0, \qquad b_0 = 1\,, \tag{6.6}$$

we agree to write $f(T) \prec g(T)$ if $a_i \leqslant b_i$ for all $i > 0$. We note that

$$f_j(T) < g_j(T); \qquad j = 1, 2 \Rightarrow f_1(T) f_2(T) \prec g_1(T) g_2(T)\,, \tag{6.7}$$

as follows from the multiplication rule for formal power series.

1.3. Lemma. *Let $f(T)$ be a formal power series of the form (6.6), and let k and l be two natural numbers such that $k < l$. Then*

$$f(T)^k \prec f(T)^l\,.$$

The *proof* is a simple consequence of implication (6.7) and the functional inequality $1 \prec f(T)$. □

1.4. Lemma. *The natural numbers* ψ_m *defined by* (6.2) *satisfy the following inequalities*:

(i) $$\frac{2^{m-1}}{m} < \psi_m \leqslant \frac{2^m}{m},$$

(ii) $\psi_{m+1} \leqslant 2\psi_m$.

Proof. (i). The upper estimate $\psi_m \leqslant 2^m/m$ follows immediately from the Witt formula (6.2). That formula also leads to the inequality

$$\psi_m \geqslant \frac{1}{m}\left(2^m - \left[\frac{m}{2}\right]2^{[\frac{m}{2}]}\right) \tag{6.8}$$

(here $[m/2]$ is the integral part of $m/2$), if all the non-zero terms on the right-hand side of (6.2) corresponding to $d > 1$ (and $d|m$) are taken with the minus sign, and we note that the number of such terms is $2^k - 1$ (that is, the number of non-empty subsets of $\{p_1, p_2, \ldots, p_k\}$ where $m = p^{\nu_1}p_2^{\nu_2} \ldots p_k^{\nu_k}$, $p_1 < p_2 < \ldots < p_k$): $2^k - 1 \leqslant p_2 p_3 \ldots p_k \leqslant [m/2]$. The lower estimate $2^{m-1}/m < \psi_m$ is now a trivial corollary of (6.8), since $[m/2] \leqslant 2^{m-1-[m/2]}$, and thus

$$\psi_m \geqslant \frac{1}{m}(2^m - 2^{m-1}) = 2^{m-1}/m.$$

(ii) Using the inequalities $\psi_{m+1} \leqslant 2^{m+1}/(m+1)$ (see (i)), and (6.8), we find that

$$\frac{\psi_{m+1}}{\psi_m} \leqslant m2^{m+1} \Bigg/ \left\{(m+1)\left(2^m - \left[\frac{m}{2}\right]2^{[\frac{m}{2}]}\right)\right\},$$

that is,

$$\frac{\psi_{m+1}}{\psi_m} \leqslant \frac{m}{m+1}\left\{2 + \frac{t2^{t+1}}{2^{2t+\omega} - t2^t}\right\} = \frac{m}{m+1}\left\{2 + \frac{2t}{2^{t+\omega} - t}\right\}, \tag{6.9}$$

where $m = 2t + \omega$, $\omega = 0$ or 1.

We can get the required inequality from (6.9). However, it is infact a formal consequence of the inclusion $\mathscr{L}_{m+1} \subseteq [\mathscr{L}_m, x] + [\mathscr{L}_m, y]$, where x and y are free generators of $\mathscr{L}$. □

1.5. Lemma. *The natural numbers* α_n *defined by the generating function* (6.4) *satisfy the inequalities*

$$\alpha_n \leqslant (n+1)2^n .$$

Proof. We note first that (6.2) is obtained by calculating the dimensions of the homogeneous components of the free 2-generator associative algebra in two different ways, and as a simple corollary of the Poincaré-Birkhoff-Witt theorem, by equating the two generating functions:

$$\frac{1}{1-2T}=\prod_{s=1}^{\infty}\left(\frac{1}{1-T^s}\right)^{\psi_s}. \tag{6.10}$$

By Lemma 1.4(ii), we have $\psi_{s+1}\leqslant 2\psi_s$. Thus, by Lemma (1.3),

$$\left(\frac{1}{1-T^s}\right)^{\psi_{s+1}}\prec\left(\frac{1}{1-T^s}\right)^{2\psi_s},$$

whence, by (6.7) and (6.10),

$$\sum_{n=0}^{\infty}\alpha_n T^n=\prod_{s=1}^{\infty}\left(\frac{1}{1-T^s}\right)^{\psi_{s+1}}\prec\prod_{s=1}^{\infty}\left(\frac{1}{1-T^s}\right)^{2\psi_s}$$

$$=\left\{\prod_{s=1}^{\infty}\left(\frac{1}{1-T^s}\right)^{\psi_s}\right\}^2=\left\{\frac{1}{1-2T}\right\}^2=\sum_{n=0}^{\infty}(n+1)2^nT^n.$$

This gives the required inequalities. □

Remark. The main lemma 1.5 can also be obtained slightly differently, by following the path used by Witt. Setting $f(T)=\sum_{n=0}^{\infty}\alpha_nT^n$ (with $\alpha_0=1$) temporarily, using (6.4) and taking logarithms, we get

$$\ln f(T)=-\sum_{s=1}^{\infty}\psi_{s+1}\ln(1-T^s),$$

whence, by formal differentiation,

$$\frac{f'(T)}{f(T)}=\sum_{s=1}^{\infty}\frac{s\psi_{s+1}}{1-T^s}T^{s-1}.$$

Thus

$$\sum_{n=0}^{\infty}n\alpha_nT^n=tf'(T)=f(T)\sum_{s=1}^{\infty}\frac{s\psi_{s+1}}{1-T^s}T^s$$

$$=\left(\sum_{k=0}^{\infty}\alpha_kT^k\right)\left(\sum_{m=1}^{\infty}N_mT^m\right),$$

where

$$N_m=\sum_{s\mid m}s\psi_{s+1}. \tag{6.11}$$

On formal multiplication of these power series, we get the recurrence relation

$$n\alpha_n=\sum_{m=1}^{n}N_m\alpha_{n-m}. \tag{6.12}$$

By Lemma 1.4(i) and 6.11),

$$N_m\leqslant\sum_{s\mid m}\frac{s}{s+1}2^{s+1}<2^{m+1}-\frac{1}{m+1}2^{m+1}+\sum_{\substack{s\mid m\\ s\geqslant 2}}2^{\frac{m}{s}+1}$$

$$< 2^{m+1} - \frac{1}{m+1} 2^{m+1} + (m+1)2^{\frac{m}{2}+1}$$

$$= 2^{m+1} - 2^{m+1-\log_2(m+1)} + 2^{\frac{m}{2}+1+\log_2(m+1)} < 2^{m+1},$$

whenever $m > 30$. For small m the proof is a matter of easy verification, so that we have in all cases:

$$N_m < 2^{m+1}. \tag{6.13}$$

Assume inductively that $\alpha_k < (k+1)2^k$, $k < n$; then, by (6.3) we get from (6.12) that

$$n\alpha_n < \sum_{m=1}^{n} 2^{m+1}(n-m+1)2^{n-m} = 2^{n+1} \sum_{k=1}^{n} k = 2^n \cdot n(n+1).$$

that is, $\alpha_n < (n+1)2^n$.

Is it true that $\alpha_n < \kappa 2^{n+1}$, where κ is some constant or slowly increasing function $\kappa(n)$ of logarithmic type? It is clear from the values for the α_n displayed in 1.2 that $\alpha_n < 2^{n+1}$ for $n < 18$, while $\alpha_{18} > 2^{19} = 524{,}288$. Furthermore, a small computer experiment using (6.12) shows that $\alpha_{20} \approx 1.111 \cdot 2^{21}$; $\alpha_{30} \approx 1.510 \cdot 2^{31}$; $\alpha_{40} \approx 1.902 \cdot 2^{41}$; $\alpha_{50} \approx 2.291 \cdot 2^{51}$, that is, the conjecture that $\kappa(n)$ grows slowly is not without foundation.

1.6. Proof of the Theorem. By Theorem 1.7.4, the free 2-generator Lie algebra $L = \mathscr{L}/E_{p-1}(\mathscr{L})$ with E_{p-1} is nilpotent. Let $c_p = p + m$ be its nilpotency class (a number which differs by 1 from the definition in 1.2.3). This means that

$$\psi_{p+m} = \dim \mathscr{L}_{p+m} = \dim \mathscr{E}_{p+m} = \phi_{p+m}.$$

From this, Lemma 1.5, inequality (6.5) and Lemma 1.4, we get a chain of inequalities:

$$p(m+1)2^m \geqslant p\alpha_m > \phi_{p+m} = \psi_{p+m} > \frac{2^{p+m-1}}{p+m}.$$

Therefore,

$$(p+m)(m+1) > \frac{2^{p-1}}{p}.$$

Even more,

$$c_p^2 = (p+m)^2 > \frac{2^{p-1}}{p},$$

so that

$$c_p > 2^{\frac{1}{2}(p-1-\log_2 p)} = 2^{(\frac{1}{2}-\varepsilon)p},$$

where

$$\varepsilon = \frac{1}{2}\frac{1+\log_2 p}{p}. \quad \square$$

§ 2. Combination of Solubility and the Engel Condition E_n

The outcome of the two-condition symbiosis mentioned in the title is determined decisively by the relative values of n and p. We consider two cases: $n > p$ and $n \leqslant p$.

2.1. Example. *There exists a Lie algebra of countable dimension over an arbitrary field of characteristic $p > 0$ having the following properties:*

(i) *L is a soluble Lie algebra with $L^{[2]} = 0$;*
(ii) *L satisfies E_{p+1};*
(iii) *L is not (globally) nilpotent.*

We shall begin with an associative algebra $A = K + W$ with multiplication $(u, v) \mapsto uv$, where $K = F[x_1, x_2, \dots]$ is the commutative algebra of truncated polynomials in countably many variables x_i (that is, $x_i^p = 0$ for all i), and $W = \langle w_\nu | \nu \in \mathbb{Z}_p^{\mathbb{N}} \rangle_F$ is the subalgebra with zero multiplication spanned by basis vectors w_ν. The indices ν and the exponent α in monomials $x^\alpha \in K$ are to be interpreted as infinite rows with components in $\mathbb{Z}_p$ that are almost all zero: $\alpha = (\alpha_1, \alpha_2, \alpha_3, \dots)$, and $\alpha_s = 0$ for all sufficiently large s. The operation of multiplication in A is completely defined by the following rules for pairs of basis vectors:

$$x^\alpha x^{\alpha'} = x^{\alpha+\alpha'}, \qquad w_\nu w_{\nu'} = 0, \qquad x^\alpha w_\nu = w_\nu, \qquad w_\nu x^\alpha = w_{\nu+\alpha} \,. \tag{6.14}$$

It can be checked without difficulty that we do indeed get an associative algebra, with elements of the form $v = f(x) + w$, where $f(x) = \sum_\alpha f_\alpha x^\alpha$, $w = \sum_\nu w^\nu w_\nu$; the coefficients $f_\alpha, w^\nu \in F$ are almost all zero. We note that W is a two-sided ideal in A. Further,

$$f(x) = f_0 + t(x), \qquad t^p = 0 \qquad \text{and} \qquad (f(x))^p = f_0^p \in F \tag{6.15}$$

(the truncated polynomial property). As vector spaces over F, L is identified with A, and we introduce the usual Lie bracket:

$$[u, v] = uv - uv, \qquad u, v \in L = A \,.$$

It is clear that $K = L + W$ is the sum of the two abelian subalgebras K and W, and that W is an ideal in L. Thus $L^{[1]} = [L, L] \subseteq W$ and $L^{[2]} \subseteq [W, W] = 0$, that is, property (i) is satisfied. To check (ii), take any elements $u, v \in L$ and form the commutator $[uv^{p+1}]$. Since $[u, v] = z \in W$ and $[W, W] = 0$,

$$[uv^2] = [zv] = [z(f + w)] = [z, f] \in W \,,$$

$$\cdots$$

$$[uv^p] = [zv^{p-1}] = [zf^{p-2}(f + w)] = [zf^{p-1}] \in W \,,$$

$$[uv^{p+1}] = [zv^p] = [zf^{p-1}(f + w)] = [zf^p] \,.$$

Recalling the enveloping algebra A and expression (6.15), we get from (1.5) that

$$[uv^{p+1}] = [zf^p] = \sum_{i=0}^{p} (-1)^i \binom{p}{i} f^i z f^{p-i} = zf^p - f^p z = zf_0^p - f_0^p z = 0 \,.$$

This means that L satisfies E_{p+1}.

Finally, we use the fact that there are infinitely many variables x_i, together with (6.14), which expresses the action of K on W and which we rewrite as

$$[w_\nu, x^\alpha] = w_{\nu+\alpha} - w_\nu .$$

According to our agreement, $x_i = x^{\varepsilon_i}$, where $\varepsilon_i = (0, \ldots, 0, 1, 0, \ldots)$ is the row with the identity in the i-th place and zeros elsewhere. Let 0 denote the zero row. Then

$$[w_0, x_1] = w_{\varepsilon_1} - w_0 ,$$

$$[w_0, w_1, x_2] = w_{\varepsilon_1+\varepsilon_2} - w_{\varepsilon_1} - w_{\varepsilon_2} + w_0 ,$$

. . . .

$$[w_0, x_1, x_2, \ldots, x_m] = w_{\varepsilon_1} + \cdots + {}_{\varepsilon_m} - \cdots + (-1)^m w_0 = \sum_\sigma \pm w_\sigma$$

(here σ runs over all rows $\varepsilon_{i_1} + \cdots + \varepsilon_{i_k}$ with $\{i_1, \ldots, i_k\} \subseteq \{1, \ldots, m\}$; the corresponding sign is $(-1)^{m-k}$). Each of the last two terms indicated shows that

$$[w_0, x_1, x_2, \ldots, x_m] \neq 0$$

for all m, that is, L is not globally nilpotent. This verifies (iii).

Remark. The fact that the Lie algebra L just constructed is locally nilpotent is obvious.

2.2. Theorem. *Let L be a Lie algebra with E_n over a field F of characteristic $p > n$ (n is arbitrary when $p = 0$). If L is soluble, then it is nilpotent. More exactly, if $L^{[r]} = 0$, then $L^s r = 0$, where*

$$s_r = \frac{n^r - 1}{n - 1} + 1 .$$

Proof. As usual (see § 1 of Chap. 5, for example), when we are discussing the nilpotency of a Lie algebra with the identity $[uv^n] = 0$, we may assume (by replacing L by $L/Z(L)$ if necessary) that L is embedded in an associative algebra A such that L generates A and $u^n = 0$ in A for all u in L. We note that this assumption has no influence on the value of the nilpotency class. We set $M = L^{[1]} = [L, L]$. We check first that

$$L^{nj+2} \subseteq M^{j+1} \tag{6.16}$$

for all $j = 0, 1, 2, \ldots$. If $n < p$ (or n is anything if $p = 0$), Proposition 1.4.6 (i) means that

$$x_1 x_2 \ldots x_n = \sum_i w_{i,1} w_{i,2} \ldots w_{i,n-1} ,$$

where one of the elements $w_{i,k}$, being a commutator in the $x_\nu \in L$, is contained in M. Splitting the product $x_1 x_2 \ldots x_{nj}$ up into j segments of length n, we arrive at an expression

$$x_1 x_2 \ldots x_{nj} = \sum \ldots w_1 \ldots w_2 \ldots w_j \ldots ,$$

in which $w_1, w_2, \ldots, w_j \in M$. If we set $w_0 = [yx_0] \in M$, then

$$[yx_0x_1x_2 \ldots x_{nj}] = \sum [w_0 \ldots w_1 \ldots w_2 \ldots w_j \ldots] \in M^{j+1}.$$

Every element of L^{nj+2} can be written as a sum of products of the form $[yx_0x_1 \ldots x_{nj}]$, so that inclusion (6.16) holds.

We proceed further by induction on r, starting from the fact that the theorem is true for $r = 1$: $s_1 = \dfrac{n^1 - 1}{n-1} + 1 = 2$ and $L^{s_1} = L^2 = L^{[1]} = 0$. Suppose that everything has been proved for solubility lengths $< r$. Since $L^{[r]} = 0 \Rightarrow M^{[r-1]} = 0$, and M satisfies all the other conditions of the theorem since it is a subalgebra of L, we have

$$M^{s_{r-1}} = 0, \qquad s_{r-1} = \frac{n^{r-1} - 1}{n-1} + 1.$$

Applying inclusion (6.16), we get

$$L^{n(s_{r-1}+1)+2} \subseteq M^{s_{r-1}} = 0.$$

Now simply note that

$$n(s_{r-1} - 1) + 2 = \frac{n^r - 1}{n-1} + 1. \quad \square$$

At first sight it would appear that this theorem merely shifts the emphasis in the problem of the global nilpotency of Lie algebras with E_n, since the solubility problem is no easier. However, we are psychologically ready to perceive the reduction thus achieved as a good sign. At least in the local situation, solubility and nilpotency were considered together during the proof of Theorem 5.3.1. Moreover, in Theorem 1.7.3 we have already made the first step in the direction of a solubility proof for an arbitrary Lie algebra L with E_n, $n < p$: like every soluble Lie algebra, L contains a non-zero abelian ideal. Of course, this is only a first approximation to the truth, since it will be clear from the results of the following section that the inequality $n < p$ is just not quite enough for solubility.

The following simple assertion, which is well-known in the general theory of varieties of Lie algebras (see [26, 2.8.3]), makes it possible to give an effective construction of ideals in L.

2.3. Proposition. *Let $f(x_1, \ldots, x_r)$ be a homogeneous associative polynomial* (*that is, an element of a free associative algebra on generators $x_1, \ldots, x_r, \ldots$*) *of degree $m_i \leqslant n < p$ in each x_i. We set*

$$J(f) = \{u \in L \mid [uf(u_1, \ldots, u_r)] = 0 \quad \textit{for all} \quad u_1, \ldots, u_r \in L\}.$$

Then $J(f)$ is an ideal of L.

Proof. We denote by $f_m(y_1, y_2, \ldots, y_m)$ the multilinear polynomial in $m = m_1 + m_2 + \cdots + m_r$ variables obtained from f by complete linearization. We may assume that the y_i are numbered in such a way that

$$f_m(\underbrace{x_1, \ldots, x_1}_{m_1}, \underbrace{x_2, \ldots, x_2}_{m_2}, \ldots, \underbrace{x_r, \ldots, x_r}_{m_r}) = m_1! \ldots m_r! \, f(x_i).$$

On the other hand, $f_m(y_1, \ldots, y_m)$ can be written as a linear combination of expressions $f\left(\sum_{i=1}^{m} \lambda_i y_i, \ldots, \sum_{j=m_1+\ldots m_{r-1}+1}^{m} \lambda_j y_j\right)$, with the λ_v running over the prime subfield F_0 of F. Our concern is about easy generalizations of the identities considered in 1.4.4. We shall not need an explicit form for these.

The remarks already made show that $J(f)$ is the same as $J(f_m) = \{u \in L | [uf_m(v_1, \ldots, v_m)] = 0$ for all $v_j \in L\}$ in all cases. By definition, $u, u' \in J(f_m) \Rightarrow \lambda u + \lambda' u' \in J(f_m)$, that is, $J(f_m)$ is a linear subspace of L. But now, it is immediately obvious from the multilinearity of f_m that

$$\begin{aligned}&[[uv]f_m(v_1, \ldots, v_m)] \\ &\quad = \{[uf_m([vv_1], \ldots, v_m)] + \cdots + [uf_m(v_1, \ldots, [vv_m])]\} \\ &\qquad + [[uf_m(v_1, \ldots, v_m)]v] = 0\end{aligned}$$

whenever $u \in J(f_m)$ and $v \in L$. Thus, $[uv] \in J(f_m) = J(f)$. □

The method for constructing ideals in L dual to Proposition 2.3 was exhibited in 5.2.2 (see assertion 2)). It relates to the case of an infinite field F, but if the degree m_i of f in each variable is strictly less than p, then F can be anything. We restrict attention to a special case.

2.4. Proposition. *Let L be a Lie algebra with E_n, $n < p$. Then, for every natural number k, the linear span J_k of the values of the Lie polynomial $[x_0 x_1^{n-1} x_2^{n-1} \ldots x_k^{n-1}]$ is an ideal of L:*

$$J_k = \langle [u_0 u_1^{n-1} u_2^{n-1} \ldots u_k^{n-1}] \quad \text{for all} \quad u_i \in L \rangle .$$

Proof. Our assertion is known for $k = 1$, since J_1 is just the ideal $E_{n-1}(L)$ (see § 1, or the beginning of § 3 in Chap. 2). For $k > 1$, one needs only transfer the factors $w \in L$ in the products $[u_0 u_1^{n-1} \ldots u_k^{n-1} w]$ to the left sufficiently often, using condition E_n, the inequality $n < p$ and formula (1.2) each time. □

2.5. Corollary. *Let L be a Lie algebra with E_n, $n < p$, contained in an associative enveloping algebra A that satisfies the (weak) identity*

$$u_1^{n-1} u_2^{n-1} \ldots u_q^{n-1} = 0, \quad u_i \in L, \quad 1 \leqslant i \leqslant q , \tag{6.17}$$

for some natural number $q = q(n)$, which we assume to be the smallest. Suppose further that we know that every Lie algebra with E_{n-1} has solubility length r_{n-1}. Then L is soluble of length $r_n \leqslant q r_{n-1}$ (and hence nilpotent, by Theorem 2.2).

Proof. We consider the chain

$$L = J_0 \supset J_1 \supset \cdots \supset J_k \supset J_{k+1} \supset \cdots \supset J_q = 0 \tag{6.18}$$

of ideals J_k of L defined in Proposition 2.4. A break in the chain at the q-th place guarantees identity (6.17). We note that $J_k \subseteq J(x_{k+1}^{n-1} \ldots x_q^{n-1})$, and we would like to study a chain of ideals of length q of type $J(f)$ in the same way (see Proposition 2.3). More essential are the inclusions

$$[J_k v^{n-1}] \subseteq J_{k+1}, \qquad 0 \leqslant k \leqslant q-1, \qquad v \in L ,$$

which are clear from the definition of the J_k and mean that the factors J_k/J_{k+1} satisfy E_{n-1}. By assumption, $(J_k/J_{k+1})^{[r_{n-1}]} = 0$, that is, $J_k^{[r_{n-1}]} \subseteq J_{k+1}$, and thus

$$L^{[r_{n-1}]} \subseteq J_1, \quad L^{[2r_{n-1}]} \subseteq J_2, \ldots, L^{[qr_{n-1}]} \subseteq J_q = 0 .$$

This proves that L is soluble and that $r_n \leqslant qr_{n-1}$. □

2.6. Example. Let L be a Lie algebra with E_4, $p > 5$. We know (see 1.2.9 and 1.8.1) that L is nilpotent since $p > 5$ (in fact $L^6 = 0$) so that $L^{[3]} = 0$ for every Lie algebra L with E_3. Thus, for $n = 4$ it is enough to try and get a weak identity of type (6.17).

Using elementary operations (permutation, left and right multiplication by elements of the enveloping associative algebra A) on identities (1.8) and (1.9), we get a system

$$\begin{aligned}
x^3y^3 &= -y^3x^3 = 13T, \\
x^2y^2x &= -y^2x^3y = -T, \\
xy^3x^2 &= -yx^3y^2 = 7T, \\
2x^2yxy^2 &= -2y^2xyx^2 = T, \\
2xyx^2y^2 &= -2yxy^2x^2 = -13T, \\
2x^2y^2xy &= -2y^2x^2yx = -25T, \\
xyxyxy &= -yxyxyx = -3T, \\
2xy^2xyx &= -2yx^2yxy = -5T, \\
2xy^2x^2y &= -2yx^2y^2x = 17T, \\
2xyxy^2x &= -2yxyx^2y = -7T,
\end{aligned} \tag{6.19}$$

where $T = -x^2y^3x$. Let $V = \langle u^3 | u \in L \rangle_F$ be the linear subspace of A spanned by the cubes of elements of L. Noting that $[xy]^3 = -21T$, we get that $21x^3y^3 = 13 \cdot 21T = -13[xy]^3$, that is, $21x^3y^3 \in V$. On the other hand, identities (1.5), (1.8) and (1.9) give that $[yx^3] = -2yx^3 - x^3y$, so that we find using (6.19):

$$\begin{aligned}
&[yx^3]y^2 + y[yx^3]y + y^2[yx^3] \\
&\quad = (-2yx^3 - x^3y)y^2 + y(-2yx^3 - x^3y)y + y^2(-2yx^3 - x^3y) \\
&\quad = x^3y^3 - 3y^2x^3y - 3yx^3y^2 = 13T - 3T + 21T = 31T .
\end{aligned}$$

By identities (1.2),

$$[yx^3]y^2 + y[yx^3]y + y^2[yx^3] = \begin{Bmatrix} [yx^3] & y \\ 1 & 2 \end{Bmatrix} \in V,$$

so that

$$31x^3y^3 = 13 \cdot 31T = 13 \begin{Bmatrix} [yx^3] & y \\ 1 & 2 \end{Bmatrix} \in V,$$

and in combination with the inclusion $21x^3y^3 \in V$ found earlier, we get $x^3y^3 \in V$.

Therefore, *the space V is an anticommutative associative algebra with the natural multiplication.* This result can be used to obtain the relation

$$tx^3y^3 = 0 \tag{6.20}$$

(details are omitted), from which there follows finally the weak identity $t^3x^3y^3 = 0$, and by Corollary 2.5, the solubility of L: $L^{[9]} = 0$. However, relation (6.20) is stronger: it enables us to deduce that $L^{[7]} = 0$. For $p = 13$, (6.19) gives at once the strong identity $x^3y^3 = 0$, so that $L^{[6]} = 0$.

If we are not aiming to get precise bounds for the characteristic of the field or for the solubility length of the Lie algebra, the weak identity necessary for application of Corollary 2.5 follows immediately from (6.19) when $p > 7$:

$$x^3y^3z^3 = -\tfrac{13}{21}[xy]^3z^3 = \tfrac{13}{21}z^3[xy]^3 = -z^3x^3y^3 = x^3z^3y^3 = -x^3y^3z^3\,.$$

However, the associative anticommutative algebra structure established above for V seems to be of independent interest.

§ 3. Insolubility for n Close to p

Is it possible to strengthen Theorem 1.7.4 by proving that a Lie algebra with E_n, $n < p$, is globally nilpotent? At least for $n = p - 2$, and thus for the group-theoretically interesting case $n = p - 1$, the answer is no.

3.1. Theorem. *There exists an insoluble, and therefore non-nilpotent, Lie algebra L with E_{p-2} over any field F of characteristic $p \geqslant 5$.*

The proof, which is divided up into a number of fairly elementary steps organised in the form of lemmas and corollaries, comes to producing a general construction which is also a vehicle for expressing the main assertion in group-theoretical language. Without loss of generality (see 5.2.2 for the reason why), we shall assume that F is algebraically closed. We denote a root of the polynomial $x^2 + 1$ by the standard symbol i, and the algebra of 2×2 matrices over F by $M = M_2(F)$. By definition, we set

$$e_1 = \begin{Vmatrix} i & 0 \\ 0 & -i \end{Vmatrix}, \qquad e_2 = \begin{Vmatrix} 0 & i \\ i & 0 \end{Vmatrix}, \qquad e_3 = \begin{Vmatrix} 0 & -1 \\ 1 & 0 \end{Vmatrix}.$$

The linear span of these elements is closed under commutation. We construct a commutative algebra K with identity on generators $x_i^{(1)}, x_i^{(2)}, x_i^{(3)}$ $(i = 1, 2, \ldots)$ and defining relations

$$x_i^{(j)}x_i^{(k)} = \delta_{jk}(x_i^{(1)})^2, \qquad x_i^{(j)}x_i^{(l)}x_i^{(t)} = 0\,, \tag{6.21}$$

where δ_{jk} is the Kronecker symbol and $i = 1, 2, \ldots$. We form the associative subalgebra A^* with identity of $M_2^* = K \otimes_F M_2$ generated by the elements

$$x_i = \sum_{j=1}^{3} x_i^{(j)} \otimes e_j, \quad i = 1, 2, \ldots .$$

We write L^* for the Lie algebra under commutation $[x, y] = xy - yx$ generated by the same elements.

For any two non-intersecting finite subsets Δ and Λ $(\Delta \cap \Lambda = \phi)$ of $\{x_1, x_2, \dots\}$, we denote by $A^*(\Delta, \Lambda)$ the subspace of A^* generated by all monomials that are of degree 1 in all the variables in Λ, degree 2 in the variables in Δ and degree 0 in all the other variables. We shall suppose that $K = A^*(\phi, \phi)$. Set $A^*_\cdot(\Lambda) = A^*(\phi, \Lambda)$, $A^*(\Delta) = A^*(\Delta, \varnothing)$.

It follows from the definition of A^* and the defining relations (6.21) og K that

$$A^* = \sum_{\Delta \cap \Lambda = \varnothing} A^*(\Delta, \Lambda)\,, \tag{6.22}$$

where the direct sum extends over all non-intersecting pairs of subsets Δ, Λ of $\{x_1, x_2, \dots\}$. Decomposition (6.22) gives A^* the structure of a graded algebra. An ideal V of A^* will be called *homogeneous* if

$$V = \sum_{\Delta \cap \Lambda = \varnothing} V \cap A^*(\Delta, \Lambda)\,.$$

We write B for the homogeneous subalgebra $\sum_\Delta A^*(\Delta)$ of A^*, where the sum extends over all subsets Δ of $\{x_1, x_2, \dots\}$.

Let J be the ideal of A^* generated by all elements of the form $x_i a x_i$, $a \in A^*$, $i = 1, 2, \dots$. It is clear that $J = \sum A^*(\Delta, \Lambda)$, where the sum extends over all pairs of disjoint subsets Δ and Λ with Δ non-empty. Let I denote the largest homogeneous ideal of A^* such that $I \cap B = 0$. We set $A = A^*/(J + I)$, $L = (L^* + I + J)/(J + I)$. Our aim is to show that *the relations*

$$y^{p-2} = 0\,, \tag{6.23}$$

$$[xy^{p-2}] = 0 \tag{6.24}$$

hold in A for all $x, y \in L$, and that L is not nilpotent when $p \geqslant 5$.

3.2. Lemma. *The following relations hold in A^* for all $x, y, z, v \in L^*$:*

$$[x \circ y, z] = 0\,; \tag{6.25}$$

$$x_i v x_i = -v\frac{x_i^2}{3}\,, \tag{6.26}$$

where $x \circ y = xy + yx$.

Proof. It follows from the definition of L^* that each of its elements has a representation in the form $\sum_{i=1}^{3} c_i e_i$, where $c_i \in K$. Since each matrix e_i has zero trace, it is enough to check relations (6.25) and (6.26) for the case $x, y, z, v \in \mathrm{sl}\,(2, K)$. Relation (6.25) then reflects the well-known fact that the square of every matrix in $\mathrm{sl}(2, K)$ lies in the centre of M_2. We note that relations (6.25) for $x = y = x_i$ imply that the elements x_i^2 lie in the centre of A^*.

It is enough to check relation (6.26) for $v = e_1, e_2, e_3$. In these cases, it can be checked without difficulty by using the defining relations (6.21) and the quaternionic relations between the e_i:

$$e_1e_2 = e_3 = -e_2e_1, \quad e_3^2 = -1 \quad (1 \text{ is the identity matrix}),$$
$$e_2e_3 = e_1 = -e_3e_2, \quad e_1^2 = -1,$$
$$e_3e_1 = e_2 = -e_1e_3, \quad e_2^2 = -1.$$

For example,

$$x_i(1 \otimes e_1)x_i = (x_i^1)^2 \otimes \sum_{k=1}^{3} e_k e_1 e_k = (x_k^{(1)})^2 \otimes (-e_1 + e_1 + e_1)$$
$$= (x_k^{(1)})^2 \otimes e_1 = \frac{x_k^2}{3}(1 \otimes e_1). \quad \square$$

Corollary. *The relations*

$$x_i^2 x_j = x_j x_i^2; \tag{6.27}$$

$$(x_i \circ v)x_i = 2v\frac{x_i^2}{3}; \tag{6.28}$$

$$[x_k x_i]ax_k = x_k a[x_i x_k] \tag{6.29}$$

hold in A^* for all $a \in A^*$, $v \in L^*$.

Proof. Relations (6.27) and (6.28) follow immediately from (6.25) and (6.26). It is enough to check (6.29) in the case where $a = v_1 \dots v_t$, the v_i being elements of L^*. We do this by induction on t. For $t = 0$, the relation is equivalent to (6.27). For $t = 1$, it follows from (6.26):

$$[x_k x_i]v_i x_k - x_k v_1[x_i x_k] = x_k[x_i v_1]x_k - x_i x_k v_1 x_k + x_k v_1 x_k x_i$$
$$= -\frac{x_k^2}{3}([x_i v_i] - x_i v_1 + v_1 x_i) = 0.$$

Thus the basis of the induction is intact, and for $t > 1$ we have

$$[x_k x_i]v_1 v_2 v_3 \dots v_t x_k = [x_k x_i]\cdot\tfrac{1}{2}(v_1 \circ v_2)v_3 \dots v_t x_k$$
$$+ [x_k x_i]\cdot\tfrac{1}{2}[v_1, v_2]v_3 \dots v_t x_k = x_k v_1 \dots v_t[x_i x_k].$$

This last equation is a corollary of the induction hypothesis and the fact that $v_1 \circ v_2$ is in the centre of A^*.

3.3. Lemma. *Suppose that $u \in A^*(\Delta, \Lambda)$. Then the equation*

$$u = \pi(u)\prod_{x_i \in \Delta}\left(\frac{x_i^2}{3}\right) \tag{6.30}$$

holds in A^*, where $\pi(u) \in A^*_\cdot(\Delta)$.

Proof. It is enough to examine the case where u is a monomial. Applying (6.25) and (6.28) in turn, we get the relation

$$x_i x_j a x_i = -x_j x_i a x_i + (x_i \circ x_j) a x_i = -x_j x_i a x_i + a(x_i \circ x_j) x_i$$
$$= -x_j x_i a x_i + 2 a x_j \frac{x_i^2}{3}.$$

We observe that the distance between the letters x_i has decreased; and we apply this process of bringing letters closer together for the word $u = \ldots x_i x_j a x_i \ldots$, for each variable x_i in Δ. At some stage we get an equation $u = u'(x_i^2/3)$, where $u' \in A^*(\Delta \backslash \{x_i\}, \Lambda)$. Relation (6.30) now follows by induction. □

Corollary A. *The monomials of the form*

$$x_{i_1}^2 x_{i_2}^2 \ldots x_{i_t}^2, \qquad i_1 < i_2 < \ldots < i_t,$$

form a basis for the algebra B.

Proof. The fact that B is the linear span of the set of monomials indicated follows from the lemma. Linear independence is proved by direct substitution of 2×2 matrices. □

Corollary B. *Let a and b be any monomials in A^* such that $ab \in B$. Then $ab = ba$.*

Proof. It is enough to consider the case where $b = x_i$: thereafter, the corollary is obtained by cyclic permutation of letters. If $b = x_i$ and $a x_i \in B$, then $a \in A^*(\Delta, \{x_i\})$, and by Lemma 3.3 we have

$$a x_i = \beta x_i \prod_{x_j \in \Delta} (x_j^2/3) x_i = x_i a,$$

since $\pi(a) = \beta x_i$ for some $\beta \in F$. □

3.4. Lemma. *The element $u \in A_*^*(\Lambda)$ lies in I if and only if*

$$uv = 0 \qquad (6.31)$$

in A^*, for every $v \in A_*^*(\Lambda)$.

Proof. If $u \in I \cap A_*^*(\Lambda)$, (6.31) follows at once from the definition. Conversely, suppose that the element u of $A_*^*(\Lambda)$ satisfies (6.31) for all $v \in A_*^*(\Lambda)$. Every homogeneous element of B contained in the ideal generated by u is a linear combination of expressions of the form cud. Here, c and d are monomials such that $dc \in A^*(\Delta_0, \Lambda)$ for some subset Δ_0. By Lemma 3.3 and Corollary B, we have

$$cud = udc = u\pi(dc) \prod_{x_j \in \Delta_0} (x_j^2/3).$$

Since $\pi(dc) \in A_*^*(\Lambda)$, (6.31) applies and we get that $cud = 0$. Thus, the ideal of A^* generated by u has non-zero intersection with B. This means that $u \in I$. □

3.5. Lemma. *Suppose that $\Lambda = \{x_1, \dots, x_l\}$, and let $f(t_1, \dots, t_l)$ be a multilinear associative polynomial. If $f(x_1, \dots, x_l) \in I \cap A^*_\cdot(\Lambda)$, then $f(v_1, \dots, v_l) = 0$ in A for arbitrary $v_1, \dots, v_l \in L$.*

Proof. Since f is multilinear, it is enough to consider the case where $v_1, \dots, v_l$ are left-normed commutators in $x_1, \dots, x_k$. But every word w in A containing two occurrences of the same letter is zero. Thus, we may assume that $f(v_1, \dots, v_l)$ is a multilinear polynomial in $x_1, \dots, x_k$. Since every permutation of the variables $x_1, x_2, \dots$. induces an automorphism of the algebras A^* and A, it is enough to check the relation $f(v_1, \dots, v_l) = 0$ in A when $v_i = [v_i', x_i]$ for each $i = 1, 2, \dots,$ where v_1' is a commutator in certain of the variables $x_{l+1}, \dots, x_k$ (the argument changes but little if some of the v_i have degree 1).

As a linear space, the algebra $A = A^*/(I + J)$ is isomorphic to $\sum_\Lambda A^*_\cdot(\Lambda)/(I \cap A^*_\cdot(\Lambda))$, where this direct sum extends over all finite subsets Λ of $\{x_1, x_2, \dots\}$. Therefore, the relation under discussion is equivalent to the inclusion $f([v_1', x_1], \dots, [v_l', x_l]) \in I$. Applying Lemmas 3.4, 3.3 and formula (6.29), we get for all $v \in A^*_\cdot(\Lambda)$:

$$f(v_1, \dots, v_l)v = f(x_1, \dots, x_l)(v|_{x_i \to [x_i v_i'],\, 1 \leqslant i \leqslant l})$$

$$= f(x_1, \dots, x_l)\pi(v|_{x_i \to [x_i v_i'],\, 1 \leqslant i \leqslant l}) \prod_{j=l+1}^{k} (x_j^2/3)\,. \qquad (6.32)$$

By the assumptions of the lemma, $f(x_1, \dots, x_l) = 0$ in A. This is equivalent to the assertion that $f(x_1, \dots, x_l) \in I$ in A^*. By Lemma 3.4, it follows from this that the element on the right-hand side of (6.32) is zero. Applying Lemma 3.4 again, we conclude from (6.32) that $f(v_1, \dots, v_l) \in I$. This proves the lemma.

3.6. Completion of the Proof of the Theorem. Applying formula (6.26) and Corollary A of Lemma 3.3, we see that the following property holds in A^*:

$$[x_1, \dots, x_k]x_k \dots x_1 = -x_k[x_1, \dots, x_{k-1}]x_k x_{k-1} \dots x_1$$
$$+ [x_1, \dots, x_{k-1}]x_k^2 x_{k-1} \dots x_1 = \tfrac{4}{3}[x_1, \dots, x_{k-1}]x_{k-1} \dots x_1 x_k^2$$
$$= (\tfrac{4}{3})^{k-1} x_1^2 x_2^2 \dots x_k^2 \neq 0\,.$$

Therefore, for every natural number k, the commutator $[x_1, \dots, x_k]$ is not contained in I, so that $[x_1, \dots, x_k] \notin I + J$. This shows that L is non-nilpotent.

Let us establish (6.23), which is equivalent in A to the linearized relations

$$\sum_{\sigma \in S_{p-2}} v_{\sigma 1} v_{\sigma 2} \dots v_{\sigma(p-2)} = 0\,, \qquad (6.33)$$

where $v_1, \dots, v_{p-2}$ are arbitrary elements of L, and the sum extends over all permutations of $\{1, \dots, p-2\}$. Applying (6.25), (6.28), (6.27) and Lemma 3.3, we get the following relation in A^*:

$$\pi\left(\left(\sum_{\sigma\in S_{p-2}} x_{\sigma 1}\dots x_{\sigma(p-2)}\right)x_{p-2}\right)$$

$$=\frac{p-3}{2}\sum_{j=1}^{p-3}\pi\left(\sum_{\tau\in S_{p-4}} x_{\tau 1}\dots x_{\tau(j-1)}x_{\tau(j+1)}\dots x_{\tau(p-3)}\cdot(x_{p-2}\circ x_j)x_{p-2}\right)$$

$$+\pi\left(\sum_{\rho\in S_{p-3}} x_{\rho 1}\dots x_{\rho(p-3)}\cdot x_{p-2}^2\right)$$

$$=\left(\frac{p-3}{2}\cdot\frac{2}{3}+1\right)\pi\left(\sum_{\rho\in S_{p-3}} x_{\rho 1}\dots x_{\rho(p-3)}x_{p-2}^2\right)\equiv 0(\operatorname{mod} p)\,.$$

(In going from the first term in this chain of equations to the second, we have used regroupings of the form $\dots x_j x_{p-2}\dots+\dots x_{p-2}x_j\dots=\dots x_j\circ x_{p-2}\dots$). Thus, $\sum_{\sigma\in S_{p-2}} x_{\sigma 1}\dots x_{\sigma(p-2)}\cdot v=0$ in A^* for all $v\in \Lambda=\{x_1,\dots,x_{p-2}\}$. By Lemma 3.4, this means that $\sum_{\sigma\in S_{p-2}} x_{\sigma 1}\dots x_{\sigma(p-2)}\in I$. But then, by Lemma 3.4, relation (6.33) holds in A for all $v_1,\dots,v_{p-2}\in L$. This proves (6.23).

Since relation (6.25) holds in A^* for all $x, y, z\in L^*$, it also holds in $A=A^*/(I+J)$ for all $x, y, z\in L$. We claim that (6.24) is a consequence of (6.23) and (6.25). Partial linearization of (6.23) leads to the relation

$$\sum_{j=0}^{p-3} y^j x y^{p-3-j}=0,\qquad x, y\in L\,,$$

whence (6.27) gives that $yxy^{p-4}=\beta xy^{p-3}$ for all $x, y\in L$ and some $\beta\in F$. In that case, we get from formula (1.5) that

$$[xy^{p-2}]=[x,\underbrace{y,\dots,y}_{p-2}]=\sum_{j=0}^{p-2}(-1)^j\binom{p-2}{j}y^j x y^{p-2-j}$$

$$=\alpha_1 xy^{p-2}+\alpha_2 yxy^{p-3}=(\alpha_1+\beta\alpha_2)xy^{p-2}=0\,.$$

Thus, we have established (6.24), and the proof of Theorem 3.1 is complete. □

3.7. The construction of A^* and L^* shows that all identities of the 2×2 matrix algebra hold in A, and all Lie identities of sl $(2, F)$ hold in L. The question arises naturally: *is it true that all the identities of some finite-dimensional Lie algebra hold in every Lie algebra with* E_{p-1} *over a field of characteristic* $p>0$?

3.8. Theorem. *We adjoin an identity to the algebra A constructed in* 3.1. *The subgroup G of the multiplicative group of the resulting algebra generated by the elements* $g_i=1+x_i$ $(i=1,2,3,\dots)$ *is an insoluble group of exponent p.*

For the proof we shall need:

3.9. Lemma. *The identity* $a^p=0$ *holds in A.*

Proof. We can represent every element a of A in the form

$$a=c+v\,, \tag{6.34}$$

where c is an element of the centre of A, expressed as a linear combination of products of the central elements $v_i \circ v_j = v_i v_j + v_j v_i$ with $v_i, v_j \in L$ (see (6.25)), and v is a linear combination of elements of L with coefficients in the centre of A. In fact,

$$v_1 v_2 v_3 \ldots v_k = \frac{1}{2}(v_1 \circ v_2) v_3 \ldots v_k + \frac{1}{2}[v_1, v_2] v_3 \ldots v_k ,$$

where the v_i come from L, and the verification of (6.34) is completed by an obvious induction on the number of commutators in each of the products. By (6.23) and its linearizations, $v^p = 0$ in A. Further, Lemma 3.5 gives that the squares $(v_j \circ v_j)^2$, where v_i and v_j are commutators in $x_1, x_2, \ldots$, are zero in A. Thus $a^p = c^p + v^p = 0$. □

Proof of Theorem 3.8. Since every element of G can be expressed in the form $g = 1 + a$, $a \in A$, by Lemma 3.9 we have $g^p = (1 + a)^p = 1 + a^p = 1$, that is, G is of exponent p.

We note that the square in A of every Lie commutator in $x_1, x_2, \ldots$ is zero. For this reason,

$$\tilde{c}(g_1, \ldots, g_k) = 1 + c(x_1, \ldots, x_k) , \tag{6.35}$$

where $\tilde{c}$ is a group commutator with some disposition of the brackets, and c is the Lie commutator with the same disposition. By Theorem 3.1, L is insoluble, and (6.35) shows that G is also insoluble. □

§ 4. Global Nilpotency for $p \gg n$

In this section we shall prove the following important result, at a level requiring heightened vigilance on the part of the reader:

4.1. Theorem. *Every Lie algebra L with E_n over a field F of characteristic zero is globally nilpotent.*

In fact what the theorem asserts is this: for every $k \geqslant c(n)$ (where $c(n)$ is a natural number depending only on n), the left-normed commutator $[x_0 x_1 \ldots x_k]$ in the absolutely free Lie algebra $\mathscr{L} = \mathrm{Lie}\langle x_0, x_1, \ldots, x_k \rangle$ over the rational field $\mathbb{Q}$ is contained in the ideal $E_n(\mathscr{L})$. In other words,

$$[x_0 x_1 x_2 \ldots x_k] = \sum_{i=1}^{r} \alpha_i \sum_{\pi \in S_n} [u_0^{(i)} u_{\pi 1}^{(i)} \ldots u_{\pi n}^{(i)}] , \tag{6.36}$$

where the α_i are in $\mathbb{Q}$, and the $u_j^{(i)}$ are basic commutators in $x_0, x_1, \ldots, x_k$. All commutators on the right-hand side of (6.36) are multilinear, and to every decomposition

$$\{x_0, x_1, \ldots, x_k\} = \bigcup_{j=0}^{n} X_j^{(i)}$$

into disjoint subsets there corresponds a set of basic commutators $u_j^{(i)}$ in the variables contained in $X_j^{(i)}$. If the α_i in expression (6.36) corresponding to $k = c(n)$

are written in reduced form $\alpha_i = q_i/p_i$ with $p_i > 0$, and we define $p_0 = \max_{1 \leqslant i \leqslant r} \{p_i\}$, then clearly *Theorem* 4.1. *remains valid over every field of finite characteristic* $p > p_0$. We write the requirement symbolically as $p \gg n$. Our argument can be formalized like this: if for every natural number m there exists a non-nilpotent Lie algebra $L(m)$ with E_n over a field $F(m)$ of characteristic greater than m, then the ultra-product $\prod_{m \in \mathbb{N}} L(m)/\mathscr{F}$ over the Frechet ultrafilter $\mathscr{F}$ is a non-nilpotent Lie algebra with E_n over the field $\prod_{m \in \mathbb{N}} F(m)/F$ of zero characteristic. Theorems 3.1 and 4.1 raise the question of the exact bound $p(n)$ for the characteristic of the ground field F, that is, the smallest function $p(n)$ such that every Lie algebra over F with E_n is globally nilpotent if the characteristic p of F is at least $p(n)$.

4.2. Reduction of the Problem. Throughout the entire section, we shall make free use of the terminology and general results in the theory of varieties as expounded in [26]. However, some definitions are given, and many will be clear from the context. The following theorem is fundamental:

Theorem A (Theorem 1.7.3). *Every Lie algebra with* E_n *over a field of characteristic zero contains a non-zero abelian ideal.*

Let F be the ground field of characteristic zero, and $\mathscr{L}$ the relatively free Lie algebra on a countable set $X = \{x_1, x_2, \ldots\}$ of generators in the variety of Lie algebras defined by E_n: $[uv^n] = 0$. Theorem A guarantees that $\mathscr{L}$ has a non-zero abelian ideal. Theorem 2.2 allows us to reduce the nilpotency proof to that of solubility. Thus, we argue by contradiction and construct a maximal ideal I of L with insoluble factor-algebra. However, we shall want this factor-algebra to be relatively free too, since it is then more convenient to handle. In this situation it suffices to prove all the basic assertions for the generators, since the latter can be mapped into any set of elements in the algebra by a suitable endomorphism. We shall often use this principle without stipulating it explicitly: if an equation is obtained between some of the generators, then the generators can be replaced by arbitrary elements. In other words, the required ideal I must be *verbal*, that is, invariant under all endomorphisms.

We shall construct I by transfinite induction, defining an ascending chain of ideals I_α indexed by ordinals α. We set $I_0 = (0)$, and proceed by induction, as follows. If α is a limit ordinal, we set $I_\alpha = \bigcup_{\nu < \alpha} I_\nu$. Otherwise $\alpha - 1$ exists, and thus $I_{\alpha - 1}$ is defined. Let $\mathscr{L} \to \bar{\mathscr{L}} = \mathscr{L}/I_{\alpha-1}$ be the natural homomorphism. Then I_α is the inverse image of the sum of the abelian ideals of $\bar{\mathscr{L}}$.

Verbality of ideals follows from three simple remarks:

1) The union of verbal ideals is verbal.

2) The sum of all the verbal ideals of a relatively free algebra of infinite rank is verbal (this follows immediately from the fact that the given sum is invariant under all automorphisms, and because that is sufficient to imply verbality, since our algebra is relatively free of infinite rank: see Theorem 4.2.9 of [26]).

3) If a homomorphism has verbal kernel, inverse images of verbal ideals under it are verbal.

We note that I_α contains $I_{\alpha-1}$ strictly if α is not a limit ordinal, by Theorem A: of course, always assuming that $\mathscr{L} \neq I_{\alpha-1}$. Therefore, cardinality arguments show that the chain of ideals must stabilize sooner or later; equivalently, $\mathscr{L}$ must be I_γ for some γ. Thus, there exists a smallest ordinal β such that $\mathscr{L}/I_\beta$ is nilpotent. Suppose that $(\mathscr{L}/I_\beta)^m = 0$, shall we say. This property is equivalent to the inclusion $[x_1, x_2, \ldots, x_m] \in I_\beta$ (as we remarked above, it is enough to establish the nilpotency requirement for the generators). The inclusion can hold only if β is a non-limit, since otherwise the left-hand side would lie in a smaller $I_{\beta'}$, contradicting the minimality of I_β. Thus, we can consider the ideal $I = I_{\beta-1}$ and the factor-algebra $L = \mathscr{L}/I$, which is a relatively free Lie algebra on the same generators (more exactly, on the generators $\bar{x}_i = x_i + I$, $i = 1, 2, \ldots$). By its definition, L is not nilpotent, while the commutator $[x_1, x_2, \ldots, x_m]$ generates a nilpotent ideal of it. This follows since every element of $I_\beta/I_{\beta-1}$ lies in the sum of finitely many abelian ideals, and consequently generates a soluble ideal; nilpotency then follows from Theorem 2.2. Recalling our remark about relatively free algebras, we can reformulate this property like this: for any m elements $u_1, u_2, \ldots, u_m$ of L, the ideal generated by the commutator $[u_1 u_2 \ldots u_m]$ is nilpotent of class not more than some fixed number independent of the u_i. Our aim now is to show that L is soluble, and thus get the desired contradiction from Theorem 2.2.

4.3. A Nilpotency Criterion for Verbal Ideals. The results of this subsection are of a general nature, and can be stated and proved for any relatively free algebra A in a variety $\mathfrak{M}$ over a field F. The algebras in $\mathfrak{M}$ need not be Lie algebras.

Thus, let $X = \{x_1, x_2, \ldots\}$ be a countably infinite set of free generators for A, and let $f = f(x_1, x_2, \ldots, x_m)$, $g = g(x_1, x_2, \ldots, x_M)$ be two multilinear elements in the generators. For any algebra $B \in \mathfrak{M}$, the polynomials f and g can be treated as maps

$$f: \underbrace{B \times \cdots \times B}_{m} \to B, \qquad g: \underbrace{B \times \cdots \times B}_{M} \to B\,.$$

The mapping means simply that elements b_i of B are substituted for the free generators. Let us agree to denote the result of the mapping by $f|_B$. In particular, f is an identity on B if and only if $f|_B = 0$.

Before coming to the statement of our main result, we must agree on terminology (this is necessary since we are working in general algebras). An ideal J of A is nilpotent of class at most q provided that the product of any q elements of it is zero, no matter what the disposition of brackets may be. Furthermore, *the verbal ideal $V_g(A)$ generated by the elements g of A* is the least verbal ideal containing g. As usual, it is the linear span of the elements $g(a_1, \ldots, a_M)$, $a_i \in A$.

Theorem B. *Let f and g be multilinear elements of the relatively free algebra $A \in \mathfrak{M}$ such that:*

(i) *for every $\mathbb{Z}_2$-graded algebra $B = B_0 + B_1$ in $\mathfrak{M}$,*

$$f|_{B_0} = 0 \Rightarrow g|_B = 0\,;$$

(ii) *f generates a nilpotent ideal of A.*
Then $V_g(A)$ is nilpotent.

The proof of Theorem B occupies a large portion of this subsection; our main Theorem 4.1 is then deduced from Theorem B, at the very end.

4.4. A Start on the Proof of Theorem B. Since the full proof is rather long, it is reasonable to split it into several steps. First of all, after restating condition (i) of Theorem B, we shall deduce from it a number of consequences in a form suitable for the subsequent technical investigation. We begin with the construction of the corresponding $\mathbb{Z}_2$-graded algebra, for which we start from A, giving each of its generators $x_1, x_2, \ldots$ a weight $\|x_i\| = 0$ or 1. We shall be interested only in gradings such that $\|x_i\| = \|x_j\|$ if $i \equiv j \pmod M$. In all, there are 2^M gradings, each of which corresponds to a set $\varepsilon = \{\varepsilon_1, \ldots, \varepsilon_M\}$, where $\varepsilon_i = 0$ or 1 and $\|x_i\| = \varepsilon_i$. By definition, the word $(\ldots(x_{i_1}x_{i_2})\ldots x_{i_k})$ has weight $\varepsilon_{i_1} + \varepsilon_{i_2} + \cdots + \varepsilon_{i_k} \pmod 2$.

Lemma 1. *For every set* $\varepsilon = \{\varepsilon_1, \ldots, \varepsilon_M\}$ *we have*

$$g(x_1, \ldots, x_M) = \sum_{j=1}^{d(\varepsilon)} \alpha_j^\varepsilon u_j^\varepsilon f(y_{1j}^\varepsilon, \ldots, y_{mj}^\varepsilon)\, v_j^\varepsilon\,,$$

where $\alpha_j^\varepsilon \in F$; *the* u_j^ε, v_j^ε, y_{ij}^ε *are words in* $x_1, \ldots, x_M$ *such that:*

1) $\|y_{ij}^\varepsilon\| = 0$ *(in the grading corresponding to* ε*)*;
2) *every summand on the right-hand side is multilinear. In particular, for each j there are no letters common to the words* u_j^ε, v_j^ε, $y_{1j}^\varepsilon, \ldots, y_{mj}^\varepsilon$.

Proof. Let $A = A_0 + A_1$ be the $\mathbb{Z}_2$-grading corresponding to ε (recall that $\|x_i\| = \varepsilon_i$). We consider the ideal J of A generated by the elements $f(y_1, \ldots, y_m)$ with $\|y_i\| = 0$ for $1 \leqslant i \leqslant m$. Since f is multilinear, we may assume that the y_j are words. Multilinearity guarantees that J is also $\mathbb{Z}_2$-graded, so that the factor-algebra $B = A/J$ is also $\mathbb{Z}_2$-graded, in accordance with the grading chosen in A; $B = B_0 + B_1$. By construction, $f|_{B_0} = 0 \Rightarrow g|_B = 0 \Rightarrow g \in J \Rightarrow g(x_1, \ldots, x_M) = \sum_{j=1}^d \alpha_j u_j f(y_{1j}, \ldots, y_{mj}) v_j$, where the u_j, v_j and y_{ij} are words. Since g depends only on $x_1, \ldots, x_M$, we can make the same proposition concerning the u_j, v_j and y_{ij}; for example, by setting all the remaining generators equal to zero.

Item 1) of the lemma is fulfilled by construction, and in order to achieve 2), we note that each word in the generators is either multilinear (when all the letters occurring in it are different), or it is not. In other words, every element a of A can be represented as a sum $a^{(P)} + a^{(N)}$, where $a^{(P)}$ is the maximal multilinear part. We have $(a+b)^{(P)} = a^{(P)} + b^{(P)}$, so that, by using the fact that g is multilinear, we can pass in the equation thus obtained to the multilinear parts and assume that each summand $u_j f(y_{1j}, \ldots, y_{mj}) v_j$ is multilinear. Since f is itself multilinear, this means that only the summands not satisfying 2) drop out, while the others remain unchanged. □

We shall not be interested in the form of the α_j, u_j and v_j (it will be enough to know that they are multilinear), so that in the interests of economy it is convenient to rewrite the equation just established in the form

$$g(x_1, \ldots, x_M) = \sum_{j=1}^{d(\varepsilon)} f(y_{1j}^\varepsilon, \ldots, y_{mj}^\varepsilon)\, W_j^\varepsilon$$

where the W_j^ε are operators in the algebra of multiplications by numbers and generators of A, and the $y_{ij}^\varepsilon = y_{ij}^\varepsilon(x_1, \ldots, x_M)$ are words of weight 0 in the grading associated with ε. We note the similar dependence $W_j^\varepsilon = W_j^\varepsilon(x_1, \ldots, x_M)$, and stress that we can substitute other generators for $x_1, \ldots, x_M$ in y_{ij}^ε and W_j^ε.

Lemma 2. *Let ε be a fixed M-set with the corresponding $\mathbb{Z}_2$-grading of A. Then*

$$g(x_{M(k-1)+1}, \quad x_{M(k-1)+2}, \ldots, x_{Mk}) = \sum_{j=1}^{d} f(y_{1j}^k, \ldots, y_{mj}^k)\, W_j^k\,,$$

for $k = 1, 2, \ldots$, where the y_{ij}^k are words and the W_j^k are operators in the algebra of multiplications depending only on $x_{M(k-1)+1}, \ldots, x_{Mk}$. Moreover,

1) *all the summands on the right are multilinear,*
2) $\| y_{ij}^k \| = 0$ *for all values of the indices.*
3) *for the various k, the summands with index j are obtained from one another by shifting the numerals of all the generators by one and the same multiple of M.*

Proof. For $k = 1$, the formula follows from Lemma 1. For larger k, it is obtained by applying the automorphism carrying $x_1, \ldots, x_M$ to $x_{M(k-1)+1}, x_{Mk}$ respectively. Since automorphisms of this type preserve the grading, property 2) is clear. Property 3) follows from the construction, and 1) from Lemma 1.

For clarity, we note that

$$y_{ij}^k = y_{ij}^\varepsilon(x_{M(k-1)+1}, \ldots, x_{Mk})\,;$$
$$W_j^k = W_j^\varepsilon(x_{M(k-1)+1}, \ldots, x_{Mk})\,,$$

emphasising that all the summands on the right depend on ε. Of course, the number d of summands also depends on ε; however, since we want to liquidate this dependence at this point, we choose d to be the largest of the 2^M numbers $d(\varepsilon)$ and add zero summands if necessary (they correspond to the operation of multiplication by 0). Thus, d is now independent of ε.

We give next a bound for the nilpotency class of a verbal ideal as in Theorem B (of course, at a purely preliminary level at the moment). Assume that the ideal of A generated by $f(x_1, \ldots, x_m)$ is nilpotent of class at most q. Our aim is to prove that the verbal ideal generated by $g(x_1, \ldots, x_M)$ is nilpotent of class not more than $N_0 = p^{2^M}$, where $p = d(q-1)+1$. If $N = MN_0$, it is enough to prove that all multilinear products of the form

$$\ldots g(x_1, \ldots, x_M) \ldots g(x_{M+1}, \ldots, x_{2M}) \ldots g(x_{N-M+1}, \ldots, x_N) \ldots$$

are zero, with any disposition of the brackets. Multilinearity means, in particular, that all the letters not occurring in g are different from $x_1, \ldots, x_M$.

Assume for a contradiction that some such product is non-zero. The main idea is to use it to obtain a more symmetrical expression, still not zero, which is a symmetrization of the zero element in the ideal generated by f. This will be done in 4.6; first we need to prove a number of combinatorial lemmas.

4.5. Symmetrization and Skew-symmetrization. This subsection is free-standing, so that the notation in it does not depend on that which has gone before. Nevertheless, the notation is coordinated as far as possible in such a way that the results obtained here apply conveniently to the objects discussed earlier.

Suppose that we are given any linear representation of the symmetric group S_N in some vector space U. The linear operator corresponding to $\sigma \in S_N$ will be denoted by $\tilde{\sigma}$. With each subgroup G of S_N we associate two natural endomorphisms H_G, $K_G: U \to U$ called *symmetrization* and *skew-symmetrization* with respect to G respectively:

$$H_G = \sum_{\sigma \in G} \tilde{\sigma}; \qquad K_G = \sum_{\sigma \in G} (\operatorname{sgn} \sigma)\tilde{\sigma} .$$

In order not to duplicate statements, it is useful to have the notation Q_G standing for either of these two endomorphisms (assertions that are true for H_G and K_G will be couched in terms of Q_G).

If $\Omega \subseteq \{1, 2, \ldots, N\}$ is any subset, it is convenient to introduce the notation $H_\Omega = H_G$, and similarly with K_Ω, Q_Ω, for the group $G = S_\Omega = \operatorname{Sym}(\Omega)$ fixing all symbols outside Ω.

For given vector spaces U, W over F, we fix some linear map $\phi: U \to W$ and an element $u \in U$ such that $\phi(u) \neq 0$. For our future needs it would be better to have an element u with the stronger property: $\phi(Q_\Omega u) \neq 0$, where Ω is a sufficiently large set and Q_Ω is either H_Ω or K_Ω. However, it is impossible to guarantee the inequality in this form (for fixed u). Nonetheless, the situation can be repaired by replacing u by $\tilde{\sigma}u$ for some $\sigma \in S_N$. Our immediate intention is to prove this. The following result explains the reason why the extra property is so advantageous.

Lemma 3. *Let G_1 be any subgroup of G. If $\phi(Q_G u) \neq 0$, then $\phi(Q_G \tilde{\sigma} u) \neq 0$ for some $\sigma \in G$.*

Proof. Let $G = \cup G_1 \sigma_i$ be the decomposition of G into right cosets modulo G_1. For example, let us consider the case $Q_G = K_G$. We have

$$K_G(u) = \sum_{\sigma \in G} (\operatorname{sgn} \sigma)\tilde{\sigma}u = \sum_i \sum_{\tau \in G_1} (\operatorname{sgn} \tau\sigma_i)\tilde{\tau}\tilde{\sigma}_i u$$

$$= \sum_i (\operatorname{sgn} \sigma_i)\left(\sum_{\tau \in G_1} (\operatorname{sgn} \tau)\tilde{\tau} \right) \tilde{\sigma}_i u = \sum_i (\operatorname{sgn} \sigma_i)\, K_{G_1}(\tilde{\sigma}_i u) .$$

Thus, by the linearity of ϕ, we have $\phi(K_{G_1}\tilde{\sigma}_i u) \neq 0$ for some i, since $\phi(K_G u) \neq 0$.

The decomposition of G into left cosets modulo G_1 would yield nothing, since extraction of some σ_i from the inequality $\phi(\sigma_i K_{G_1} u) \neq 0$ is not always possible.

Lemma 4. *There exist a subset Ω of $\{1, 2, \ldots, N\}$ and an element σ of S_N such that*

(i) $\phi(Q_\Omega \tilde{\sigma} u) \neq 0$;

(ii) $|\Omega| = [\sqrt{N}]$.

Proof. As is well known (see for example [26, 3.2.7]), the identity element of the group algebra $F[S_N]$ can be represented in the form $1 = \sum \alpha_D e_D$, where the sum extends over all Young tableaux D of order N. Here α_D is a rational number, and

$$e_D = \sum_{\pi \in R(D),\, \kappa \in C(D)} (\operatorname{sgn} \kappa)\kappa\pi ,$$

where $R(D)$ is the subgroup fixing the rows of D, and $C(D)$ that fixing the columns. Since $\phi(u) \neq 0$, we have $\phi(\tilde{e}_D u) \neq 0$ for some tableau D. This tableau has either a row of length $t \geqslant \sqrt{N}$, or else a column of height $t \geqslant \sqrt{N}$. We shall consider the two cases separately.

a) D contains a row $D_{(i)} = (i_1, \ldots, i_t)$. Then, for some $\pi \in R(D)$ we have

$$\phi\left(\sum_{\kappa \in C(D)} (\operatorname{sgn} \kappa)\tilde{\kappa}(\tilde{\pi} u)\right) \neq 0$$

that is, $\phi(K_{C(D)}\tilde{\pi} u) \neq 0$. Let Ω be a subset of $\{i_1, \ldots, i_t\}$ of cardinal $[\sqrt{N}]$. Then $S_\Omega \subseteq S_{\{i_1, \ldots, i_t\}} \subseteq C(D)$. Thus, by Lemma 3, $\phi(K_\Omega \tilde{\sigma}\tilde{\pi} u) \neq 0$ for some $\sigma \in C(D)$, as required.

b) D contains a column $D^{(i)} = {}^T(i_1, \ldots, i_t)$. Again, we have $\phi\left(\sum_{\pi \in R(D)} \tilde{\kappa}\tilde{\pi} u\right) \neq 0$ for some $\kappa \in C(D)$, that is,

$$\phi(\tilde{\kappa} H_{R(D)} u) \neq 0 \Rightarrow \phi(\tilde{\kappa} H_{R(D)} \tilde{\kappa}^{-1} \tilde{\kappa} u) \neq 0 .$$

But $\tilde{\kappa} H_{R(D)} \tilde{\kappa}^{-1} = H_{R(D')}$ for some tableau D' of the same dimension. Thus $\phi(H_{R(D')}\tilde{\kappa} u) \neq 0$, and all we need do now is choose Ω inside the column conjugate to $D^{(i)}$ under κ, and complete the discussion as in a). □

The next step is to construct subsets $\Omega_1, \Omega_2, \ldots, \Omega_M$ such that

$$\phi(Q_{\Omega_1} Q_{\Omega_2} \cdots Q_{\Omega_M} \tilde{\sigma} u) \neq 0 .$$

The subsets must be pairwise disjoint and obtained from one another by successive shifts through one and the same number, and finally all elements of Ω_i must be congruent to i modulo M. The sets have the same cardinal; if the cardinal is p, then N must be taken large enough.

Lemma 5. *Set $M = MN_0$, where $N_0 = p^{2M}$. Then there exist an element σ of S_N, numbers $i_1, \ldots, i_p$ with $1 \leqslant i_1, \ldots, i_p \leqslant N_0$ and sets*

$$\Omega_r = \{M(i_1 - 1) + r, \ldots, M\{i_p - 1) + r\}, \qquad r = 1, 2, \ldots, M ,$$

such that

$$\phi(Q_{\Omega_1} \cdots Q_{\Omega_M} \tilde{\sigma} u) \neq 0 .$$

(As before, every endomorphism Q_{Ω_i} is H_{Ω_i} or K_{Ω_i}).

Proof. We have, of course, to apply Lemma 4 several times over; not always to the original situation, however, but with different N and ϕ. At the first step, for N we take $N_0 = p^{2M}$ and consider the subgroup isomorphic to S_{N_0} acting non-trivially only on the symbols congruent to 1 modulo M. For our map we choose the original ϕ of Lemma 4. In this situation, Lemma 4 means that there exist a permutation σ_1

in S_{N_0} and a set Ω of cardinal $N_1 = \sqrt{N_0} = p^{2^{M-1}}$, that is, a set of integers $k_1, \ldots, k_{N_1}$ with $1 \leqslant k_1 < \cdots < k_{N_1}$ such that $\phi(Q_{P_1}\tilde{\sigma}_1 u) \neq 0$, where

$$P_1 = \{(k_1 - 1)M + 1, (k_2 - 1)M + 1, \ldots, (k_{N_1} - 1)M + 1\} .$$

The transition from Ω to P_1 requires that we use not S_{N_0} itself but the group isomorphic to it acting on the indices that are congruent to 1 modulo M; P_1 is not the desired Ω_1, but a considerably larger set. The intermediate steps concerned with the choice of the k_j suggest how it can be diminished.

For the second step, we consider the subgroup isomorphic to S_{N_1} acting on the set

$$\{(k_1 - 1)M + 2, (k_2 - 1)M + 2, \ldots, (k_{N_1} - 1)M + 2\} ,$$

obtained by a shift of 1 to P_1. For ϕ we take the map $v \mapsto \phi(Q_{P_1}\tilde{\sigma}_1 v)$. By Lemma 4, there exists $\sigma_2 \in S_{N_1}$ and a set P_2 of cardinal $N_2 = \sqrt{N_1} = p^{2^{M-2}}$ corresponding to a choice of index-set $\{l_1, \ldots, l_{N_2}\} \subseteq \{k_1, k_2, \ldots, k_{N_1}\}$ such that $\phi(Q_{P_2}Q_{P_1}\tilde{\sigma}_1\tilde{\sigma}_2 u) \neq 0$, where

$$P_2 = \{(l_1 - 1)M + 2, (l_2 - 1)M + 2, \ldots, (l_{N_2} - 1)M + 2\} .$$

This process repeats by analogy with what has gone before; at the m-th step we arrive at the required index-set

$$\{i_1, \ldots, i_{p = N_M}\} \subseteq \cdots \subseteq \{l_1, \ldots, l_{N_2}\} \subseteq \{k_1, \ldots, k_{N_1}\} .$$

For the corresponding P_i of cardinal $N_i = p^{2^{M-i}}$, all of whose elements are congruent to i modulo M, we have

$$\phi(Q_{P_{M-1}}Q_{P_{M-1}} \cdots Q_{P_1}\tilde{\sigma}_1\tilde{\sigma}_2 \ldots \tilde{\sigma}_M u) \neq 0 .$$

In particular, the endomorphism Q_{P_M} corresponds to the set

$$\begin{aligned} \Omega_M &= \{(i_1 - 1)M + M, \ldots, (i_p - 1)M + M\} \\ &= \{i_1 M, \ldots, i_p M\} . \end{aligned}$$

It is natural in this context to set

$$\Omega_r = \{(i_1 - 1)M + r, \ldots, (i_p - 1)M + r\} .$$

Note that the Ω_i are pairwise disjoint, so that the corresponding endomorphisms Q_{Ω_i} permute in pairs; thus, by Lemma 3, the required inequality $\phi(Q_{\Omega_M} \cdots Q_{\Omega_1}\tilde{\sigma}u) \neq 0$ follows immediately from the similar inequality $\phi(Q_{P_M} \cdots Q_{P_1}\sigma_0 u) \neq 0$ just obtained. □

The rule for constructing the Ω_i can be made clearer by using the following picture. If the set $\{1, 2, \ldots, N\}$ is divided into N_0 collections of M numbers in a consecutive fashion, one considers the i_1-th collection, the i_2-th, and so on, up to the i_p-th:

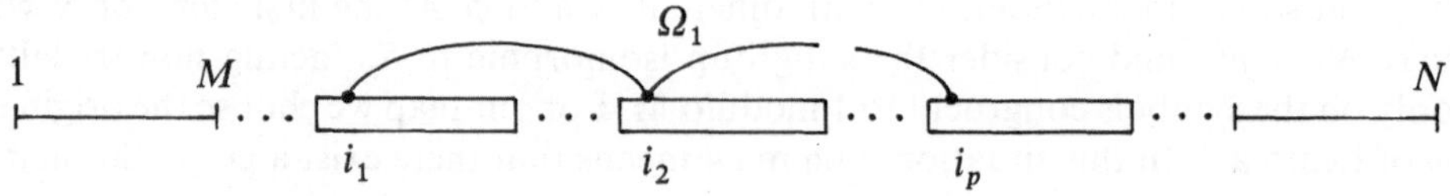

Then Ω_1 is the set of first members in the collections, Ω_2 the set of second members, *etc.*

We note the following fact. Under the action of elements of S_{Ω_i}, that is, under the action of permutations τ of the indices standing in the i-th positions of the distinguished collections of numbers just described, an element $v = Q_{\Omega_1} \ldots Q_{\Omega_M} \tilde{\sigma} u$ will either remain unchanged (if $Q_{\Omega_i} = H_{\Omega_i}$), or else be multiplied by sgn τ, since $\tilde{\tau} K_{\Omega_i} = (\operatorname{sgn} \tau) K_{\Omega_i}$. If S_N also acts on the image of ϕ, of course in a way that is consonant with the original action on u, the same assertion holds in the case $\phi(v) \neq 0$.

4.6. Conclusion of the Proof of Theorem B. The time has come to combine the information obtained in the preceding subsections. Thus, the natural numbers $p = d(q-1)+1$, $N = Mp^{2M}$, $N_0 = p^{2M}$ have the same meanings as in 4.4; they are used as they were in 4.5. The group S_N acts on A by permuting the generators; however, for the proper construction of ϕ, it is more convenient to replace the U in 4.5 by the tensor power $A^{\otimes N}$. The action is defined in the natural way: if $\sigma \in S_N$, then $\tilde{\sigma}(v_1 \otimes \cdots \otimes v_N) = v_{\sigma(1)} \otimes \cdots \otimes v_{\sigma(N)}$. We note simply that the action is defined by the position of the vectors v_i, not their suffices. For example, if $N = 3$ and $\sigma = (1, 2)$, then $\tilde{\sigma}(v_2 \otimes v_3 \otimes v_1) = v_3 \otimes v_2 \otimes v_1$, and not $v_1 \otimes v_3 \otimes v_2$ (naturally, since the action must not depend on the notation used for the elements).

We consider now the expression

$$\ldots g(x_1, \ldots, x_M) \ldots g(x_{N-M+1}, \ldots, x_N) \ldots,$$

assumed to be non-zero, and the mapping $\phi : A^{\otimes N} \to A$ corresponding to it, which we define by substitution:

$$\phi(v_1 \otimes \cdots \otimes v_N) = \ldots g(v_1, \ldots, v_M) \ldots g(v_{N-M+1}, \ldots, v_N) \ldots,$$

(all the other letters are fixed). If $u = x_1 \otimes \cdots \otimes x_N$, then $\phi(u) \neq 0$. In a case like this, Lemma 5 guarantees the existence of σ in S_N and sets $\Omega_1, \ldots, \Omega_M$ such that $\phi(Q_{\Omega_1} \ldots Q_{\Omega_M} \tilde{\sigma} u) \neq 0$. As always, all elements of Ω_i are congruent to i modulo M, and Ω_{i+1} is obtained from Ω_i by a shift (see the end of 4.5).

We introduce a small change in the notation for the generators of A, which is allowable since A is relatively free. Firstly, having chosen the notation so that $\tilde{\sigma} u = x_1 \otimes \cdots \otimes x_N$, we get rid of σ from the given inequality, and thus of $\tilde{\sigma}$: $\phi(Q_{\Omega_1} \ldots Q_{\Omega_M} u) \neq 0$. But ϕ is linear and S_N acts on A, so that

$$Q_{\Omega_1} \ldots Q_{\Omega_M}(\ldots g(x_1, \ldots, x_M) \ldots g(x_{N-M+1}, \ldots, x_N) \ldots) \neq 0 .$$

The endomorphisms Q_{Ω_i} are viewed here as linear combinations of automorphisms $\tilde{\sigma}$ corresponding to the permutations σ of the generating set. This also gives a slight nuance to $\tilde{\sigma}$, in that we have moved over from action on positions in $A^{\otimes N}$ to action on suffices in A. However, since action on suffices as applied to $x_1 \otimes \ldots \otimes x_N$ is the same as action on position, this interpretation is legal. To work with A and suffices right away would be more difficult, because of the need to ensure that the corresponding linear map ϕ is well-defined.

Let us go on with the argument. Among the N_0 occurrences in the given expression, only p pieces undergo the action of symmetrization or skew-symmetrization. It is reasonable to ignore the remaining places and introduce new notation for the generators in such a way that the first p numbers go to the given p occurrences. In other words, the new enumeration must reduce the left-hand side of the last inequality to the form

$$Q_{\Omega_1}\dots Q_{\Omega_M}(\dots g(x_1,\dots,x_M)\dots g(x_{M+1},\dots,x_{2M})\dots g(x_{(P-1)M+1},\dots,x_{PM})\dots,)\,,$$

where only the given p occurrences are distinguished. However, as well as the way they are written, the Ω_i themselves are changed, since the action takes place not on position, but on suffix:

$$\Omega_i = \{i, M+i, \dots, (p-1)M+i\}, \qquad 1 \leqslant i \leqslant M\,.$$

Thus

$$Q_{\Omega_1}\dots Q_{\Omega_M}(\dots g(x_1,\dots,x_M)\dots g(x_{(P-1)M+1},\dots,x_{PM})\dots) \neq 0(*)$$

Consider the set $\varepsilon = \{\varepsilon_1, \dots, \varepsilon_M\}$, where $\varepsilon_i = 0$ if $Q_{\Omega_i} = H_{\Omega_i}$ and $\varepsilon_i = 1$ if $Q_{\Omega_i} = K_{\Omega_i}$. This ε corresponds to a completely determined grading of A. By Lemma 2,

$$g(x_{(k-1)M+1},\dots,x_{kM}) = \sum_{j=1}^{d} f(y_{ij}^k,\dots,y_{mj}^k)\,W_j^k\,,$$

where all the summands are multilinear and $\|y_{ij}^k\| = 0$. By substituting these equations in (*), we deduce that

$$Q = Q_{\Omega_1}\dots Q_{\Omega_M}(\dots f(y_{1j_1}^1,\dots y_{mj_1}^1,)\,W_{j_1}^1,\dots f(y_{1j_p}^p,\dots,y_{mj_p}^p)\,W_{j_p}^p) \neq 0\,.$$

for some $j_1, \dots, j_p$ such that $1 \leqslant j_1, \dots, j_p \leqslant d$. Because of the choice of $p = d(q-1)+1$, there are q coincidences among the numbers $j_1, \dots, j_p$; $j_{\nu_1} = j_{\nu_2} \dots = j_{\nu_q} = j$, shall we say. By Lemma 2, the expressions

$$f(y_{ij}^{\nu_i},\dots,y_{mj}^{\nu_i})\,W_j^{\nu_i}, \qquad i = 1, 2, \dots, q\,,$$

are obtained from one another by shifts of the suffices on the generators by one and the same multiple of M.

We claim now that Q is symmetric in $y_{1j}^{\nu_1}, \dots, y_{1j}^{\nu_q}$. For example, let us show that $y_{1j}^{\nu_1}$ and $y_{1j}^{\nu_2}$ can be interchanged. Let τ be the permutation corresponding to a shift of the generators occurring in $y_{1j}^{\nu_1}$ to those occurring in $y_{1j}^{\nu_2}$. It can be expressed as a product of transpositions $\tau_k \in S_{\Omega_k}$ (τ_k rearranges the suffices congruent to k modulo M). Consider the corresponding automorphisms $\tilde{\tau}$ and $\tilde{\tau}_k$. On the one hand, action of $\tilde{\tau}$ is the interchange of $y_{1j}^{\nu_1}$ and $y_{2j}^{\nu_2}$ (by multilinearity: see Lemma 2, and Lemma 1 if necessary), while the remaining words and the multiplication operators W_j remain unchanged. On the other hand, $\tau Q = Q$, since $\tau_k Q_{\Omega_k} = (-1)^{\|x_k\|} Q_{\Omega_k}$ (see the end of 4.5), $\|y.:\| = 0$, and the operators Q_{Ω_k} commute. Therefore, the expression for Q is unaltered when $y_{1j}^{\nu_1}$ and $y_{1j}^{\nu_2}$ are interchanged. Symmetry of Q in all $y_{1j}^{\nu_1}, \dots, y_{1j}^{\nu_2}$ is proved in the same way; and similarly, Q is symmetric in $y_{kj}^{\nu_1}, \dots, y_{kj}^{\nu_q}$, for every $k = 1, 2, \dots, m$.

However, this is the desired contradiction, since by symmetry the multiple expression

$$\sum_{\sigma_1 \in S_q}\dots\sum_{\sigma_m \in S_q}(\dots f(y_{1j}^{\nu_{\sigma_1(1)}},\dots,y_{mj}^{\nu_{\sigma_m(1)}})\,W_j^{\nu_1}\dots f(y_{1j}^{\nu_{\sigma_m(q)}},\dots,y_{mj}^{\nu_{\sigma_m(q)}})\,W_j^{\nu_q}\dots)\,,$$

which is non-zero by construction, must in fact be zero, since it is obtained on linearizing the zero expression

$$\ldots f(z_1, \ldots, z_m) W_j^{\nu_1} \ldots f(z_1, \ldots, z_m) W_j^{\nu_q} \ldots . \quad \square$$

4.7. Proposition. *Let $L = L_0 + L_1$ be a $\mathbb{Z}_2$-graded n-Engel Lie algebra. If L_0 is nilpotent of class at most m, then L is nilpotent of class at most $M = c_n(m)$.*

Proof. Write $A(L_0)$ for the associative algebra generated by the set $\{\operatorname{ad} x | x \in L_0\}$. Nilpotency of L_0 implies that of $A(L_0)$ (in the associative sense: see Corollary 1.4.7). Thus $L_1(A(L_0))^c = 0$ for sufficiently large c. On the other hand,

$$L^{[1]} = (L^{[1]})_0 + (L^{[1]})_1 = ([L_0, L_0] + [L_1, L_1]) + [L_1, L_0] ,$$

and induction on k proves that $(L^{[k]})_1 \subseteq L_1(A(L_0))^k$. In fact,

$$(L^{(k+1)})_1 = [L^{[k]}, L^{[k]}]_1 = [(L^{[k]})_1, (L^{[k]})_0]$$
$$\subseteq [(L^{[k]})_1, L_0] \subseteq [L_1(A(L_0))^k, L_0] \subseteq L_1(A(L_0))^{k+1} .$$

This means that $L^{[c]} \subseteq L_0$, whence it follows that L is soluble. Nilpotency of L follows from Theorem 2.2. □

4.8. Proof of Theorem 4.1. Let $A = L$ be the Lie algebra constructed at the end of 4.2. The ideal of A generated by the element $f = [x_1 x_2 \ldots x_m]$ is nilpotent. The element f, the variety $\mathfrak{M}$ defined by A, and the commutator $[x_1 x_2 \ldots x_M]$, $M = c_n(m)$ (which is suitable to play the part of multilinear element g—just as in the proof of Proposition 4.7), can all be put into the statement of Theorem B. By that theorem, the verbal ideal $V_g(A)$ of A generated by g is nilpotent. The factor-algebra $A/V_g(A)$ is also nilpotent, of class at most M. Thus A is soluble, which is the contradiction promised at the end of 4.2. □

§ 5. Commentary

5.1. Theorem 1.1 was preceded by results of Bomshik Chang [41] and F. M. Malyshev [178], from which it followed that the nilpotency class c_p of the free 2-generator Lie algebra $L = \text{Lie}\langle x, y | (\text{ad } u)^{p-1} = 0 \rangle$ with E_{p-1} cannot be a linear function of p (in fact, for every $\alpha \in R$, $c_p > \alpha p$ for large enough p). Malyshev's bound related to the function $c(2, n)$, for arbitrary n. These results were proved by complicated combinatorial reasoning using the dual grading in L relative to the generators x and y. S. I. Adian and N. N. Repin [9] have established that there is an exponential lower bound, namely $c(2, n) > 2^{n/15}$; they did this by applying some arguments of Malyshev but ignoring the grading, by bringing all the homogeneous elements of the ideal $\mathscr{E} = \mathscr{E}_n(L)$ into the discussion. The ground field F in their theorem is quite arbitrary. While it is the case that, contrary to their assertion, the ideal $\mathscr{E}$ is not homogeneous for $|F| < n$, this is of no consequence: one has to go from $\mathscr{E}$ to a larger homogeneous ideal I_n containing all the components of $\mathscr{E}_{p+m}$.

The fairly subtle Malyshev-Adian-Repin analytical bounds are connected, after a finite calculation, with the non-triviality of the Witt formula (6.2).

In our proof of Theorem 1.1 (see [145]), we started from the fact that the Witt formula is best understood etymologically—*via* the generating function (6.10). The generating function method, the effectiveness of which has been confirmed in the most diverse branches of mathematics, has allowed us to manage with extremely rough arguments. Formal power series and relation (6.12) introduce the totality of the numbers α_m reflecting bounds for ϕ_{p+m} into the discussion all at once. A more exact result has thus been achieved. How much it can be strengthened (yes, and whether it needs to be?) is not clear at present; however, the question of an explicit upper bound for c_p is now even more complex than it was thirty years ago (see Appendix 1 in this connection).

5.2. Theorem 1.1 refutes a strange conjecture to be found in the survey article [142], which arose under the influence of relation (1.9), apparently. The conjecture that the associative enveloping algebra A of a Lie algebra with E_n, n arbitrary, over a field of characteristic zero satisfies the (weak) identity

$$u_1^{n-1}u_2^{n-1}\dots u_{n-1}^{n-1}=0$$

would be more than answered by the contents and results of the first two sections. According to results of Higgins [106] (Theorem 2.2 and Corollary 2.5), this would give that L is globally nilpotent, with effective upper bounds for the class. Regrettably, almost no approaches to this conjecture exist. Even the case $n = 4$ is quite non-trivial: the interesting arguments of A. A. Zolotykh [288] strengthening the Higgins calculations are employed in Example 2.6. Some of the material of § 2 can be found in [26], including Example 2.1, which is due to Cohn [43], although we have preferred an independent treatment in this case.

5.3. The same applies to the fundamentally important Theorem 3.1 (see [26], Chap. 8, § 8); the proof given here is from a manuscript of Razmyslov, which was courteously placed at the author's disposal. The original 1971 proof (see [220]), as well as Razmyslov's remarkable result [221] about the insolubility of the variety of groups defined by the identical relation $x^4 = 1$, are at the root of the method he developed, which is of very great generality (see [224]). An example of an insoluble group of exponent 5 was given in 1970 (see [21]); a convenient account is to be found in [22].

Assertions like those in Theorems 3.1 and 3.8 point up the complexity of Lie algebras with E_{p-1} and of groups satisfying the identical relaxation $x^p = 1$. The problem of how $c(d, p-1)$ grows for fixed p arose as long ago as the fifties. If Theorem 3.1 has been proved in 1952, it would perhaps have detracted from the desire to engage in the search for a proof of Theorem 1.7.4.

5.4. Theorem 4.1, which is due to Zel'manov (see [284]), answers an old and very natural question in the theory of Lie algebras, one which touches directly on a similar question in group theory: must a torsion-free group be nilpotent if it satisfies the n-th Engel condition $x^{(n)}\circ y = 1$? Here $x, y \in G$, $x^{(1)}\circ y = x\circ y$

$= x^{-1}y^{-1}xy, x^{(k+1)} \circ y = x \circ (x^{(k)} \circ y)$. For $n = 2$, the positive answer is fairly easily obtained; however, for $n = 3$ it is already a serious problem, and it was solved by Heineken (see [99] and the literature there cited). More exactly, he proved that a group G satisfying the third Engel condition is nilpotent of class 4 if G does not contain elements of order 2 or 5. In these last two cases the group is merely locally nilpotent (see Example 2.9!), so that the torsion freeness condition (or rather, the absence of elements whose orders are small in comparison with n) is essential. For $n \geqslant 4$, the question of the nilpotency of G remains open; however, Theorem 4.1 about Lie algebras increases its chances of having a positive solution. We recall that the Engel condition $x^{n(x, y)} \circ y = 1$, with $n(x, y)$ depending on the elements x and y of G, does not even guarantee local nilpotency—the achievement of Golod [55] has not yet been outshone for elegance.

It must be stated that in the summer of 1984, when the plan of the book was constructed, Theorem 4.1 figured as a serious problem, thrown out as a challenge to mathematicians with a combinatorial cast of thought. A very little earlier, Zel'manov had remarked (in the course of some lectures that he was delivering) on the importance of Theorem 1 in the survey [142], which was given there without proof and is essentially equivalent to Theorem 1.7.3. The transfinite construction in § 4.2 was realized precisely as a consequence of the assertion. It served as the starting-point for the later brilliant arguments of Zel'manov. A first sketch of these was made on the 18th of December 1984 in the school for Lie algebras and their applications in mathematics and physics. Our treatment is somewhat different in detail, and rather too short in § 4.2, so that it is appropriate to supplement it by a few detailed remarks.

5.5. Everything relating to Theorem 4.1 has been written in fresh symbols. Zel'manov simplified the proof considerably towards the end of 1986; we have reproduced here this new proof, following the exposition delivered by Ufnarovskij to the Third School on Lie algebras in December 1987. The principal feature of the new approach is a much more active use of graded $\mathbb{Z}_2$-algebras. It is a response to the tendency now appearing to rely on the philosophy of superalgebras in the theories of very different varieties of linear algebras: associative, alternative, Lie, Jordan. The method enables one to avoid long calculations.

5.6. An endomorphism ϕ of the associative enveloping algebra A of a Lie algebra L is said to be an *endomorphism of the pair* (L, A) if $\phi(L) \subseteq L$. An ideal I of A is a *verbal ideal of the pair* (L, A) if I is invariant under all endomorphisms of the pair (L, A). The pair $(L(X), A(X))$ constructed in § 4 is *relatively free* in the sense that $L(X)$ is generated by X and every mapping $X \to L$ extends to an endomorphism of the pair $(L(X), A(X))$. This follows from the arguments developed below, in which $(L_0(X), A_0(X))$ is any relatively free pair (in particular, absolutely free).

5.7. Lemma. *Let* $(L_0(X), A_0(X))$ *be a relatively free pair, I an ideal of* $A_0(X)$ *and* $A_0(X) \to \bar{A}_0 = A_0(X)/I$ *the natural homomorphism. The following assertions are equivalent*:

(i) *I is a verbal ideal of the pair* $(L_0(X), A_0(X))$;

(ii) *the pair* $(\bar{L}_0, \bar{A}_0)$ *is relatively free on the free generating set* $\bar{X}$ *(here* $\bar{L}_0 = (L_0(X) + I)/I$).

Proof. (i) ⇒ (ii). For simplicity we set $L_0 = L_0(X)$, $A_0 = A_0(X)$, and we consider any map $\bar{\phi}: \bar{x}_i \mapsto \bar{a}_i$, where $x_i \in X$, $a_i \in L_0$ for $i \in \mathbb{N}$. By assumption, the map $\phi: x_i \mapsto a_i$, $i \in N$, extends to an endomorphism $\tilde{\phi}$ of the pair (L_0, A_0). Since I is a verbal ideal, we have $\tilde{\phi}(I) \subseteq I$. This means that $\tilde{\phi}$ induces an endomorphism $\bar{\tilde{\phi}}$ of the pair $(\bar{L}_0, \bar{A}_0)$. It is easy to see that $\bar{\tilde{\phi}}$ extends $\bar{\phi}$.

(ii) ⇒ (i). Assume that $\bar{X} = \{\bar{x}_i | i \in \mathbb{N}\}$ is a free generating set for the pair $(\bar{L}_0, \bar{A}_0)$. We have to show that $\phi(I) \subseteq I$. Let f be a non-commutative polynomial in X such that $f(x_i) \in I$, that is, $f(\bar{x}_i) = 0$. The map $\bar{x}_i \mapsto \overline{\phi(x_i)}$, $i \in \mathbb{N}$, extends to an endomorphism of the pair (L_0, A_0). Therefore

$$f(\overline{\phi(x_i)}) = \overline{f(\phi(x_i))} = \overline{\phi(f(x_i))} = 0 ,$$

so that $\phi(f(x_i)) \in I$. □

5.8. Lemma. *Let* $(L_0(X), A_0(X))$ *be a relatively free pair with generating set* X, *and let* M *be an abelian ideal of* $L_0(X)$. *Then, for every endomorphism* ϕ *of the pair* $(L_0(X), A_0(X))$, $\phi(M)$ *generates an abelian ideal in* $L_0(X)$.

Proof. Let a, b be any elements of M. Our problem is to prove that the relation

$$g_0 \cdot \operatorname{ad} \phi(a) \cdot \operatorname{ad} g_1 \cdot \ldots \cdot \operatorname{ad} g_m \cdot \operatorname{ad} \phi(b) = 0$$

holds for all $g_0, g_1, \ldots, g_m \in L_0(X)$. The elements a, b, $\phi(a)$, $\phi(b)$ can be expressed as linear combinations of commutators in X. Suppose that only the generators $x_1, \ldots, x_{d-1} \in X$ occur in these expressions. We consider the map

$$\psi: \begin{cases} x_i \mapsto \phi(x_i) & \text{for} \quad 1 \leqslant i \leqslant d-1 , \\ x_i \mapsto g_{i-d} & \text{for} \quad d \leqslant i \leqslant m+d , \\ x_i \mapsto x_i & \text{for} \quad m+d < i \end{cases}$$

and extend it to an endomorphism $\tilde{\psi}$ of the pair $(L_0(X), A_0(X))$. Application of $\tilde{\psi}$ to the equation

$$x_d \operatorname{ad} a(x_1, \ldots, x_{d-1}) \cdot \operatorname{ad} x_{d+1} \cdot \ldots \cdot \operatorname{ad} x_{d+m} \cdot \operatorname{ad} b(x_1, \ldots, x_{d-1}) = 0$$

gives the required relation. □

5.9. Corollary. *The sum* $I_{ab}(L_0)$ *of all the abelian ideals of the algebra* $L_0 = L_0(X)$ *is invariant under all the endomorphisms of the pair* (L_0, A_0), *and the ideal* $I_{A_0}(I_{ab}(L_0))$ *of* A_0 *generated by* $I_{ab}(L_0)$ *is verbal.*

5.10. Lemma. $(I_{A_0}(M))^n = 0$ *for every abelian ideal* M *of* L_0 *if the pair* (L_0, A_0) *satisfies the weak identity* $x^n = 0$.

Proof. Use the reasoning of 2) in 5.1.2. □

5.11. The Ideals $\mathbf{I}_\alpha$ in 4.2 are Verbal. Starting from the free pair $(\mathscr{L}(X), \mathfrak{A}(X))$ and the verbal ideal $V = I_{\mathfrak{A}(X)}([xy^n], x^n|x, y \in \mathscr{L}(X))$, we get from Lemma 5.7 a relatively free pair $(L(X), A(X))$. Set $I_0 = 0$. Since a union of verbal ideals is a verbal ideal, the ideal $I_\alpha = \bigcup_{\beta<\alpha} I_\beta$ is verbal when α is a limit ordinal. If α is not a limit ordinal, and the verbal ideal $I_{\alpha-1}$ is already defined, Lemma 5.7 as applied to the homomorphism $\pi: A(X) \to \bar{A}_\alpha = A(X)/I_{\alpha-1}$ gives that the pair $(\bar{L}_\alpha, \bar{A}_\alpha)$ is relatively free. Corollary 5.9 shows that $I_{\bar{A}_\alpha}(I_{ab}(\bar{L}_\alpha))$ is verbal. By Lemma 5.7 once more, this means that the pair obtained from $(\bar{L}, \bar{A})$ on factoring by $I_{\bar{A}_\alpha}(I_{ab}(\bar{L}_\alpha))$ is also relatively free. Let I_α be the inverse image of $I_{\bar{A}_\alpha}(I_{ab}(\bar{L}_\alpha))$ under π. Since the pair $((L(X) + I_\alpha)/I_\alpha, A(X)/I_\alpha)$ is relatively free, it follows from Lemma 5.7 that I_α is verbal in $(L(X), A(X))$.

5.12. By Lemma 5.10, $I_{\bar{A}}(I_{ab}(\bar{L}))$ is the sum of nilpotent ideals of $\bar{A}$. In that case, every finite set of elements of I_α generates a nilpotent ideal of $A(X)$ modulo $I_{\alpha-1}$.

In the (assumed) non-nilpotent algebra $A(X) = \bigcup_\alpha I_\alpha$, we choose the minimal ordinal α for which the factor-algebra $A(X)/I_\alpha$ is nilpotent. Then α is not a limit ordinal. For, we have $x_1 \ldots x_m \in I_\alpha$ for the free generators $x_i \in X$, where m is some natural number. Let β be the smallest ordinal such that $x_1 \ldots x_m \in I_\beta$. Clearly, β is not a limit ordinal, and $\beta \leqslant \alpha$. Since I_β is verbal, it now follows that $A(X)^m \subseteq I_\beta$, that is, $\beta = \alpha$.

5.13. To check that the definition $\mathscr{K}_G(f) = \sum \alpha_v \mathscr{K}_G(v)$ from 4.5 is really a definition, we need to check: whenever $\sum w_i = 0$, where the w_i are monomials, every skew-symmetrization of this expression is also zero.

With this aim, we call the word $y_{j_1} \ldots y_{j_r}$ the *type* of the monomial $v_0 y_{j_1} v_1 \ldots y_{j_r} v_r$, and we group the monomials w_i according to types. It follows easily from the definition of free product that every isotypic component is zero. Thus, without loss of generality, we may assume straightaway that all the monomials w_i have one and the same type $y_{j_1} \ldots y_{j_r}$, that is,

$$w_i = v_{i0} y_{j_1} v_{i_1} y_{j_2} \ldots y_{j_r} v_{i_r}, \qquad v_{i_k} \in A \cup 1 .$$

From the properties of free products, it also follows that

$$\sum_i v_{i0} \otimes v_{i_1} \otimes \cdots \otimes v_{i_r} = 0 .$$

For arbitrary $a_1, \ldots, a_r \in A$, we now have

$$\sum_i v_{i0} a_1 v_{i_1} \ldots a_r v_{i_r} = 0 ,$$

from which it follows that $\sum \mathscr{K}_G(w_i) = 0$.

5.14. On the Proof of Lemma 4 in 4.5. We dwell in detail on the concluding step. Assume that the Young tableau D contains a row $D_{(i)} = (i_1, \ldots, i_r)$ of length

$r \geqslant [\sqrt{l}]$. We consider the subgroup $G_R \subseteq S_l$ consisting of all permutations which leave all the points in $\{1, \ldots, l\} \backslash \{i_1, \ldots, i_r\}$ fixed. Let $\sigma_1, \ldots, \sigma_k$ be a right transversal of $R(D)$ mod G_R. Then $e_D = \sum_i \sum_{\substack{p \in G_R \\ q \in C(D)}} \varepsilon_p p \sigma_i q$, so that there exist permutations $q \in C(D)$, σ_i such that

$$\Phi\left(\sum_{p \in G_R} p\sigma_i q v_1 \otimes \cdots \otimes v_l\right) \neq 0 .$$

However, the expression $\sum_{p \in G_R} p \sigma_i q v_1 \otimes \cdots \otimes v_l$ is symmetric in $v_{i_1}, \ldots, v_{i_r}$.

We assume now that D contains a column $D_{(i)} = {}^t(i_1, \ldots, i_r)$, and that $\sigma_1, \ldots, \sigma_k$ form a right transversal of $C(D)$ mod G_C. Then, for some $p \in R(D)$ and σ_i we have

$$\Phi\left(\sum_{q \in G_C} \varepsilon_q p q \sigma_i v_1 \otimes \cdots \otimes v_l\right) \neq 0 .$$

The expression $\sum_{q \in G_C} \varepsilon_q p q p^{-1} p \sigma_i v_1 \otimes \cdots \otimes v_l$ is skew-symmetric in $v_{p(i_1)}, \ldots, v_{p(i_r)}$.

5.15. Theorem 4.1 is a further, and perhaps more convincing, argument in favour of the "extended" solution of RBP, as was done in 1958 *via* Theorem 1.7.1.

The traditional remark on the desirability of a constructive proof of Theorem 4.1 remains—alas!—outside the limits of reasonable opinion: such a proof is still out of reach even for comparatively small values of n. Formula (6.36) can be written easily in explicit form for $n = 2$, when $c_2 = 2$:

$$[xyz] = [x(yz + zy)] + \tfrac{1}{3}[y(xz + zx)] + \tfrac{2}{3}[z(xy + yx)] .$$

one has to consider the longer product $[xyzt]$). For $n = 3$ and $c_3 = 5$, formula (6.36) would take a menacingly complicated form, while for $n = 4$ and $c_4 = 29$ (? See [99]), we would hit what is essentially the general solution, where it is necessary to seek an intrinsic motivation for the shape of the desired formula. In these circumstances, would it not be better to revert to the conjecture in 5.2?

In connection with this problem, it is perhaps relevant to recall the almost idyllic picture in the theory of associative nil-algebras: the efforts of Nagata [194], Higman [107], Kuz'min [154] and Razmyslov [223] have shown that

$$\frac{n(n+1)}{2} \leqslant f(n) \leqslant n^2 ,$$

where $f(n)$ is the nilpotency class of the free countably generated associative algebra satisfying the identity $x^n = 0$ over a field of characteristic zero. Kuz'min has conjectured that $f(n)$ is in fact $\dfrac{n(n+1)}{2}$. Theorem 4.1 gives the nilpotency of every associative algebra A over $\mathbb{Q}$ with countable generating set X that satisfies the weak identity $u^n = 0$ (this for all $u \in L(X)$). What can be said about the nilpotency class of such an algebra A?

5.16. *Remark.* According to L. A. Bokut', E. Formanek recalled a striking fact during a lecture at a conference in Oberwolfach in May 1986. It appears that there were precursors to Nagata [194] and Higman [107]: Yu. S. Dubnov and V. K. Ivanov [47] obtained the same results in different language. Their wartime paper went unnoticed.

Chapter 7

Finite p-Groups and Lie Algebras

The character of this chapter is almost entirely that of a survey, and in it we shall outline the shape of the linear methods in finite group theory that have a direct bearing on our principal theme, the restricted Burnside problem. The only subject treated in any detail in M. R. Vaughan-Lee's elegant approach to describing the multilinear identities of the Lie algebra $L(B(d, p))$ associated with the Burnside group $B(d, p)$. In particular, it is proved by elementary methods (see Corollary 2.3 in § 3) that $L(B(d, p))$ satisfies E_{p-1}. The solution of RBP provided here is therefore independent of other sources.

§ 1. Fundamental Relations Between Groups and Lie Algebras

1.1. We shall not touch on all aspects of the connection between the structure of groups and of Lie algebras. We note simply that at the root of the matter there lie the formal identities forming a bridge between the multiplication and commutation in groups on the one hand, and addition and multiplication in Lie algebras on the other.

We agree at once that a Lie algebra in this section is understood to be a Lie ring with a commutative ring K as domain of operators. In other words, a Lie K-algebra is a K-module L equipped with a bilinear operation $[x, y]$ of multiplication with the usual properties (see § 2 of Chap. 1). As a rule, our K will be the ring of integers $\mathbb{Z}$ or the residue class ring $\mathbb{Z}_q$, $q = p^\alpha$. The concepts of subalgebra $N \subset L$ and ideal $N \lhd L$ have the same sense as previously, except that N has to be a K-module. As usual,

$$L = L^1 \supset L^2 \supset L^3 \supset \cdots$$

is the lower central series and

$$L = L^{[0]} \supset L^{[1]} \supset L^{[2]} \supset \cdots$$

the derived series of L. Finally, a Lie K-algebra L is said to be *graded* if it is graded as a K-module, that is, when

$$L = \bigoplus_{i=1}^{\infty} L_i, \qquad [L_i, L_j] \subseteq L_{i+j}.$$

In a multiplicatively written group G, the symbol (x, y) means the *commutator* of two elements $x, y \in G$: $(x, y) = x^{-1}y^{-1}xy$. Further, if $x^y = y^{-1}xy$ is the conjugate of x by y (the result of applying the inner automorphism induced by y to x), then obviously

$$x^y = x(x, y), \qquad xy = yx^y = yx(x, y)\,.$$

Also,

$$(xy, z) = (x, z)^y(y, z), \qquad (x, yz) = (x, z)(x, y)^z\,. \tag{7.1}$$

Finally, the following identity discovered by P. Hall holds in groups:

$$((x, y), z^x)((z, x), y^z)((y, z), x^y) = 1\,; \tag{7.2}$$

it is remarkably reminiscent of the Jacobi identity in Lie algebras. The same idea is expressed in the Witt identity:

$$((x, y^{-1}), z)^y((y, z^{-1}), x)^z((z, x^{-1}), y)^x = 1\,. \tag{7.3}$$

A *central series* in a group G is any decreasing sequence of subgroups $G = H_1 \supset H_2 \supset H_3 \supset \cdots$ such that $(H_i, H_j) \subset H_{i+j}$ for all $i, j \geqslant 1$. Here, (H_i, H_j) is the *mutual commutator subgroup*, that generated by the commutators of all elements $x \in H_i$, $y \in H_j$. The most important example is the lower central series, with terms $\gamma_1(G) = G$, $\gamma_i(G) = (\gamma_{i-1}(G), G)$. The factor-groups H_i/H_{i+1} of a central series are abelian, since $(H_i, H_i) \subseteq H_{2i} \subset H_{i+1}$. Following Lazard [155], we write the group H_i/H_{i+1} additively and denote an element of H_i/H_{i+1} by $\tilde{x}_i$, that is, the coset mod H_{i+1} of the element x_i of H_i. Clearly, $\widetilde{x_i x_i'} = \tilde{x}_i + \tilde{x}_i'$.

We form the direct sum $L(G; (H_i)) = \bigoplus_{i \geqslant 1}(H_i/H_{i+1})$, which is abelian group, and introduce a grading by agreeing that elements of H_i/H_{i+1} are homogeneous of degree i. We set

$$[\tilde{x}_i, \tilde{x}_j] = (x_i, x_j)\,,$$

and then extend the definition of the bracket multiplication

$$[\tilde{x}, \tilde{y}] = \sum_{s,t} [\tilde{x}_s, \tilde{y}_t]$$

by linearity to all pairs of elements $\tilde{x} = \sum \tilde{x}_s$, $\tilde{y} = \sum \tilde{y}_t$ of $L(G; (H_i))$. It follows immediately from identities (7.1)–(7.3) that $[\tilde{x}, \tilde{y}]$ is a bilinear operation giving $L(G; (H_i))$ the structure of a Lie algebra with operator domain K depending on the original group G and the choice of the central series (H_i). By varying (H_i), we can hope to get varying information about G, assuming of course, that the necessary condition $\bigcap_i H_i = 1$ is satisfied (that is, G is nilpotent, or at least ω-nilpotent). If this is so, then Card G = Card $L(G; (H_i))$.

In what follows, we shall consider only the lower central series $(\gamma_i(G))$ of G, and the symbol $L(G) = \oplus L_i$, $[L_i, L_j] \subset L_{i+j}$, will denote the corresponding graded Lie algebra. It is the existence of the richer and more easily-studied structure on $L(G)$ that enables one, in principle at least, to simplify the solution of certain group-theoretical questions. Some good examples can be found in the monograph [116]. We are interested here only in the one line, that relating to RBP for prime-power exponent $q = p^\alpha$. To that end, we consider the group

$$G = B_0(d, q) = B(d, q) \Big/ \Big(\bigcap_i \gamma_i(B(d, q)) \Big) ,$$

where $B(d, q) = \mathscr{F} / \mathscr{F}^q$ is the free Burnside group ($\mathscr{F}$ is the absolutely free group on d generators, and $\mathscr{F}^q = \langle f^q | f \in \mathscr{F} \rangle$). Whenever Card $G < \infty$, G is the signal that RBP has a positive solution for exponent q; all the information on the Burnside group $B(d, q)$ ignored by Lie algebras is "hidden" in the intersection $\bigcap_i \gamma_i (B(d, q))$. In every case, G is ω-nilpotent (that is, $\bigcap_i \gamma_i(G) = 1$), d-generator and of exponent q. In its turn, the associated Lie algebra $L = L(G)$ is d-generator and has operator domain $K = \mathbb{Z}_q$, so that $qu = 0$ for all $u \in L$. But this is merely an elementary and very superficial corollary of the identical relation $x^q = 1$.

Using arguments like the "commutator collecting process" of P. Hall (see [83, 115]), one can get new and more interesting relations. One of these, mentioned by Higman in [111], has the form of an identical relation "mod p" in $L(G)$:

$$\sum_{\sigma \in S_{q-1}} [u_0 u_{\sigma 1} u_{\sigma 2} \ldots u_{\sigma(q-1)}] \equiv 0 (\bmod p) . \tag{7.4}$$

For $q = p$ (that is, $\alpha = 1$), so that $pL(G) = 0$, we get the identical relation

$$\sum_{\sigma \in S_{p-1}} [u_0 u_{\sigma 1} u_{\sigma 2} \ldots u_{\sigma(p-1)}] = 0 ,$$

which is equivalent to the Engel condition E_{p-1}: $[uv^{p-1}] = 0$. Other routes have lead to this identity: the Zassenhaus group-theoretical identity

$$(\ldots((x, \underbrace{y), y), \ldots, y)}_{p-1} \equiv 1 (\bmod \gamma_{p+1}(G))$$

valid in a group G of prime exponent p (see [83, 115]), and the approaches of Magnus [171–175], Sanov [231, 232] and Higman [108] are among them. We shall see in § 3 that E_{p-1} (which is our alpha and omega in investigating RBP for prime exponent) arises as the very start of an infinite chain of relations, whose derivation does not require any further apparatus. Moreover, for many other purposes it is useful to have available one of the realizations of a free group in a ring of formal power series.

1.2. Let $\mathfrak{A} = \mathfrak{A}(X)$ be the free associative $\mathbb{Q}$-algebra on the generating set $X = \{x_1, \ldots, x_d\}$, and $\mathscr{L} = \mathscr{L}(X)$ the free Lie $\mathbb{Z}$-algebra generated in $\mathfrak{A}$ by the same generators under the bracket operation $[u, v] = uv - vu$. Let $\mathfrak{A}_r$ be the $\mathbb{Z}$-module freely spanned by the monomials $x_{i_1} x_{i_2} \ldots x_{i_r}$ of degree r, and set $\mathscr{L}_r = \mathfrak{A}_r \cap \mathscr{L}$. Then

$$\mathfrak{A} = \bigoplus_{r=0}^{\infty} \mathfrak{A}_r, \qquad \mathfrak{A}_r \mathfrak{A}_s \subseteq \mathfrak{A}_{r+s} \qquad (r, s \geqslant 0) ,$$

$$\mathscr{L} = \bigoplus_{i=1}^{\infty} \mathscr{L}_r, \qquad [\mathscr{L}_r, \mathscr{L}_s] \subseteq \mathscr{L}_{r+s} \qquad (r, s \geqslant 1) ,$$

so that $\mathfrak{A}$ and $\mathscr{L}$ are graded algebras.

By the completed associative algebra $\hat{\mathfrak{A}} = \prod_{r=0}^{\infty} \mathfrak{A}_r$ we mean the algebra of formal power series

$$f = f_0 + f_1 + f_2 + \cdots \qquad (f_r \in \mathfrak{A}_r) .$$

$\hat{\mathfrak{A}}$ contains the Lie $\mathbb{Z}$-algebra $\hat{\mathscr{L}}$ of formal Lie power series $f = \sum f_r$ with $f_r \in \mathscr{L}_r$.

For every power series $u \in \hat{\mathfrak{A}}$ without free term we define the exponential

$$\mathrm{e}(u) = \exp u = \sum_{i=0}^{\infty} \frac{u^i}{i!} .$$

By the Campbell-Hausdorff formula (see [26, 31]), we have for all $x, y \in \hat{\mathscr{L}}$:

$$\mathrm{e}(x)\mathrm{e}(y) = \mathrm{e}(z) ,$$

where

$$z = z(x, y) = x + y + \frac{1}{2}[x, y] + \cdots$$

is an element of $\hat{\mathscr{L}}$.

The group of invertible elements of $\hat{\mathfrak{A}}$ contains the subgroup $\mathscr{F}$ generated by the exponentials $\mathrm{e}(x_1), \ldots, \mathrm{e}(x_d)$. As is well known (see [175], for example), $\mathscr{F}$ is a free group on the free generators $\mathrm{e}(x_1), \ldots, \mathrm{e}(x_d)$. Every element w of $\gamma_n(\mathscr{F})$ not contained in $\gamma_{n+1}(\mathscr{F})$ is of the form $\mathrm{e}(u)$, where $u = a_n + a_{n+1} + \cdots$ and $0 \neq a_n \in \mathscr{L}_n$, $a_m \in \mathbb{Q}\mathscr{L}_m$ for $m > n$. Setting $\mathrm{gr}(w) = a_n$ (and $\mathrm{gr}(w) = 0$ if $w = 1$), we define a map

$$\gamma_{n+1}(\mathscr{F}) \mapsto 0, \qquad f\gamma_{n+1}(\mathscr{F}) \mapsto \mathrm{gr}(f) \quad \text{for} \quad f \in \gamma_n(\mathscr{F}) \backslash \gamma_{n+1}(\mathscr{F}) ,$$

which, by the Magnus-Witt theorem, is an isomorphism from $\gamma_n(\mathscr{F})/\gamma_{n+1}(\mathscr{F})$ to $\mathscr{L}_n$, $n = 1, 2, \ldots$ If $0 \neq a = \mathrm{gr}(f) \in \mathscr{L}_n$, $0 \neq b = \mathrm{gr}(h) \in \mathscr{L}_m$, then $[a, b] \in \mathscr{L}_{n+m}$. Moreover, every element $[a, b] \neq 0$ is of the form $[a, b] = \mathrm{gr}((f, h))$, and $(f, h) \in \gamma_{n+m+1}(\mathscr{F})$ if $[a, b] = 0$. Therefore, $\mathscr{L}$ is the Lie $\mathbb{Z}$-algebra associated with $\mathscr{F}$ in the sense of 1.1.

The Lie algebra $L = L(B_0(d, p)) = L(B(d, p))$ associated with $B_0(d, p)$ (see 1.1) and with the free Burnside group $B(d, p) = \mathscr{F}/\mathscr{F}^p$ can be written as $\mathscr{L}/J$, where J is some homogeneous ideal of the free Lie $\mathbb{Z}$-algebra $\mathscr{L}$. This means that $J = \bigoplus_{n=1}^{\infty} J_n$, where $J_n = J \cap \mathscr{L}_n$. Since

$$\gamma_n(\mathscr{F}/\mathscr{F}^p) = \gamma_n(\mathscr{F})\mathscr{F}^p/\mathscr{F}^p \cong \gamma_n(\mathscr{F})/(\gamma_n(\mathscr{F}) \cap \mathscr{F}^p) ,$$

$J_n = \mathscr{L}_n \cap \{\mathrm{gr}(f) | f \in \mathscr{F}^p\}$ by definition. We know that $p\mathscr{L} \subseteq J$. The next lemma gives some preliminary information on other elements contained in J.

1.3. Lemma. *Let $T_n(x_1, \ldots, x_n)$ be the homogeneous component of the polynomial*

$$((1 + x_1)(1 + x_2) \ldots (1 + x_n))^p ,$$

which has degree 1 in each of the variables $x_1, x_2, \ldots, x_n$ $(n = 1, 2, \ldots)$. Then every multilinear element of J is contained in the subalgebra of $\mathfrak{A}$ generated by elements of the form $T_n(y_1, \ldots, y_n)$, where $y_i \in \{x_1, \ldots, x_d\}$ for $1 \leqslant i \leqslant n$.

Proof. The free group $\mathscr{F}$ consists of elements $e(\alpha_1 y_1)e(\alpha_2 y_2)\dots e(\alpha_k y_k)$, where $\alpha_i \in \mathbb{Z}$, $y_i \in \{x_1, \dots, x_d\}$ for $1 \leqslant i \leqslant k$. Thus $\mathscr{F}^p$ is generated by the elements

$$(e(\alpha_1 y_1)e(\alpha_2 y_2)\dots e(\alpha_k y_k))^p$$

$$= 1 + \sum \alpha_1^{n_1} \dots \alpha_k^{n_k} \frac{y_1^{n_{11}} \dots y_k^{n_{k1}} \dots y_1^{n_{1p}} \dots y_k^{n_{kp}}}{n_{11}! \dots n_{k1}! \dots n_{1p}! \dots n_{kp}!}$$

$$= 1 + \sum \frac{\alpha_1^{n_1} \dots \alpha_k^{n_k}}{n_1! \dots n_k!} \binom{n_1}{n_{11}, \dots, n_{1k}} \cdots \binom{n_p}{n_{1p}, \dots, n_{kp}}$$

$$\times y_1^{n_{11}} \dots y_k^{n_{k1}} \dots y_1^{n_{1p}} \dots y_k^{n_{kp}},$$

where the summation runs over all decompositions $n_i = \sum n_{ij}$ with $\sum n_i > 0$. Redistributing terms and using the definition of T_n, we get an expression for the right-hand side in the form

$$1 + \sum \frac{\alpha_1^{n_1} \dots \alpha_k^{n_k}}{n_1! \dots n_k!} T_{n_1 + \dots + n_k}(\underbrace{y_1, \dots, y_1}_{n_1}, \underbrace{y_2, \dots, y_2}_{n_2}, \dots, \underbrace{y_k, \dots, y_k}_{n_k}).$$

If now $w \in \mathscr{F}^p$, then $w = 1 + a$, where a is an infinite sum of products of elements like

$$\frac{1}{n_1! \dots n_k!} T_{n_1 + \dots + n_k}(y_1, \dots, y_1, \dots, y_k, \dots, y_k).$$

It is clear that $gr(w)$ is expressible as a finite sum, and if $gr(w)$ is a multilinear element, the coefficient $1/n_i!$ occurring in its expression must be 1. Since every multilinear element of J is of the form $gr(w)$ for some $w \in \mathscr{F}^p$, the proof of the lemma is complete. □

We shall use this lemma later; right now we shall exhibit a panorama that has been in slow development over a period of forty or fifty years. The most succulent colours have been added very recently indeed.

§ 2. The Ideal of Relations (a General Survey)

2.1. Set $G = B_0(d, p)$. As we saw in 1.2, $L = L(G) = \mathscr{L}/J$, where $\mathscr{L} = \bigoplus_n \mathscr{L}_n$ is the free Lie $\mathbb{Z}$-algebra (free Lie ring) on d generators $x_1, \dots, x_d$, and $J = \bigoplus_{n=1}^{\infty} J_n$ is the graded *ideal of relations*. Since $L_n = \mathscr{L}_n / J_n$, the problem comes to a description of the homogeneous components J_n. Witt's formula

$$\psi_n(d) = \frac{1}{n} \sum_{\delta | n} \mu\left(\frac{n}{\delta}\right) d^\delta$$

gives an expression for the dimension $\psi_n(d)$ of the homogeneous component L_n of L. If we calculate the dimension $\phi_n(d)$ of the component J_n of J, we get

$$\dim L_n = \psi_n(d) - \phi_n(d)$$

which is the rank of $\gamma_n(G)/\gamma_{n+1}(G)$.

We make great use of the fact that $J \supset p\mathscr{L} + \mathscr{E}$, where $\mathscr{E} = E_{p-1}(\mathscr{L})$ is the ideal considered in § 3 of Chap. 2; it had already arisen in pre-war days in the study of the correspondence between groups and Lie algebras. Sanov [232] proved that

$$J_n = (p\mathscr{L} + \mathscr{E})_n, \qquad 1 \leqslant n \leqslant 2p-2\,.$$

The homogeneous component $\mathscr{E}_n$ (do not confuse this with the Engel condition E_n!) is spanned by the elements

$$\langle u_0, u_1, \ldots, u_{p-1} \rangle = [u_0 S(u_1, \ldots, u_{p-1})] = \sum_{\sigma \in S_{p-1}} [u_0 u_{\sigma 1} \ldots u_{\sigma(p-1)}] \tag{7.5}$$

which are symmetric mod $p\mathscr{L}$ (not to be confused with the linear span $\langle u_0, \ldots, u_{p-1} \rangle_{\mathbb{Z}}$!), where the $u_i = u_i(x_1, \ldots, x_d)$ are homogeneous basic commutators and $\sum_{i=0}^{p-1} \deg u_i = n$. Although the equation $J_{2p-1} = (p\mathscr{L} + \mathscr{E})_{2p-1}$ holds for $d = 2$ (see [134], where it was also asserted that $J_{2p} = (p\mathscr{L} + \mathscr{E})_{2p}$, but that was an error), and Sanov [232] expressed the general opinion that J and $p\mathscr{L} + \mathscr{E}$ are the same, in 1973 G. E. Wall [267, 268] proved that the following inclusion holds:

$$J \supset p\mathscr{L} + \mathscr{E} + \mathscr{W}\,.$$

Here $\mathscr{W}$ is the ideal of $\mathscr{L}$ generated by the elements of the form

$$\langle\langle u_0, u_1, \ldots, u_{2p-2} \rangle\rangle = \frac{1}{p!} \sum_{j_1 < \cdots < j_{p-1}} \; \sum_{i_0 < i_1 < \cdots < i_{p-1}} [[u_{i_0} S(u_{i_1}, \ldots, u_{i_{p-1}})] S(u_{j_1}, \ldots, u_{j_{p-1}})]$$

$$\{j_1, \ldots, j_{p-1}; i_0, \ldots, i_{p-1}\} = \{0, 1, \ldots, 2p-2\}\,. \tag{7.6}$$

It is known that the elements (7.6) are symmetric mod $(p\mathscr{L} + \mathscr{E})$. Moreover, Wall [267] proved the following non-inclusion by hand for $p = 5$, while Cannon did it for $p = 5$ and 7 using a computer:

$$\langle\langle \underbrace{x_1, \ldots, x_1}_{p-1}, \underbrace{x_2, \ldots, x_2}_{p-1}, x_3 \rangle\rangle \notin p\mathscr{L} + \mathscr{E}\,.$$

Thus $\mathscr{W} \nsubseteq p\mathscr{L} + \mathscr{E}$, at least when $d \geqslant 3$. In view of the result in [134] mentioned above and the fact that $J = p\mathscr{L} + \mathscr{E}$ for $d = 2$, $p = 5$ (see [131, 98 and 151]), it was possible to hope that the case $d = 2$ is exceptional. However, Khukhro [128] used a computer to establish that $\mathscr{W}_{2p} \nsubseteq p\mathscr{L} + \mathscr{E}$ for $p = 7$. More exactly,

$$\langle\langle [x_1, x_2], x_1, \ldots, x_1, x_2, \ldots, x_2 \rangle\rangle \notin p\mathscr{L} + \mathscr{E}\,.$$

Wall also proved in [269] that

$$J_n = (p\mathscr{L} + \mathscr{E} + \mathscr{W})_n, \qquad 1 \leqslant n \leqslant 3p-3\,.$$

True, for components of weight $\geqslant p$ in any of the variables, this was done under the assumption that the conjecture (see [134]) about the independence of elements (7.5) of degree $\leqslant 2p-2$ is correct (it is easy to see that the elements of degree $2p-1$ and higher are dependent). Havas, Newman and Vaughan-Lee [96] confirmed this conjecture (which is obviously true for $d = 2$, $p = 5$) in the cases $d = 2$, $p = 7$ and $d = 3$, $p = 5$.

Khukhro [129] has discovered a (formally) new relation of degree $3p - 2$. It turns out that $\mathscr{L}$ contains the ideal (which we denote provisionally by $\mathscr{X}$) generated by the following elements, which are homogeneous and symmetric mod $(p\mathscr{L} + \mathscr{E} + \mathscr{W})$:

$$\langle\langle\langle u_0, u_1, \ldots, u_{3p-3}\rangle\rangle\rangle$$

$$= \frac{1}{(p!)^2}\Big\{\textstyle\sum_1 [[[u_{i_0} S(u_{i_1}, \ldots, u_{i_{p-1}})]\, S(u_{j_1}, \ldots, u_{j_{p-1}})]$$

$$\times S(u_{k_1}, \ldots, u_{k_{p-1}})] + \frac{1}{2}\textstyle\sum_2 [[u_{i_0} S(u_{i_1}, \ldots, u_{i_{p-1}})]$$

$$\times S([u_{j_0} S(u_{j_1}, \ldots, u_{j_{p-1}})], u_{k_2}, \ldots, u_{k_{p-1}})]\Big\}.$$

In this, $\sum_1$ denotes summation over all collections

$$i_0 < i_1 < \cdots < i_{p-1}, j_1 < \cdots < j_{p-1}, k_1 < \cdots < k_{p-1},$$

$$\{i_0, i_1, \ldots, i_{p-1}, j_1, \ldots, j_{p-1}, k_1, \ldots, k_{p-1}\} = \{0, 1, \ldots, 3p-3\},$$

and $\sum_2$ that over collections

$$i_0 < i_1 < \cdots < i_{p-1}, j_0 < j_1 < \cdots < j_{p-1}, k_2 < \cdots < k_{p-1},$$

$$\{i_0, i_1, \ldots, i_{p-1}, j_0, \ldots, j_{p-1}, k_2, \ldots, k_{p-1}\} = \{0, 1, \ldots, 3p-3\}.$$

It was proved that the homogeneous components of J of weight $\leqslant p - 1$ in every variable and total weight $3p - 2$ are generated mod $(p\mathscr{L} + \mathscr{E} + \mathscr{W})$ by the elements $\langle\langle\langle x_{i_0}, x_{i_1}, \ldots, x_{i_{3p-3}}\rangle\rangle\rangle$. Khukhro made the natural conjecture that, for $p \geqslant 5$ (and maybe $d \geqslant 4$), not all of these elements lie in $(p\mathscr{L} + \mathscr{E} + \mathscr{W})$; however, no computer experiments have been carried out. A concrete 'suspicious" element is

$$\langle\langle\langle \underbrace{x_1, \ldots, x_1}_{p-1}, \underbrace{x_2, \ldots, x_2}_{p-1}, \underbrace{x_3, \ldots, x_3}_{p-1}, x_4\rangle\rangle\rangle,$$

which resembles the Wall element mentioned above.

It is the necessity for increasing the generating number at each stage to guarantee new relations in $L(G)$ that motivates the assumption that d should be taken large enough. It is sometimes expedient to start at the very beginning with the free Lie $\mathbb{Z}$-algebra $\mathscr{L} = \mathrm{Lie}\langle x_1, x_2, \ldots\rangle$ on a countable infinity of generators, as Wall did in [271, 272]. Let $\mathfrak{B}_q$ be the Burnside variety consisting of all groups of exponent dividing $q = p^\alpha$. Further, let $\mathscr{F}(\mathfrak{B}_q)$ be the free group of this variety of countably infinite rank, and $L(\mathscr{F}(\mathfrak{B}_q))$ the graded Lie $\mathbb{Z}_q$-algebra associated with the lower central series of $\mathscr{F}(\mathfrak{B}_q)$. There is a natural epimorphism $\mathscr{L} \to L(\mathscr{F}(\mathfrak{B}_q))$ with kernel $J = J(q)$. The elements of J are called the *Lie relations* of $\mathfrak{B}_q$ (we use the same letters $\mathscr{L}$ and J regardless of the cardinal of the generating set X). Wall used a more capacious term—he called a Lie polynomial

$f(x_1, \ldots, x_r) \in J$ a *Lie relator* for $\mathfrak{B}_q$. It is not hard to see that if the relator $f(x_1, \ldots, x_r)$ is multilinear with respect to some subset $\{x_1, \ldots, x_r\}$ of the free generators of L, then $f(u_1, \ldots, u_r) \in J$ for every choice of elements $u_1, \ldots, u_r \in L$, and obviously $f(\bar{u}_1, \ldots, \bar{u}_r) = 0$ for $\bar{u}_1, \ldots, \bar{u}_r \in L(\mathscr{F}(\mathfrak{B}_q))$.

2.2. In his interesting paper [262], Vaughan-Lee introduces certain multilinear polynomials $K_{n+1}(x_1, \ldots, x_{n+1})$, $n \geqslant p-1$, in the free generators $x_1, x_2, \ldots, x_{n+1}$ of the free Lie algebra $\mathscr{L}$ over $\mathbb{Z}$. If $S = \{i, j, \ldots, k\}$ is a non-empty subset of $\{1, 2, \ldots, n\}$ with $i < j < \cdots < k$ and $u \in \mathscr{L}$, we set

$$[u, x_S] = [u, x_i, x_j, \ldots, x_k] .$$

Then, by definition,

$$K_{n+1}(x_1, \ldots, x_n, x_{n+1}) = \sum [x_{n+1}, x_{S(1)}, x_{S(2)}, \ldots, x_{S(p-1)}] , \tag{7.7}$$

where the summation extends over all decompositions of $\{1, 2, \ldots, n\}$ as unions of disjoint *non-empty* subsets $S(1), S(2), \ldots, S(p-1)$. It is convenient to set $K_1(x) = px$.

Further, let

$$I_1 \subset \cdots \subset I_p \subset I_{p+1} \subset \cdots$$

be the ascending chain of ideals of $\mathscr{L}$ in which I_m is the ideal generated by the set

$$\{K_n(u_1, u_2, \ldots, u_n) | u_1, \ldots, u_n \in \mathscr{L}; n \leqslant m\} .$$

Then $I_1 = \cdots = I_{p-1} = p\mathscr{L}$. This introduces the following constructively-defined ideal:

$$I = I(p) = \bigcup_{m=1}^{\infty} I_m .$$

It turns out that $I \subseteq J(p) = J$, which is more than the inclusion $p\mathscr{L} \subset J$, that is:

2.3. Theorem. *Let G be any group of exponent p. Then its associated Lie $\mathbb{Z}_p$-algebra $L(G)$ satisfies the identities*

$$K_n = 0, \qquad n = p, p+1, \ldots .$$

Corollary. *The Lie $\mathbb{Z}_p$-algebra $L(G)$ associated with a group of prime exponent satisfies the Engel condition E_{p-1}.*

Proof. By (7.7),

$$K_p(x_1, x_2, \ldots, x_p) = \sum_{\sigma \in S_{p-1}} [x_p, x_{\sigma 1}, x_{\sigma 2}, \ldots, x_{\sigma(p-1)}] .$$

By Theorem 2.3 with $x_p = u$, $x_1 = x_2 = \cdots = x_{p-1} = v \in L(G)$, we have

$$0 = K_p(v, v, \ldots, v, u) = (p-1)! [uv^{p-1}] ,$$

so that $[uv^{p-1}] = 0$.

Remark. Account has been taken in the statement of Theorem 2.3 of the isomorphism

$$L(G) = \mathscr{L}/J \cong (\mathscr{L}/p\mathscr{L})/(J/p\mathscr{L}) = \bar{\mathscr{L}}/\bar{J}\,,$$

where $\bar{\mathscr{L}}$ is a free Lie algebra over $\mathbb{Z}_p$ and $\bar{J}$ is an ideal in it. Consequently, we can sharpen things up by noting the ideal equation $I_p = p\mathscr{L} + E_{p-1}(\mathscr{L})$. The results stated here acquire special significance because of our next theorem.

2.4. Theorem. *Every multilinear identity of $L(\mathscr{F}(\mathfrak{B}_p))$ is a consequence of the identities $K_n = 0$, $n \geqslant p$, and $px = 0$.*

In other words, every multilinear element of J is contained in I. The proofs of Theorems 2.3 and 2.4 are given in § 3.

Wall developed the above results and confirmed Vaughan-Lee's conjecture on the possibilities for jumps in the chain $I_1 \subset \ldots \subset I_p \subset I_{p+1} \subset \ldots$.

2.5. Theorem. *Suppose that $q = p^\alpha$, where p is prime. The following assertion holds for the ideal $I = I(q)$:*

$$I_r = I_{r-1} \quad \text{whenever} \quad r \not\equiv 1 \,(\mathrm{mod}\,(p-1))\,. \tag{7.8}$$

In other words, the only jumps possible are those of the type

$$I_{1+s(p-1)} \neq I_{s(p-1)}\,, \tag{7.8'}$$

while $I_{1+s(p-1)} = I_{(s+1)(p-1)}$ for all s.

The reader is referred to the original paper [271] for the proof of this result.

What is known in fact about the jumps (7.8′)? We get the following by reformulating the results in 2.1 in the obvious way:

$$p\mathscr{L} = I_{p-1} \neq I_p = \cdots = I_{2p-2} = p\mathscr{L} + \mathscr{E} \qquad \text{(I. N. Sanov)}\,,$$

$$I_{2p-2} \neq I_{2p-1} = \cdots = I_{3p-3} = p\mathscr{L} + \mathscr{E} + \mathscr{W} \qquad \text{(G. E. Wall)}\,,$$

$$I_{3p-3} \overset{?}{\neq} I_{3p-2} = \cdots = I_{4p-4} \overset{?}{\neq} p\mathscr{L} + \mathscr{E} + \mathscr{W} + \mathscr{X} \qquad \text{(E. I. Khukhro)}\,.$$

2.6. If (7.8′) is true for $s = 1, 2, \ldots, s_0$, the problem arises about the largest value possible for s_0 with fixed $q = p^\alpha$ (and sufficiently large generating number d). It is known that $s_0 = 1$ for $p = 2$ and 3, and $s_0 \geqslant 2$ for $p = 5$ and 7. Khukhro's "suspicious" relation $\langle\langle\langle\cdots\rangle\rangle\rangle = 0$ suggests the possibility that $s_0 \geqslant 3$. However, this is already in the region where the arbiter is (still) a computer. We note only that jumps of the type (7.8′) are directly related to the Hughes problem in group theory. For every finite p-group P and every exponent $q = p^\alpha$, the Hughes subgroup $H_q(P)$ of P is defined to be the subgroup generated by the elements of order $> q$ of P. What values can the index $|P : H_q(P)|$ take when $H_q(P) \neq 1$?

Theorem (see [271]). *Let s_0 be a positive integer, and suppose that the jumps* (7.8′) *occur in the chain of ideals of multilinear relations for the Burnside variety $\mathfrak{B}_q$ when*

$s = 1, 2, \ldots, s_0$. *Then there is a finite p-group P with* $H_q(P) \neq 1$ *such that* $|P:H_q(P)| = p^{s_0}$.

Next, the Hughes problem has links with delicate inequalities for the nilpotency classes of some "extremal" p-groups. Thus, in its more exact formulation, including that of describing $J(q)$, RBP turns out to have direct connections with the other deep questions that saturate the theory of finite p-groups.

2.7. Reverting to the results of his computer tests, Vaughan-Lee observes in [262] that it is possible in principle to compute $\mathscr{L}_n/(\mathscr{L}_n \cap I_m)$ directly for any given values of the parameters d, p, m and n. However, the dimensions obtained can be very impressive and one generally restricts to small d and p. In the case $d = 2$, $p = 5$, Havas, Wall and Wamsley [98] used the so-called *nilpotent-quotient algorithm* for Lie rings in calculating $\mathscr{L}/I_p$. For $(d, p) = (2, 5)$, the factor-algebra $\mathscr{L}/I_p$ has order 5^{34} and nilpotency class 12. Indeed, these were the upper bounds obtained by hand (see [131]). However, the nilpotent quotient algorithm for groups was also used in [98]. It turns out that $B_0(2, 5)$ has the very same order 5^{34}. It follows from these results that $J = I_p$ for $d = 2$, $p = 5$. Havas, Newman and Vaughan-Lee [96] have described a nilpotent quotient algorithm for graded Lie rings and used it in the case $d = 3$, $p = 5$ to show that $\mathscr{L}/I_p$ has order 5^{2292} and class 17, while $\mathscr{L}/I_{2p-1}$ has order 5^{2282} and class 17. The same authors used Higman's combinatorial method [108] to prove that $B_0(d, 5)$ has nilpotency class at most $6d$. Vaughan-Lee has informed me also of a proof that $I_9 = I_{13}$ for $p = 5$ and all d. Altogether, for $d = 3$ this gives that $\mathscr{L}/I_{2p-1} = \mathscr{L}/I$, so that $|B_0(3, 5)| \leqslant 5^{2282}$. This inequality is sharp provided that the following general conjecture is true for $d = 3$, $p = 5$:

2.8. Conjecture (Vaughan-Lee). *For all d and p,*

$$J = I\,,$$

that is, every identity of $L(\mathscr{F}(\mathfrak{B}_p))$ *is a consequence of the* (*constructively computed*) *multilinear identities.*

No means have yet been suggested for confirming the important conjecture 2.8, which is a significant sharpening of the connection between finite p-groups and Lie algebras. The only authentic case is $d = 2$, $p = 5$, and it should be possible to confirm the conjecture for $d = 3$, $p = 5$; but even here, as Vaughan-Lee feels, one has to rest one's hopes on the use of the nilpotent quotient algorithm for groups on a computer with a larger memory.

§ 3. Multilinear Relations: Proofs of Theorems 2.3 and 2.4

Changing the notation of § 5 in Chapter 6 slightly, we set

$$(x, y; 1) = (x, y) = x^{-1}y^{-1}xy, \qquad (x, y; n+1) = ((x, y; n), y)$$

for arbitrary elements x, y of a group G. As with Lie algebras, the usual left-norming convention operates for repeated commutation:

$$(x, y_1, y_2, \ldots, y_k) = (\ldots((x, y_1), y_2), \ldots, y_k)\,.$$

The smallest normal subgroup containing x, that is, the *normal closure* of x, is denoted by $\langle x^G \rangle$.

3.1. Let G be any group of prime exponent p with identity element denoted by e; and let x, y be any elements of G. For some time we shall be assuming that the normal closure $\langle x^G \rangle$ is abelian. Then $\langle x^G \rangle$ is an elementary abelian p-group on which G acts by conjugation. We make $\langle x^G \rangle$ into a $\mathbb{Z}_pG$-module by setting

$$b^a = (g_1^{-1} b g_1)^{n(1)} (g_2^{-1} b g_2)^{n(2)} \ldots (g_k^{-1} b g_k)^{n(k)}$$

as usual, for arbitrary $b \in \langle x^G \rangle$, $a = n(1)g_1 + n(2)g_2 + \cdots + n(k)g_k \in \mathbb{Z}_pG$ (we identify $\mathbb{Z}_p$ with the numbers $0, 1, \ldots, p-1$). In that case,

$$e = y^{-p}(yx)^p = (y^{-(p-1)} x y^{(p-1)}) \ldots (y^{-2} x y^2)(y^{-1} x y) x = x^a\,,$$

where $a = e + y + \cdots y^{p-1} = (y - e)^{p-1} \in \mathbb{Z}_pG$. However, since $b^{y-e} = (b, y)$ is a commutator,

$$e = x^a = (x, y; p-1)\,. \tag{7.9}$$

We recall that this Engel condition in groups of exponent p is valid for the time being only for those x whose normal closures are abelian.

Suppose now that $y_1, y_2, \ldots, y_n \in G$ and $n \geqslant p-1$. We take $y = y_1 y_2 \ldots y_n$. If $S = \{i, j, \ldots, k\}$ is a non-empty subset of $\{1, 2, \ldots, n\}$ with $i < j < \cdots < k$ and $b \in \langle x^G \rangle$, we set

$$(b, y_S) = (b, y_i, y_j, \ldots, y_k)\,.$$

The second of the relations (7.1), rewritten in the form

$$(b, y_1 y_2) = (b, y_2)(b, y_1)(b, y_1, y_2)\,,$$

can be carried over without difficulty to more factors:

$$\begin{aligned}(b, y_1 y_2 y_3) &= (b, y_1(y_2 y_3)) = (b, y_2 y_3)(b, y_1)(b, y_1, y_2 y_3)\\ &= (b, y_3)(b, y_2)(b, y_2, y_3)(b, y_1)(b, y_1, y_3)(b, y_1, y_2)(b, y_1, y_2, y_3)\,.\end{aligned}$$

A simple induction on n leads to the expression

$$(b, y) = \prod_S (b, y_S)\,, \tag{7.10}$$

where the product is taken over all non-empty subsets of $\{1, 2, \ldots, n\}$. No care is necessary over the orders of the factors since all the (b, y_S) lie in $\langle x^G \rangle$ and therefore commute with each other. Comparing (7.9) and (7.10), we get the identity

$$\prod (x, y_{S(1)}, y_{S(2)}, \ldots, y_{S(p-1)}) = e\,, \tag{7.11}$$

in which the product is taken over all sequences $S(1), S(2), \ldots, S(p-1)$ of non-empty subsets of $\{1, 2, \ldots, n\}$. The elements $y_1, y_2, \ldots, y_n$ are completely arbitrary, so that (7.11) remains true if y_1 is replaced by the identity element e of G. However, if $1 \in S(1) \cup S(2) \cup \cdots \cup S(p-1)$ and y_1 is replaced by e, we have $(x, y_{S(1)}, \ldots, y_{S(p-1)}) = e$, the trivial commutator. At the same time, the analogous commutator with $1 \notin \bigcup_i S(i)$ remains unchanged under this substitution. Thus (7.11) contains a factor of the form

$$\prod_{1 \notin \bigcup_i S(i)} (x, y_{S(1)}, y_{S(2)}, \ldots, y_{S(p-1)}) = e\,,$$

removal of which gives

$$\prod_{1 \in \bigcup_i S(i)} (x, y_{S(1)}, y_{S(2)}, \ldots, y_{S(p-1)}) = e\,.$$

Applying similar arguments to this identity (replacing y_2 by e and removing the factor $\prod_{2 \notin \bigcup_i S(i)} (x, y_{S(1)}, \ldots, y_{S(p-1)}) = e$) shows that

$$\prod_{\{1,2\} \subseteq \bigcup_i S(i)} (x, y_{S(1)}, y_{S(2)}, \ldots, y_{S(p-1)}) = e\,.$$

We repeat these arguments for the indices $3, 4, \ldots, n$, and get the following identity after finitely many steps:

$$\prod_{\{1,2,\ldots,n\} \subset \bigcup_i S(i)} (x, y_{S(1)}, y_{S(2)}, \ldots, y_{S(p-1)}) = e\,. \tag{7.12}$$

3.2. Assume now that $G = B(n+1, p)$ is the Burnside group of exponent p on the generators $x, y_1, \ldots, y_n$. We isolate in G the set of *special commutators*, meaning x, x^{-1}, $y_1, y_1^{-1}, \ldots, y_n, y_n^{-1}$ and all (a, b), where a and b are special commutators.

Since every element g of G is a product of special commutators and $x^g = x(x, g)$, it follows from the identities (7.1) that the normal closure $\langle x^G \rangle$ is generated by special commutators each of which has at least one occurrence from $\{x, x^{-1}\}$ (that is, x or x^{-1} occurs in each). The commutator subgroup $\langle x^G \rangle'$ of $\langle x^G \rangle$ is then generated by special commutators with at least two occurrences from $\{x, x^{-1}\}$.

If we waive the commutativity of $\langle x^G \rangle$ in the arguments of 3.1 that led to identity (7.12), what happens is that (7.12) is expressed as an inclusion

$$\prod_{\{1,2,\ldots,n\} \subset \bigcup_i S(i)} (x, y_{S(1)}, y_{S(2)}, \ldots, y_{S(p-1)}) \in \langle x^G \rangle'\,.$$

Using the remark about the generating set for $\langle x^G \rangle'$, we get a relation

$$\prod_{\{1,2,\ldots,n\} \subset \bigcup_i S(i)} (x, y_{S(1)}, y_{S(2)}, \ldots, y_{S(p-1)}) = c_1 c_2 \ldots c_k\,, \tag{7.13}$$

in which $c_1, c_2, \ldots, c_k$ are special commutators in G, each of which has at least two occurrences from $\{x, x^{-1}\}$. We have thus proved part of the following result.

3.3. Lemma. *Every element c_i, $1 \leqslant i \leqslant k$, in (7.13) has at least two occurrences from $\{x, x^{-1}\}$ and at least one from $\{y_j, y_j^{-1}\}$ for each $j = 1, 2, \ldots, n$.*

Proof. We proceed by induction on j, starting from the trivially true case $j = 0$. Assume that all the c_i, $1 \leqslant i \leqslant k$, have an occurrence from $\{y_j, y_j^{-1}\}$ for each $j = 1, 2, \ldots, r-1$ but that not all the c_i have occurrences from $\{y_r, y_r^{-1}\}$. By introducing a larger number of special commutators, we may assume that the c_i not having occurrences from $\{y_r, y_r^{-1}\}$ stand to the left of those that do. For, if c_i does not have an occurrence from $\{y_r, y_r^{-1}\}$ but c_{i-1} does, we replace $c_{i-1}c_i$ in the product $c_1c_2 \ldots c_k$ by $c_ic_{i-1}(c_{i-1}, c_i)$. The additional special commutator (c_{i-1}, c_i) thus introduced has not less than two (in fact, not less than four) occurrences from $\{x, x^{-1}\}$ and one from $\{y_j, y_j^{-1}\}$ for $j = 1, 2, \ldots, r$. Operating in this way, we replace $c_1c_2 \ldots c_k$ by a product $d_1d_2 \ldots d_s$, where $d_1, d_2, \ldots, d_s$ are special commutators each of which has at least two occurrences from $\{x, x^{-1}\}$ and at least one from $\{y_j, y_j^{-1}\}$ for $j = 1, 2, \ldots, r-1$, and $d_1, \ldots, d_t$ have no occurrences from $\{y_r, y_r^{-1}\}$ while $d_{t+1}, \ldots, d_s$ do.

Let δ_r be the endomorphism of the relatively free group G taking x to x, y_j to y_j for $j \neq r$, and y_r to e. Application of δ_r to both sides of relation (7.13), with $c_1c_2 \ldots c_k$ replaced by $d_1d_2 \ldots d_td_{t+1} \ldots d_s$, gives $e = d_1 \ldots d_t$. Thus

$$\prod_{\{1, 2, \ldots, n\} \subset \bigcup_i S(i)} (x, y_{S(1)}, y_{S(2)}, \ldots, y_{S(p-1)}) = d_{t+1} \ldots d_s .$$

This establishes the inductive step from $r-1$ to r. □

3.4. Proof of Theorem 2.3. Relation (7.13), which was proved for the case where $x, y_1, \ldots, y_n$ are free generators of the relatively free group G of exponent p, remains true under all endomorphisms; so it holds for all elements $x, y_1, \ldots, y_n$ of G. Suppose that $x \in \gamma_m(G)$ and $y_j \in \gamma_{m(j)}(G)$ for $j = 1, 2, \ldots, n$, and set

$$l = m + m(1) + m(2) + \cdots + m(n) .$$

It is clear that every special commutator having at least two occurrences from $\{x, x^{-1}\}$ and one from $\{y_j, y_j^{-1}\}$ for $j = 1, 2, \ldots, n$ must lie in $\gamma_{l+1}(G)$. Similarly, if $S(1), S(2), \ldots, S(p-1)$ are non-empty subsets of $\{1, 2, \ldots, n\}$ such that $\{1, 2, \ldots, n\} = \bigcup_i S(i)$, then $(x, y_{S(1)}, \ldots, y_{S(p-1)}) \in \gamma_{l+\nu}(G)$, $\nu \geqslant 0$, where $\nu \geqslant 1$ for any case where $S(i) \cap S(i') \neq \phi$ for some indices $i \neq i'$. On the basis of these arguments and Lemma 3.3, we conclude that

$$\prod (x, y_{S(1)}, y_{S(2)}, \ldots, y_{S(p-1)}) \in \gamma_{l+1}(G) ,$$

where the product extends over all decompositions of $\{1, 2, \ldots, n\}$ as disjoint unions of nonempty subsets $S(1), S(2), \ldots, S(p-1)$.

Turning to the associated Lie $\mathbb{Z}_p$-algebra $L(G)$ and recalling Definition (7.7) of the polynomials K_{n+1}, we see that

$$K_{n+1}(u_i, u_2, \ldots, u_{n+1}) = 0 , \tag{7.14}$$

where $u_{n+1} = \tilde{x}_{n+1} = x_{n+1}\gamma_m(G)$, $u_i = \tilde{x}_i = x_i\gamma_{m(i)}(G)$ for $1 \leqslant i \leqslant n$. Since relation (7.14) is linear in $u_1, u_2, \ldots, u_{n+1}$, it holds for all elements $u_1, u_2, \ldots, u_{n+1} \in L(G)$, not simply the homogeneous ones. We have thus arrived at the multilinear identities $K_{n+1} = 0$ in $L(G)$. □

3.5. Before embarking on the proof of Theorem 2.4, we shall establish a number of auxiliary results. We observe that the obvious inclusion $p\mathscr{L} \subset J$ enables us to assume henceforth without loss of generality that $\mathfrak{A}$ and $\mathscr{L}$ are free algebras over $\mathbb{Z}_p$. As before, J is a graded ideal in $\mathscr{L}$. Every multilinear identity in the Lie algebra $L = L(B(d, p)) = \mathscr{L}/J$ over a field of characteristic p lifts in a natural way to a multilinear identity with rational coefficients. This means that Lemma 1.3 holds in characteristic p.

Following [270], we denote by Λ the algebra of all linear transformations on $\mathfrak{A}$, and for $v \in \mathfrak{A}$ we define $\operatorname{ad} v \in \Lambda$: $u \operatorname{ad} v = [u, v]$. Since $x_1, x_2, \ldots$ are free generators of A, there is a uniquely defined homomorphism $\phi: \mathfrak{A} \to \Lambda$ such that $\phi(x_i) = \operatorname{ad} x_i$, $i = 1, 2, \ldots$. For $u, v \in \mathfrak{A}$ we set

$$[u|v] = u\phi(v) ,$$

so that

$$[u|1] = u, \ [[u|v]|w] = [u|vw] .$$

Although

$$[u, x_i] = u \operatorname{ad} x_i = [u|x_i] ,$$

it by no means follows from the definition that $[u|v] = [u, v]$. The elements v such that this is true for all u comprise a Lie subalgebra of $\mathfrak{A}$.

Remembering the advantages of the symbolism $[u, y_S]$ from 2.2 (which we shall not use any more, in fact), we introduce the following very similar notation. For every non-empty subset $S = \{i, j, \ldots, k\}$ of integers with $i < j < \cdots < k$, we set $x_S = x_i x_j \ldots x_k$, and $x_\varnothing = 1$. Further, it is convenient to agree on the notation

$$\bar{x}_S = (-1)^{|S|} x_k \ldots x_j x_i, \qquad \bar{x}_\varnothing = 1 .$$

The defining relation (7.7) now takes the form

$$K_n(x_1, x_2, \ldots, x_n) = [x_n|\textstyle\sum x_{S(1)}x_{S(2)} \ldots x_{S(p-1)}] , \tag{7.15}$$

where, as before, the sum extends over all decompositions of $\{1, 2, \ldots, n\}$ as unions of disjoint subsets $S(1), S(2), \ldots, S(p-1)$. It is natural to set $K_n = 0$ for $n < p$, since the characteristic of the ground field is p.

The polynomial $T_n(x_1, \ldots, x_n)$ in the statement of Lemma 1.3 can be written in the convenient and explicit form

$$T_n = \textstyle\sum x_{S(1)}x_{S(2)} \ldots x_{S(p)} , \tag{7.16}$$

where the summation is over all decompositions of $\{1, 2, \ldots, n\}$ into disjoint subsets $S(i)$, where this time some may be empty.

Since by definition T_n is the homogeneous component of $((1 + x_1) \ldots (1 + x_n))^p$ of degree 1 in $x_1, \ldots, x_n$, the relation

$$((1 + x_1) \ldots (1 + x_n))^p = ((1 + x_1) \ldots (1 + x_n) - 1)^p + 1$$

holds in characteristic p, and the polynomial $((1+x_1)\dots(1+x_n)-1)^p$ has no terms of degree $< p$, then, as with K_n, we have $T_n = 0$ for $n < p$.

It is appropriate to replace expression (7.15) for K_n by one in which empty subsets $S(i)$ are allowed. With this aim, we note that the sum $\sum_{S(i)\neq\varnothing} x_{S(1)}x_{S(2)}\dots x_{S(p-1)}$ is the homogeneous component of degree 1 in $x_1,\dots,x_{n-1}$ of the polynomial $((1+x_1)\dots(1+x_{n-1})-1)^{p-1}$. However, in characteristic p, the latter is $\sum_{r=0}^{p-1}((1+x_1)\dots(1+x_{n-1}))^r$, and every summand $((1+x_1)\dots(1+x_{n-1}))^r$ has for its homogeneous component of degree 1 in $x_1,\dots,x_{n-1}$ a sum $\sum x_{S(1)}x_{S(2)}\dots x_{S(r)}$, where empty subsets $S(i)$ are allowed. We have arrived at the required expression

$$K_n = \sum_{r=1}^{p-1}[x_n|\sum x_{S(1)}x_{S(2)}\dots x_{S(r)}]\,. \tag{7.17}$$

As before (see 2.2), by I_m we mean the ideal of L generated by the set

$$\{K_n(a_1,\dots,a_n)|a_i\in\mathscr{L},\qquad 1\leqslant i\leqslant n, n\leqslant m\}\,.$$

By its very definition, I_m is a graded ideal. It is clear that $I_{p-1}=0$, and $I_m\subseteq J$ for all m, by Theorem 2.3. Further, let $V_{\mathfrak{A}}(I_m)$ be the ideal of the associative algebra $\mathfrak{A}$ generated by I_m.

Lemma (see [268]). $V_{\mathfrak{A}}(I_m)\cap\mathscr{L} = I_m$.

Proof. For fixed m, we can assume that $\mathscr{L}$ is generated by m elements. By Corollary 2.3 and Theorem 7.4, the factor-algebra $\bar{\mathscr{L}} = \mathscr{L}/I_m$ is nilpotent of class $\leqslant c$. Since $\bar{\mathscr{L}}$ is a graded algebra over $\mathbb{Z}_p$, $\bar{\mathscr{L}}$ is embedded in the universal enveloping algebra $U = U(\bar{\mathscr{L}})$; the latter is a graded associative algebra whose grading is consonant with that in $\bar{\mathscr{L}}$. We denote by $U^{(k)}$ the module over $\mathbb{Z}_p$ spanned by all monomials $u_1u_2\dots u_k$ with $u_i\in\bar{\mathscr{L}}$. The lemma can then be expressed in the form

$$\bar{\mathscr{L}}\cap(U^{(c+1)}+U^{(c+2)}+\cdots)=0\,.$$

But this is clear, since the homogeneous components of L of degree $> c$ are zero. □

3.6. Proof of Theorem 2.4. 1). If we could verify that

$$K_n \equiv T_n \bmod V_{\mathfrak{A}}(I_{n-1}),\qquad n\geqslant p\,, \tag{7.18}$$

this would be enough. Because, in that case, $T_n\in V_{\mathfrak{A}}(I_n)$ for all n (recall that $T_n = 0$ for $n < p$). By Lemma 1.3, we conclude from this that every multilinear element of J is contained in $V_{\mathfrak{A}}(I_n)$ for some n. Enlisting the aid of Lemma 3.5, we now get Theorem 2.4.

2) We turn to the proof of congruence (7.18), which we do by induction on n. Revising the decomposition of left-normed commutators (see 1.4.1) in the new notation of 3.5, we get

$$[x_n|x_1x_2\dots x_{n-1}] = \sum \bar{x}_{S(1)}x_nx_{S(2)}\,, \tag{7.19}$$

where the summation extends over all disjoint decompositions $\{1, 2, \dots, n-1\} = S(1) \cup S(2)$, including $S(1) = \varnothing$ and $S(2) = \varnothing$. An r-fold application of this argument to K_n in the form (7.17) leads us to an expression

$$K_n = \sum_{r=1}^{p-1} \{\sum \bar{x}_{S(1)}\bar{x}_{S(2)} \dots \bar{x}_{S(r)}x_nx_{S(r+1)}x_{S(r+2)} \dots x_{S(2r)}\}\,,$$

where the sum in curly brackets extends over all decompositions of $\{1, 2, \dots, n-1\}$ into disjoint subsets $S(1), S(2), \dots, S(r)$, $1 \leqslant r \leqslant p-1$. Further, we consider the multilinear element

$$W = \sum_{r=1}^{p-1} \{\sum x_{S(1)} \dots x_{S(p)}\bar{x}_{S(p+1)} \dots \bar{x}_{S(p+r)}x_nx_{S(p+r+1)} \dots x_{S(p+2r)}\}\,,$$

where the inner sum extends over all decompositions of $\{1, 2, \dots, n-1\}$ into disjoint subsets $S(1), S(2), \dots, S(p+2r)$. If $S = \{i, j, \dots, k\}$ is a non-empty subset of $\{1, 2, \dots, n-1\}$, we gather together all terms with $S(1) \cup \dots \cup S(p) = S$ in W to give an element

$$\sum_{r=1}^{p-1} \{T_{|S|}(x_i, x_j, \dots, x_k) \sum \bar{x}_{S(p+1)} \dots \bar{x}_{S(p+r)}x_n \dots x_{S(p+2r)}\}\,,$$

which, by induction, lies in $V_{\mathfrak{A}}(I_{n-1})$.

On the other hand, the sum of the remaining terms of W with $S(1) \cup \dots \cup S(p) = \varnothing$ is just K_n. Thus, $W \equiv K_n \bmod V_{\mathfrak{A}}(I_{n-1})$.

3) We prove now that $W \equiv T_n (\bmod V_{\mathfrak{A}}(I_{n-1}))$. The formal substitution $x_n = 1$ in (7.19) leads us to the relation

$$\sum \bar{x}_{S(1)}x_{S(2)} = 0\,,$$

where, as before, the sum extends over all decompositions of $\{1, 2, \dots, n-1\}$ into disjoint (possibly empty) subsets $S(1)$ and $S(2)$. Clearly, $\sum x_{S(1)}\bar{x}_{S(2)} = 0$. Therefore, if S is a non-empty subset of $\{1, 2, \dots, n-1\}$, then all terms in W such that $S(p) \cup S(p+1) = S$ together give zero, that is,

$$W = \sum_{r=1}^{p-1} \{\sum x_{S(1)} \dots x_{S(p-1)}\bar{x}_{S(p+2)} \dots \bar{x}_{S(p+r)}x_n \dots x_{S(p+2r)}\}\,.$$

Repeating this argument with the pairs of subsets $S(p-1)$, $S(p+2)$; $S(p-2)$, $S(p+3)$; $\dots$; $S(p-r+1)$, $S(p+r)$, we arrive at the identity

$$W = \sum_{r=1}^{p-1} \{\sum x_{S(1)} \dots x_{S(p-r)}x_nx_{S(p+r+1)} \dots x_{S(p+2r)}\}\,,$$

or, after redistributing terms in the sum and renumbering subsets,

$$W = \sum_{r=1}^{n-1} \{\sum x_{S(1)} \dots x_{S(r)}x_nx_{S(r+1)} \dots x_{S(p)}\}\,.$$

On the other hand, recalling (7.16), we get

$$\begin{aligned}T_n &= \sum_{\cup S(i)=\{1,\ldots,n\}} x_{S(1)}x_{S(2)}\cdots x_{S(p)}\\ &= \sum_{r=1}^{p}\Bigg\{\sum_{\substack{\cup S(i)=\{1,\ldots,n\}\\ n\in S(r)}} x_{S(1)}x_{S(2)}\cdots x_{S(p)}\Bigg\}\\ &= \sum_{r=1}^{p}\Bigg\{\sum_{\cup S(i)=\{1,\ldots,n-1\}} x_{S(1)}\cdots x_{S(r)}x_n\cdots x_{S(p)}\Bigg\}\\ &= W + T_{n-1}(x_1,\ldots,x_{n-1})x_n\,.\end{aligned}$$

By induction, $T_{n-1}\in V_{\mathfrak{A}}(I_{n-1})$. We get congruence (7.18) by combining the deductions of parts 2) and 3). □

3.7. The arguments of 3.6 lead to the following useful property of the K_n when supplemented by a minor analysis of expression (7.16).

Proposition. *K_n is symmetric modulo I_{n-1}, that is,*

$$K_n(x_1,\ldots,x_n)\equiv K_n(x_{\sigma 1},\ldots,x_{\sigma n}) \bmod I_{n-1}$$

for every permutation $\sigma\in S_n$.

Proof. Let i, $1\leqslant i\leqslant n-1$, be a fixed index. We write $T_n(x_1,\ldots,x_n)=\sum x_{S(1)}\cdots x_{S(p)}=A+B$, where A is the sum of the terms $x_{S(1)}\cdots x_{S(p)}$ such that i and $i+1$ lie in different subsets $S(j)$, $S(k)$, $j\neq k$, and B the sum of the terms such that $i, i+1\in S(j)$ for some j. Clearly, $B=T_{n-1}(x_1,\ldots,x_{i-1},x_ix_{i+1},x_{i+2},\ldots,x_n)$. Interchanging the positions of x_i and x_{i+1} changes B into $T_{n-1}(x_1,\ldots,x_{i+1}x_i,\ldots,x_n)$ and leaves A unchanged—all that happens is that summands of A are redistributed. Thus,

$$\begin{aligned}&T_n(x_1,\ldots,x_ix_{i+1},\ldots,x_n)-T_n(x_1,\ldots,x_{i+1}x_i,\ldots,x_n)\\ &\quad= T_{n-1}(x_1,\ldots,x_ix_{i+1},\ldots,x_n)-T_{n-1}(x_1,\ldots,x_{i+1}x_i\ldots,x_n)\\ &\quad= T_{n-1}(x_1,\ldots,x_ix_{i+1}-x_{i+1}x_i,\ldots,x_n)\\ &\quad= T_{n-1}(x_1,\ldots,[x_i,x_{i+1}],\ldots,x_n)\,,\end{aligned}$$

which gives, by congruence (7.18),

$$K_n(x_1,\ldots,x_i,x_{i+1},\ldots,x_n)-K_n(x_1,\ldots,x_{i+1},x_i,\ldots,x_n)\in V_{\mathfrak{A}}(I_{n-1})\,.$$

But $K_n\in\mathscr{L}$ by definition, while $V_{\mathfrak{A}}(L_{n-1})\cap\mathscr{L}=I_{n-1}$ by Lemma 3.5. Hence,

$$K_n(x_{\tau 1},\ldots,x_{\tau n})\equiv K_n(x_1,\ldots,x_n) \bmod I_{n-1}$$

where τ is the transposition in S_n interchanging i and $i+1$. Our assertion now follows. □

§ 4. Commentary

4.1. The Campbell-Hausdorff formula (see 1.2), which plays a fundamental rôle in the theory of Lie groups, has been a source of much investigation in abstract group theory. The formal group law

$$x \circ y = \ln(e(x)e(y)) = z(x, y) = x + y + \frac{1}{2}[xy] + \cdots$$

on $\hat{\mathscr{L}}$ (here 0 is the identity, and $-a$ the inverse of a for each a in $\hat{\mathscr{L}}$) enables one to express relations between group elements in terms of properties of the series $z(x, y)$. It is not an accident that the calculation of the terms in this series is really difficult, albeit that there is a formal explicit expression for each homogeneous component $z_{r,s}$ of $z(x, y)$ of degree r in x and s in y (see [31], Chapters 1–3, for example). Various recurrence formulae are more convenient in many situations. If we set

$$z_n = \sum_{r+s=n} z_{r,s}, \qquad z_{(r)} = \sum_{s=0}^{\infty} z_{r,s} = \frac{\omega_r(x, y)}{r!},$$

then we have

$$z(x, y) = \sum_{n=1}^{\infty} z_n = \sum_{r=0}^{\infty} \frac{\omega_r(x, y)}{r!},$$

where

$$\omega_r(x, y) = \underset{y \to \omega(x, y)}{D} \omega_{r-1}(x, y), \qquad r \geqslant 1,$$

$$\omega_0 = y, \qquad \omega_1 = \omega(x, y) = x + \frac{1}{2}[x, y] + \sum_{i=1}^{\infty} \pi_i [xy^{2i}]. \tag{7.20}$$

Here

$$\pi_r = \frac{(-1)^{r-1} b_{2r}}{(2r)!};$$

$b_2 = 1/6$, $b_4 = -1/30$, $b_6 = 1/42$, ... are the Bernoulli numbers; D is the linearization operator ("non-commutative differentiation"):

$$f(x, y + \lambda u) = f(x, y) + \lambda \underset{y \to u}{D} f(x, y) + \cdots$$

Hausdorff's formula (see [92]) is also a recurrence:

$$\sum_{n=1}^{\infty} n z_n(x, y) = x + y + \frac{1}{2}[zy] - \frac{1}{2}[zx] + \sum_{i=1}^{\infty} \pi_i [z_1 z^{2i}]. \tag{7.21}$$

Baker's formula (see [25]) gives an ideal recurrence relation for the even components $z_{2k}(x, y)$:

$$z_{2k} = -\frac{1}{2} \sum_{s=1}^{2k-1} \frac{1}{s!} [z_{2k-s} x^s]. \tag{7.22}$$

Easy calculations show that

$$z(x, y) = x + y + \frac{1}{2}[xy] + \frac{1}{12}[xy^2] + \frac{1}{12}[yx^2] + \frac{1}{24}[yx^2y] - \frac{1}{720}[xy^4]$$

$$- \frac{1}{720}[yx^4] + \frac{1}{360}[xy^3x] + \frac{1}{360}[yx^3y] - \frac{1}{120}[xy^2xy]$$

$$- \frac{1}{120}[yx^2yx] + \frac{1}{1440}[xy^4x] - \frac{1}{1440}[yx^4y] - \frac{1}{720}[xy^3x^2]$$

$$+ \frac{1}{240}[xy^2xyx] + \cdots$$

It seems quite impossible to use this. However, based on (7.20)–(7.22), arithmetical properties of the Bernoulli numbers (Staudt's theorem, among others), and on the elementary formula

$$\mathrm{e}(-y)\mathrm{e}(x)\mathrm{e}(y) = \mathrm{e}\left(\sum_{k=0}^{\infty} \frac{[xy^k]}{k!}\right),$$

fairly deep information can be obtained about the normal subgroup $\mathfrak{F}^q$ of the free group $\mathfrak{F}$ (and about the ideal $J(q)$, $q = p^\alpha$). Some results are noted in 4.8 below.

In addition, active use is made in Wall's papers dealing with the correspondence between groups and Lie algebras (see [269, 271] for example) of the Artin-Hasse p-exponential function

$$\mathrm{e}_p(t) = \mathrm{e}\left(\sum_{r=0}^{\infty} \frac{1}{p^r} t^{p^r}\right),$$

whose power series decomposition contains only p-integral rational coefficients. The truncated exponential $\sum_{k=0}^{p-1} \frac{1}{k!} t^k$ is also useful on transition to characteristic p.

4.2. In accordance with the notation introduced in § 1 of Chapter 6, let $\phi^*_{p+n}(d)$ be the number of different elements of the form (7.5) with $\sum \deg u_i = p + n$ in the free Lie algebra $\mathscr{L}$ over $\mathbb{Z}_p$. This number is obtained if some ordered basis B of $\mathscr{L}$ consisting of homogeneous elements is taken and all possible forms of type (1.7) are constructed:

$$\begin{Bmatrix} u_1 & u_2 & \dots & u_s \\ j_1 & j_2 & \dots & j_s \end{Bmatrix}$$

with $u_i \in B$, $u_1 < u_2 < \cdots < u_s$, $\sum_{k=1}^{s} j_k = p$, $j_k \neq p$ for all k, $\sum_j j_k \deg u_k = p + n$. By definition, the ideal $\mathscr{E} = E_{p-1}(\mathscr{L})$ is spanned by these forms, so that the conjecture from [134] mentioned in 2.1 takes the form

$$\phi_{p+n}(d) = \phi^*_{p+n}(d), \qquad 0 \leqslant n \leqslant p-2.$$

This is true for $n \leqslant 4$ (see [134], Theorem 5). Separation of the homogeneous forms of degree r in x and $p+n-r$ in y for $d=2$ also leads to the equality $\phi_{r,\,p+n-r} = \phi^*_{r,\,p+n-r}$ for $r \leqslant 3$ and $n \leqslant p-2$. The conjecture is still open in the general case.

4.3. The arguments in § 3 lead fairly easily to the following assertion.

The ideal I_n is generated mod I_{n-1} *by the elements $K_n(u_1, \ldots, u_n)$ with homogeneous basic $u_1 \in B$, $u_1 \leqslant \cdots \leqslant u_n$ such that $u_i \neq u_{i+p-1}$ for $1 \leqslant i \leqslant i+p-1 \leqslant n$.*

Proof (see [262]). Since

$$[K_n(x_1, \ldots, x_i, \ldots, x_n), x_{n+1}] = \sum_{i=1}^{n} K_n(x_1, \ldots, [x_i, x_{n+1}], \ldots, x_n)\,,$$

I_n is generated mod I_n by the $K_n(u_1, \ldots, u_n)$ with homogeneous basic u_i, $1 \leqslant i \leqslant n$. By Proposition 3.7, we may assume that $u_1 \leqslant \cdots \leqslant u_n$. We claim that $K_n(u_1, \ldots, u_n) \in I_{n-1}$ if $u_i = u_{i+1} = \cdots = u_{i+p-1}$ for some i. This is obvious for $n = p$. If $n > p$, then by Proposition 3.7,

$$\begin{aligned} K_n(u_1, \ldots, u_n) &= K_n(u_i, \ldots, u_{i+p-1}, u_{i+p}, \ldots, u_n, u_1, \ldots, u_{i-1}) \operatorname{mod} I_{n-1} \\ &= K_n(u_i, \ldots, u_i, u_{i+p}, \ldots, u_n, u_1, \ldots, u_{i-1}). \end{aligned}$$

We consider now the image of the commutator $[x_n | x_{S(1)} \ldots {}_{S(p-1)}]$ from expression (7.15) for K_n in which u_i is substituted for $x_1, \ldots, x_p$ and u_{i+p} for $x_{p+1}, \ldots, u_{i-1}$ for x_n. If all the variables $x_1, \ldots, x_p$ lie in $S(j)$ for some j, the image of $[x_n | x_{S(1)} \cdots x_{S(p-1)}]$ lies in $E_{p-1}(\mathscr{L}) = I_p$. If this is not so, we consider all terms $[x_n | x_{T(1)} \cdots x_{T(p-1)}]$ satisfying the two conditions:

$$\begin{aligned} &|S(j)| = |T(j)|, \qquad 1 \leqslant j \leqslant p-1\,, \\ &S(j) \cap \{x_{p+1}, \ldots, x_n\} = T(j) \cap \{x_{p+1}, \ldots, x_n\}, \qquad 1 \leqslant j \leqslant p-1\,. \end{aligned}$$

The number of such terms is divisible by p, and when u_i is substituted for $x_1, \ldots, x_p$, all terms take the same value. Thus, we get 0 mod p on adding the images of all terms of this form, and thus $K_n(u_i, \ldots, u_{i+p-1}, u_{i+p}, \ldots, u_n, u_1, \ldots, u_{i-1}) \in I_p$. □

A significant portion of the results contained in Theorem 2.3 and 2.4 (but not in Corollary 2.3) carry over to the case $q = p$ (see [271]).

4.4. The description of the ideal $I(p)$ of multilinear relations, suitably refined to the case of a Lie algebra $\mathscr{L}$ with a fixed finite number d of generators, will in all probability call to life investigations on the properties of the universal group $B_0(d, p)$. However, it will certainly not be easy to accomplish this, since an explicit description of all—not merely the multilinear—relations still requires the aid of a computer, and of course the possibilities for the latter are limited.

Something has been achieved in this direction without a computer. For instance, very recently Adian and Repin [10] have shown that the nilpotency class of $B_0(2, p)$ is greater than $2^{p/15.4}$; this came about from a refinement of their earlier arguments (see 6.5.1), and rests on the results of Vaughan-Lee [262] describing identities. Consequences of their results are that $\dim L(B_0(2, p)) > 2^{2^{p/15.5}}$ and

$$|B_0(2, p)| > p^{2^{2^{p/15.5}}}$$

for sufficiently large primes p. Note that *realistic* explicit upper bounds of this type for $B_0(2, p)$ are out of reach at the moment, despite the existence of recursive bounds (Appendix I).

Thus we see that the free Engel Lie algebra $L(2, p-1)$ and the Lie algebra $L(B_0(2, p))$ associated with $B_0(2, p)$ are asymptotically very similar; the constants 1/15 and 1/15.5 (which could perhaps be improved to $1/2 - \varepsilon$ and 1/3 if our proofs from 1 of Chap. 6 are used), differ but little from each other. This would lead us to expect any additional identities holding in $L(B_0(2, p))$ to be of too high a degree for us to worry about their independence.

The best understood of all these groups is $B_0(2, 5)$ (see § 8 in Chap. 1): as we have already seen, it has order 5^{34}, nilpotency class 12 and Engel index 6. The multiplication table for its basic commutators is essentially known (see [98]). All this information provoked the interesting approach of M. Hall Jr. and C. S. Sims [90] to the elucidation of the cardinal, and still open, question of whether $B_0(2, 5)$ and $B(2, 5)$ are the same; that is, whether $B(2, 5)$ is finite. Some very specific difficulties of a computational nature have arisen in the path of this question.

Earlier, the author [138, 142] suggested a simple comparison of the Engel properties of the Burnside groups. Since $(a, b; 6) = 1$ in $B_0(2, 5)$, one can ask: is it true that the equation

$$(x, y; 6) = \prod_{i=1}^{m} w_i(x, y)^5$$

has a solution in the free group F on two generators x, y? If not, $B(2, 5)$ is infinite.

It is not difficult to check the identities

$$(x, y) = x^{-2}(xy^{-1})^2 y^2, \qquad (x, (x, y)) = (x^{-1}y^{-1})^3(yxy)^3(y^{-1}(x, y))^3,$$

$$((x, y), y) = (xy)^{-3(x, y)}(yx^2)^{3xy}x^{-3y}.$$

The properties of $B(2, 4)$ are such that the expression

$$(x, y; 5) = \prod_{i=1}^{m} w_i(x, y)^4$$

is valid only for the fifth Engel word. The computer analysis carried out by Havas [94] showed that $m \leqslant 250$ for the smallest number m in such an expression. A. V. Korlyukov (Grodno: unpublished) has shown quite recently that $m \leqslant 28$. How many fifth powers will be necessary to express $(x, y; 6)$?

4.5. The cohomological properties of $B_0(2, p)$ have been investigated by Venkov [264] for arbitrary p. There are many obscurities here.

4.6. We have remarked already (see § 5 of Chap. 6) that there is a very extensive literature devoted to $B(d, 4) = B_0(d, 4)$. The insolubility of $B(\infty, 4)$, established by

Razmyslov [221] (see also [26], Chap. 8), manisfests itself in properties of subgroups isomorphic to $B(d, 4)$ embedded in it. Thus, after the publication of Razmyslov's paper, the very subtle (albeit conditional) results about nilpotency classes (see [78, 75]) lead to the definitive conclusion: for $d \geqslant 3$, the nilpotency class of $B(d, 4)$ is $3d - 2$, and 5 for $d = 2$. More exactly, $\gamma_{3d-2}(B(d, 4))$ is an elementary abelian group of rank $d(d + 1)/2$. Vaughan-Lee [261] proved that the derived length of $B(d, 4)$ is exactly k, where $2^{k-1} < 3d - 2 \leqslant 2^k$. It is proved in [180] that

$$4^d \leqslant 2\log_2|B(d, 4)| \leqslant (4 + 2\sqrt{2})^d .$$

Persistent attempts have been undertaken to prove that $B(2, 8)$ is finite (see [66–68]), simply by feeling the way; however, as yet it has been possible to find only preliminary information concerning subgroups of this group generated by elements of special form. On the other hand, in 1978 M. Hall Jr. informed the author that he expected $B(2, 8)$ to be infinite; this happened again in 1987. Here is his more precise conjecture: "Let $G = \langle a, b, c | a^2 = b^2 = c^2 = 1, x^8 = 1 \rangle$, and take $A = (bc)^4$, $B = (ac)^4$, $C = (ab)^4$. Then $\langle A, B, C \rangle \cong \langle a, b, c \rangle$. In particular, G is infinite." Incidentally, G' is 5-generator and $(G')^2$ is 32-generator.

At the moment there is no reason to give any preference to such a conjecture, but perhaps it is appropriate to work on RBP for exponent 8. Sanov [231] showed in 1951 that all groups of exponent 8 satisfy the 23rd Engel congruence. Krause [149] reduced this to 14. There is an example in [66] of a group of exponent 8 not satisfying the 11-th Engel congruence. It is known that a study of groups with the 14-th Engel congruence is incredibly difficult; other identities have to be brought in.

4.7. Machine experimentation on the groups $B(d, q)$ is pretty impressive. The leaders in this area of computation are groups of mathematicians in Australia and Great Britain. The following tables are extracted from [262]; in them there appears the numeral n and the rank (more exactly, the p-rank) of the factor $\gamma_n(B(d, q)/\gamma_{n+1}(B(d, q))$ of the lower central series of $B(d, q)$. The p-rank of an abelian group A of order p^r is simply the exponent r.

$B(2, 4)$

n	Rank
1	2
2	3
3	2
4	3
5	2
Σ	12

$B(3, 4)$

n	Rank
1	3
2	6
3	8
4	17
5	21
6	8
7	6
Σ	69

$B(4, 4)$

n	Rank
1	4
2	10
3	20
4	55
5	99
6	84
7	80
8	40
9	20
10	10
Σ	422

$B(d, 4), d > 4$

n	Rank
1	d
2	$d + \binom{d}{2}$
3	$2\binom{d}{2} + 2\binom{d}{3}$
4	$3\binom{d}{2} + 8\binom{d}{3} + 5\binom{d}{4}$
5	$2\binom{d}{2} + 15\binom{d}{3} + 27\binom{d}{4} + 14\binom{d}{5}$
6	?
.	?
.	?
$3d - 3$	$(d-1)d(d+1)/3$
$3d - 2$	$d(d+1)/2$

$B_0(2, 5)$

n	Rank
1	2
2	1
3	2
4	3
5	2
6	4
7	4
8	4
9	6
10	3
11	2
12	1
$\sum$	34

$B_0(3, 5)$

n	Rank
1	3
2	3
3	8
4	18
5	30
6	71
7	132
8	240
9	411
10	486
11	453
12	278
13	$\leqslant 87$
14	$\leqslant 48$
15	$\leqslant 8$
16	$\leqslant 3$
17	$\leqslant 3$
$\sum$	$\leqslant 2282$

$B_0(2, 7)$

n	Rank
1	2
2	1
3	2
4	3
5	6
6	9
7	12
8	23
9	36
10	61
11	94
12	159
13	260
14	406
15	640
16	985
17	1510
18	2157
19	?
.	?
.	?
$\sum$	>6366

It is expected that $B_0(2, 7)$ has nilpotency class about 30 and order about 7^{10000}.

$B_0(2, 8)$

n	Rank
1	2
2	3
3	5
4	5
5	9
6	11
7	18
8	30
9	56
10	99
11	177
12	307
13	558
14	960
15	?
.	?
.	?
Σ	>2240

$B_0(2, 9)$

n	Rank
1	2
2	3
3	3
4	3
5	6
6	9
7	18
8	30
9	56
10	98
11	176
12	320
13	572
14	1042
15	?
.	?
.	?
Σ	>2338

4.8. Let us look at the general situation of an arbitrary group G of prime-power exponent p^k, $k > 1$. Using the Campbell-Hausdorff formula, Sanov [231] showed that

$$(x_m, y; s(p^s - p^{s-1}))^{p^{k-s}} \in \gamma_r(G), \qquad s = 1, 2, \ldots, k\,, \tag{7.23}$$

for all $x_m \in \gamma_m(G)$ and $y \in G$, where

$$r = \min(pm + 1, m + s(p^s - p^{s-1}) + 1)\,.$$

Result (7.23) is non-trivial when $m \geqslant sp^{s-1}$, and if $x_m = (x, y; sp^{s-1} - 1)$, the inclusion

$$(x, y; sp^s - 1)^{p^{k-s}} \in \gamma_{sp^s+1}(G)\,, \tag{7.24}$$

holds; for $s = k$, it is equivalent to the $(kp^k - 1)$-th Engel congruence. Inclusion (7.23) is improved slightly in [53] to give

$$r = \min(pm + p - 1, m + s(p^s - p^{s-1}) + 1)\,;$$

the proof is more elementary and uses group algebras.

Even earlier, Bruck [33] had formulated the philosophical question: "What makes a group finite?", and stated the useful conjecture that (7.24) can be replaced when $r = k$ by the Engel congruence

$$(x, y; f(k) - 1) \equiv 1 \bmod \gamma_{f(k)+1}(G) \tag{7.25}$$

with $f(k) = kp^k - (k-1)p^{k-1} - 1)$. Krause proved it for $p^k = 8$, when he obtained the 14-th Engel congruence mentioned above. The translations of (7.24) or (7.25) into the language of the associated algebra $L(G)$ seem to give rather little, but it is more essential that the injection of other multilinear identities will improve the mechanics of the connection between groups and Lie algebras. The results of E. I. Zel'manov (1987; see also [286]) are very convincing in this respect. Let us pay some attention to them.

4.9. The following question arises in connection with (7.4) and (7.24):

Problem ($G_{r,s}$). *Assume that the Lie algebra L over* $\mathbb{Z}_p$ *generated by the set* $X = \langle x_1, \ldots, x_d, \ldots \rangle$

a) *satisfies the linearized Engel identity of degree r:*

$$\sum_{\sigma \in S_r} [u_0 u_{\sigma 1} u_{\sigma 2} \ldots u_{\sigma r}] = 0\,;$$

b) *has the property that* $(\mathrm{ad}\ w)^s = 0$ *for some natural number* $s \geqslant r$ *and every commutator* $w \in L$ *(that is, every Lie word in X).*

Is L locally nilpotent?

It is comparatively easy to show that a positive solution to (G_{p^k-1,kp^k-1}) yields that *every Lie algebra over the ring* $\mathbb{Z}_pk$ with a) $r = p^k_{-1}$, and b) $s = kp^k - 1$ is locally nilpotent. However, the Lie algebra $L = L(G)$ of a group G satisfying the identical relation $x^q = 1$, $q = p^k$, is of this sort; that is,

$$(G_{p^k-1,\,kp^k-1}) \Rightarrow \text{ RBP for exponent } p^k$$

(implication here means the *positive* solution).

By identity (1.1) of Chap. 1, conditions a), b) in problem $(G_{r,s})$ are satisfied in every Lie algebra satisfying the identity $[uv^r] = 0\,(s \geqslant r)$. Therefore, we have another restatement of our main problem:

Problem (E_n). Is every Lie algebra with E_n, that is, $[uv^n] = 0$, locally nilpotent?

Finally, we can impose a stronger condition on L, namely that L and its p-envelope $\tilde{L}$ satisfy E_n and all its partial linearizations:

$$\left[u_0 \begin{Bmatrix} u_1 & \ldots & u_s \\ j_1 & \ldots & j_s \end{Bmatrix}\right] = 0, \qquad j_1 + \cdots + j_s = n\,.$$

Problem ($\tilde{E}_n$). *Is every Lie p-algebra over* $\mathbb{Z}_p$ *satisfying* E_n *and all its partial linearizations locally nilpotent?*

The implications $(G_{r,s}) \Rightarrow (E_n) \Rightarrow (\tilde{E}_n)$ in a plan for a positive solution of these problems can in fact be put into a circle (this is the main achievement of Zel'manov (see [286])):

$$\begin{array}{ccc} & (G_{r,s}) & \\ \swarrow & & \nwarrow \\ (E_n) & \Rightarrow & (\tilde{E}_n) \end{array}$$

Deduction: *If* $(\tilde{E}_p k_{-1})$ *has a positive solution, so does* RBP *for exponent* p^k.

However, we must make an essential sharpening of all this. Suppose that the symbol PS(T; d) means: "*the positive solution of problem T for groups (or algebras) on d generators*". Then, what has been established is this:

$$PS(\tilde{E}_{p^k-1}; \tilde{d}) \Rightarrow PS(RBP\ p^k; d)\ ;$$

however, $\tilde{d}$ has to be chosen significantly larger than d (in symbols, $\tilde{d} \gg d$). We will come across a similar phenomenon a little later (see 4.10).

The question arises here about possible methods for solving $(\tilde{E}_n)$. The reason why p-algebras come into the picture is that (E_n) implies the identity $x^{[p^l]} = 0$ for $p^l \geqslant n$ (by definition, $x^{[p^l]} = \pi^l(x)$, where $\pi : x \mapsto x^{[p]}$ is the p-map). It is natural to select an element $a \neq 0$ such that $a^{[p]} = 0$, that is, $(\text{ad } a)^p = 0$.

As usual, we say that an element $c \neq 0$ is a (*thin*) *sandwich* if

$$(\text{ad } c)^2 = 0, (\text{ad } c)(\text{ad } u)(\text{ad } c) = 0, \qquad u \in L$$

(the second condition is not superfluous when $p = 2$). The descent from a with $(\text{ad a})^p = 0$ to a thin sandwich—even if this is possible—is a very delicate operation, about which we say nothing. However, if a sandwich $c = f(y_1, y_2, \ldots, y_m)$ can be found, the problem is solved. This is because at the very outset we can assume that the locally nilpotent radical $R(L)$ of L is zero. In addition, since all partial linearizations are implied in $(\tilde{E}_n)$, the ground field $F \supset \mathbb{Z}_p$ can be assumed infinite. If $f(L)$ is the set of values of $f(y_1, y_2, \ldots, y_m)$ in L, then as we remarked in 5.2.2, the F-subspace V_f of L spanned by $f(L)$ is an ideal of L. In our case, V_f is generated by thin sandwiches. It is locally nilpotent (by Theorem 3.1 of Chap. 5, and the improved version of it for all $p>0$ in [287]), in contradiction to our original premiss.

Thus, to solve $(\tilde{E}_n)$, all that remains to do is to overcome the problem of "smallness"—that is, to construct at least one sandwich in our Lie algebra L over a field F of characteristic $p > 0$. Zel'manov has shown that a Lie algebra with E_n is locally nilpotent as long as $n \leqslant p^2 - p$. This has been done independently for $n = 6$ and arbitrary $p > 0$ by Zel'manov; and for $n = 7$, $p = 3$ by Chanyshev. Unfortunately, RBP for exponents $p^k = 8$ and 9 is associated with the pairs $n = 7$, $p = 2$ and $n = 8$, $p = 3$, and what is in store for us here appears to be in the future!

The same methods have been used to prove that there exists a function $e(n)$ of a natural argument such that every nilpotent n-Engel torsion-free group (see 5.5.4) has nilpotency class at most $e(n)$. Furthermore, every m-generator nilpotent 4-Engel group is nilpotent of class at most $e'(m)$.

In conclusion, we note that A. D. Chanyshev [42] has given an effective proof that every 2-generator Lie algebra L with E_5 over Z_2 is nilpotent: in fact, $L^{31} = 0$.

4.10. Theorem (G. Higman and P. Hall [91]). *Suppose that*

$$n = p_1^{k_1} \ldots p_r^{k_r}, \qquad 2 \leqslant p_1 < \cdots < p_r$$

where the p_i *are primes. Suppose further that*

(i) *the maximal finite group* $B_0(d_i, p_i^{k_i})$ *exists for* $i = 1, \ldots, r$ *and all* d_i;

(ii) *the number of pairwise non-isomorphic finite simple groups of exponent m dividing n is finite.*
Then $B_0(d, n)$ exists for every d.

Thus, because of the classification of finite simple groups (which is now said to be complete), RBP for arbitrary exponent n reduces to RBP for primary exponents dividing n, with $d_i \gg d$ as always.

In the case of odd exponent n (that is, $p_1 > 2$), condition (ii) is unnecessary, because of the Feit-Thompson theorem on the solubility of groups of odd order. It is also unnecessary when $r = 2$ because of Burnside's classical $p^\alpha q^\beta$-theorem.

M. Hall Jr [83, 86] has proved that groups of exponent 6 are locally finite (see also the simplified proof of M. F. Newman [203]). Thus, $B(d, 6) = B_0(d, 6)$. It follows from this and from results in [91] that

$$|B(d, 6)| = 2^a \cdot 3^{b + \binom{b}{2} + \binom{b}{3}},$$

where $a = 1 + (d-1)3^{d + \binom{d}{2} + \binom{d}{3}}$ and $b = 1 + (d-1)2^d$.

4.11. Khukhro solved RBP for group in the variety $\mathfrak{M}_p$(see [130]): $G \in \mathfrak{M}_p$ if and only if G has a splitting automorphism ϕ of order p, that is,

$$x x^\phi x^{\phi^2} \ldots x^{\phi^{p-1}} = 1$$

for all $x \in G$. Groups of exponent p and finite groups with regular automorphisms of prime order p lie in $\mathfrak{M}_p$. It is proved that the locally nilpotent groups in $\mathfrak{M}_p$ comprise a variety of groups with operators. More exactly, there is a natural number $f(d, p)$ such that the nilpotency class of every d-generator nilpotent group possessing a splitting automorphism of order p is not more than $f(d, p)$. In particular, every finite d-generator splitting p-group has a subgroup of index p and nilpotency class not more than $f(d-1, p)$. This is another interpretation of the results about p-groups not coinciding with their Hughes subgroup $H_p(G)$ (see 2.6). The proofs are based on Theorem 1.7.5 and the Thompson-Higman-Kostrikin-Kreknin theorem about the existence of an upper bound $f(p)$ for the nilpotency class of a finite group with a regular automorphism of prime order p.

Appendix 1

An Effective Version of the Proof of Theorem 1.7.4 in Terms of Recursive Functions (due to E. I. Zel'manov)

We denote by $\phi(m, n)$ the nilpotency class of the m-generator free Lie algebra $L = L(m, n)$ with E_n over a field of characteristic $p > n$. For convenience, here and below nilpotency class in the usually-accepted sense (see Definition 1.2.3) is increased by 1, so that $L^{\phi(m,n)-1} \neq 0$, $L^{\phi(m,n)} = 0$. Theorem 1.7.4 establishes the existence of $\phi(m, n)$. Here we shall construct a primitive recursive function $\tilde{\phi}(m, n)$ majorizing $\phi(m, n)$. It is possible to give "effective" versions of the various proofs contained in the book; however, we shall do this only for the shortest one—the geodesic. We note, however, that $\tilde{\phi}(m, n)$ cannot be regarded as a reasonable estimate for $\phi(m, n)$. For example, it gives unrealistically large estimates for the known value of $\phi(2, 4)$, namely 13, and for the conjectured value $\phi(2, 6) \approx 30$. More than that, in our opinion no effective version of the proofs in this book can give reasonable estimates for $\phi(m, n)$. For this reason, we have not been ashamed to choose the very worst bounds anywhere below, provided that the choice was motivated by considerations of simplicity and brevity.

A.1. Sandwich Algebras (effective version of 5.1.3, pp. 111, 112). We consider a pair (L, A), where L is a Lie algebra and A its associative enveloping algebra; L is generated by thin sandwiches $a_1, \dots, a_m$ of the pair (L, A). We assume further that $u^n = 0$ in A for all $u \in L$, and $n < p$. Denote the greatest lower bound of the nilpotency classes of the associative enveloping algebras in all such pairs by $\psi(m, n)$. Clearly, $\psi(m, 1) = 1$ and $\psi(1, n) = n$. Since the arguments proceed by induction on m and n, we consider the Lie subalgebra L_0 of L generated by $a_1, \dots, a_{m-1}$. It is clear that $\dim L_0 < (m-1)\psi^{(m-1,n)}$. In this section we construct a primitive recursive function majorizing $\psi(m, n)$, using the parameters

$$q = \psi((\dim L_0)^2, n-1),$$

$$r_0 = [4q + 2\psi(m-1, n)]\psi(m-1, n) + 2q + 1,$$

$$r = 2^{n-3}r_0 + (2^{n-3} - 1)(n-1)\psi(m, n-1).$$

The space spanned by the commutators $[a_m, x, x']$ with $x, x' \in L_0$ has dimension at most $(\dim L_0)^2$. It was proved at the end of 5.1.3 that the following relation holds for arbitrary commutators $x_i, x'_i \in L_0$, $1 \leqslant i \leqslant q$:

$$\left(\prod_{i=1}^{q} [a_m, x_i, x'_i]\right) a_m = 0.$$

It follows from this (see pages 109–110) that the following relation holds for arbitrary commutators $x_0, x_1, x'_1, \ldots, x_{2q+1}, x'_{2q+1}$ in L_0:

$$[x_0 a_m x_1 x'_1 a_m \ldots a_m x_{2q+1} x'_{2q+1} a_m] = 0 .$$

Now by 3.5.3, L contains a non-zero *thick* sandwich of the pair (L, A) of one of the following types:

$$x_0 ;$$

$$[x_0 a_m x_1 x'_1 a_m \ldots x_t x'_t a_m] ;$$

$$[x_0 a_m x_1 x'_1 a_m \ldots x_t x'_t a_m x''_1 \ldots x''_k],$$

where $t \leqslant 2q$; $x_i, x'_i, x''_i \in L_0$. It is easy to see that $k < 2\psi(m-1, n) - 1$. In all cases, the weight of this element in $a_1, \ldots, a_m$ does not exceed

$$(2t + 1 + k)\psi(m-1, n) + t + 1 \leqslant r_0 .$$

For a subset M of A, the ideal of A generated by M is denoted by $I_A(M)$. When $M = \{x\}$, we write $I_A(x)$.

Lemma 1. *The Lie algebra L contains a non-zero commutator c of weight less than r such that $I_A(c)^n = 0$.*

Proof. It has been proved earlier that L contains a sandwich of the pair (L, A) of thickness 2 and weight at most r_0. Denote it by c_2. If $I_A(c_2)^n \neq 0$, there exist words $w_i = a_{j_1} \ldots a_{j_{d_i}}$, $1 \leqslant i \leqslant n-1$, such that

$$c_2 w_1 c_2 \ldots c_2 w_{n-1} c_2 \neq 0 .$$

Here we may assume that the set of words $w_1, \ldots, w_{n-1}$ has length vector $(d_1, \ldots, d_{n-1})$ which is lexicographically smallest among those of the words with the stated property. In that case, the length of each word w_i is less than $\psi(m, n-1)$. Otherwise (see the end of the proof of 5.1.3 on p. 112), we have

$$c_2 w_i c_2 = \sum_j c_2 w'_{ij} c_2 w''_{ij} ,$$

where the words w''_{ij} are non-empty, and the w'_{ij} have length less than d_i.

Furthermore, transfer of unnecessary words to the left in the usual way leads to the equality

$$c_2 w_1 c_2 \ldots c_2 w_{n-1} c_2 = \sum \pm v_i c_{i1} \ldots c_{in} ,$$

where the v_i are words in $a_1, \ldots, a_m$; the c_{ij} are commutators in $c_2, a_1, \ldots, a_m$; and every summand on the right-hand side has exactly the same composition as on the left. This means that there is an index i such that $c_{i1} \ldots c_{in} \neq 0$. It follows from this that $[c_{ij}, c_{ik}] \neq 0$ for some j, k with $1 \leqslant j < k \leqslant n$. Otherwise

$$c_{i1} \ldots c_{in} = \frac{1}{n!} \begin{Bmatrix} c_{i1} & \ldots & c_{in} \\ 1 & \ldots & 1 \end{Bmatrix} = 0 .$$

We set $c_3 = [c_{ij}, c_{ik}]$. As was proved in § 3 of Chap. 5, c_3 is a sandwich of the pair (L, A) of thickness at least 3. Its weight in $a_1, \ldots, a_m$ does not exceed $2r_0 + (n-1)\psi(m, n-1)$. If $I_A(c_3)^n \neq 0$, we use similar methods to construct a commutator c_4 of weight less than $2(2r_0 + (n-1)\psi(m, n-1)) + (n-1)\psi(m, n-1)$, which is a sandwich of thickness at least 4, *etc.* At the $(n-1)$th step we get a commutator c_{n-1} of weight less than $2^{n-3}r_0 + (2^{n-3} - 1)(n-1)\psi(m, n-1) = r$, which obviously has the property that $c_{n-1}Ac_{n-1} = 0$. This proves the lemma. □

Lemma 2. *The following inequality holds*:

$$\psi(m, n) < n^{m^r} .$$

Proof. By Lemma 1, there is a commutator w_1 in $a_1, \ldots, a_m$ of weight less than r such that $I_1^n = 0$, where $I_1 = I_A(w_1)$. Applying Lemma 1 to the factor-pair $(L + I_1/I_1, A/I_1)$, we find that there exists a commutator w_2 of weight less than r such that $w_2 \notin I_1$ and $I_2^n \subseteq I_1$, where $I_2 = I_A(w_1, w_2)$, *etc.* This gives rise to an ascending chain of ideals $I_1 \subset I_2 \subset \cdots \subset I_k$. The ideal I_k contains all commutators of weight less than r in $a_1, \ldots, a_m$ so that $I_k = A$. Since there are not more than $m + m^2 + \cdots + m^{r-1} < m^r$ different commutators of weight less than r, we have $k < m^r$. Furthermore, $I_{i+1}^n \subseteq I_i$, so that $I_k^{n^{m^r}} = 0$. This proves the lemma. □

Next, we introduce the function $\tilde{\psi}(m, n)$ recursively, replacing r at every point by an explicitly defined parameter:

$$\tilde{\psi}(m, 1) = 1, \qquad \tilde{\psi}(1, n) = n ;$$

$$\tilde{q} = \tilde{\psi}(m^{2\tilde{\psi}(m-1, n)}, n-1) ;$$

$$\tilde{r}_0 = (4\tilde{q} + 2\tilde{\psi}(m-1, n))\tilde{\psi}(m-1, n) + 2\tilde{q} + 1 ;$$

$$\tilde{r} = 2^{n-3}\tilde{r}_0 + (2^{n-3} - 1)(n-1)\tilde{\psi}(m, n-1) ;$$

$$\tilde{\psi}(m, n) = n^{m^{\tilde{r}}} .$$

We have shown that $\psi(m, n) < \tilde{\psi}(m, n)$ (see Lemma 2).

A.2. The Locally Nilpotent Radical from an Effective Point of View. We introduce once more the free m-generator Lie algebra $L = L(m, n)$ with E_n over a field of characteristic $p > n$ into the discussion. Let $a_1, \ldots, a_m$ be a set of free generators for it. As was proved in § 2 of Chap. 5, the ground field can be assumed to be infinite. Then, for every homogeneous Lie polynomial f, the linear span of the set $f(L)$ of values of f is an ideal in L, which we denote by $V_f(L)$.

Suppose that $f = f(x_1, \ldots, x_k)$ has degree d_i in x_i, $1 \leqslant i \leqslant k$, and set $d = \sum_{i=1}^k d_i$.

Lemma 3. *Assume that the factor-algebra $L/V_f(L)$ is nilpotent of class N. Then the algebra $V_f(L)$ has a generating set consisting of not more than $m^{nNd} \cdot d!$ elements of $f(L)$.*

Proof. Any commutator v in $a_1, \ldots, a_m$ of weight nN can be represented in the form $v = \sum_i \pm [v_{i1}, v_{i2}]$, where v_{i1}, v_{i2} are commutators in $a_1, \ldots, a_m$ of weights at least N. In fact, suppose that $v = [a_{i_1}, \ldots, a_{i_{nN}}]$, $1 \leqslant i_1, \ldots, i_{nN} \leqslant m$. By Proposition 1.4.6 (see p. 15), $v = \sum_j \pm [a_{i_1}, \ldots, a_{i_N}, w_{j,1}, \ldots, w_{j,n-1}]$, where the $w_{j,v}$ are commutators in $a_1, \ldots, a_m$. If the weights of all the $w_{j,v}$ are at most $N-1$ for some j, then $nN - N \leqslant (n-1)(N-1)$. This is a contradiction, and it means that $w_{j,v}$ has weight at least N for some v. The assertion concerning the weights of the commutators v_{i1} and v_{i2} follows from this.

Let $\tilde{f} = \tilde{f}(y_1, \ldots, y_d)$ be the complete linearization of the polynomial f. Assume that the variables y_j with $1 \leqslant j \leqslant d_1$ appear on linearization with respect to x_1, and the y_j with $d_1 < j \leqslant d_1 + d_2$ on linearization with respect to x_2, *etc.* Let $u_1, \ldots, u_d$ be any collection of elements of L. There may be coincidences among the u_j with $d_1 + \cdots d_i < j \leqslant d_1 + \cdots + d_i + d_{i+1}$. Split the elements u_j of this form into equivalence classes with respect to coincidence, and let $d_{ii'}$, $1 \leqslant i' \leqslant q_i$, be the cardinals of the equivalence classes. If similar terms connected with just the one equivalence class of cardinal $d_{ii'}$ are included in $\tilde{f}(u_1, \ldots, u_d)$, the factor $d_{ii'}!$ will arise. Thus the normalized element

$$f_{(n)}(u_1, \ldots, u_d) = \tilde{f}(u_1, \ldots, u_d) \Big/ \prod_{\substack{i \leqslant i \leqslant k \\ 1 \leqslant i' \leqslant q_i}} (d_{ii'}!)$$

is well-defined. We must emphasise once more that, formally speaking, the numerator and denominator on the right-hand side may be zero. However, if we bring all similar terms into the numerator, the denominator cancels, and the object we arrive at is the one we want.

The ideal $I = V_f(L)$ is homogeneous. Its homogeneous component I_t of weight t is spanned by elements $\tilde{f}(u_1, \ldots, u_d)$, where $u_1, \ldots, u_d$ are commutators in $a_1, \ldots, a_m$ of total weight t. Every commutator of weight N in $a_1, \ldots, a_m$ lies in I. It follows from what was proved above that the algebra I is generated by the subspace $\sum_{t<nN} I_t$. In fact, for $t \geqslant nN$, a commutator v of weight t can be represented in the form $v = \sum_i [v_{i1}, v_{i2}]$, where the commutators v_{i1} and v_{i2} lie in homogeneous components of I of smaller weights.

The number of different cortèges $(b_1, \ldots, b_d)$, where the b_i are non-empty associative words in $a_1, \ldots, a_m$ of total degree t, is $m^t \cdot \binom{m^t - 1}{d-1}$. For, there are m^t ways of choosing words $b = b_1 \ldots b_d$ of degree t. Thereafter, it remains for us to choose $d-1$ different positions in b at which the subwords $b_1, \ldots, b_{d-1}$ end (the initial position is excluded).

Since $m^t \cdot \binom{m^t - 1}{d-1} < m^{td}$, the algebra I has a generating set consisting of not more than $\sum_{t=d}^{nN-1} m^{td} < m^{nNd}$ elements of $\tilde{f}(L)$. However, every element of $\tilde{f}(L)$ is a linear combination of not more than $d_1! \ldots d_k! < d!$ elements of $f(L)$. The lemma follows from this. □

A.3. Completion of the Proof (construction of the primitive recursive function $\tilde{\phi}(m, n)$). It was proved in § 2 of Chap. 5 that for every n-Engel Lie algebra L (with $n < p$), there exists a Lie polynomial f that is not the identity on L and such that $f(L)$ consists of nil-elements of index at most 3. Combining Chap. 2, § 2 and Chap. 5, § 2, we can strengthen this result by proving: *for every n-Engel Lie algebra $L(n < p)$ there exists a Lie polynomial f that is not the identity on L and such that its set $f(L)$ of values consists of thin sandwiches.*

With this aim, we consider the set of integral cortèges

$$\alpha = (k_1, \ldots, k_s | r_1, \ldots, r_t) = (\alpha' | \alpha''),$$

such that $0 < s, t$; $3 \leqslant k_s < k_{s-1} < \cdots < k_1 \leqslant n-1$; $r_1 \leqslant n-2$, $2r_{i+1} \leqslant r_i$, and α'' does not end with more than two zeroes. The descent to thin sandwiches in § 2 of Chap. 2 was effected using polynomials of the form

$$g_s(b, u) = [u[bu]^s b^2].$$

We set

$$f_\alpha(x_0, \ldots, x_s, u_1, \ldots, u_t)$$
$$= g_{r_t}(g_{r_{t-1}}(\ldots g_{r_1}([x_s \ldots [x_2[x_1 x_0^{k_1}]^{k_2} \ldots]^{k_s}, u_1), \ldots, u_{t-1}), u_t),$$

Clearly, $f_{(\phi|\phi)}(x_s) = x_s$. The polynomial f_α is homogeneous, of degree

$$(r_t + 2)((r_{t-1} + 2)(\ldots (r_1 + 2)[k_s(\ldots (k_2(k_1 + 1) + 1) \ldots) + 1]$$
$$+ (r_1 + 1)) + (r_2 + 1)) \ldots) + (r_t + 1),$$

which is obviously less than $(n!)^2$.

We order cortèges lexicographically. If $\alpha = (\alpha'|\alpha'')$, $\beta = (\beta'|\beta'')$ and α' is a left end of β', then $\beta > \alpha$; if $\alpha' = \beta'$ and α'' is a left end of β'', then $\beta > \alpha$. It is easy to see that every cortège other than $(\phi|\phi)$ has an immediate predecessor, namely $\beta = \max\{\gamma | \gamma < \alpha\}$. We denote it by $\alpha - 1$.

It follows from the results of § 2 of Chap. 2 and § 2 of Chap. 5 that every element of $f_{\alpha_1}(L)$, where $\alpha_1 = (n-1, n-2, \ldots, 3 | n-2, [\frac{n-2}{2}])$, is a thin sandwich. Moreover, if the identity $f_{\alpha-1} = 0$ holds in an n-Engel Lie algebra L, then every element of $f_\alpha(L)$ is a thin sandwich.

We consider now the sequence starting with α_1 and consisting of terms $\alpha_{i+1} = \alpha_i - 1$, $i \geqslant 1$. At some point we get $\alpha_k = (\phi|\phi)$. Of course, the number k is less than the total number of cortèges, which is less than $(n!)^2$.

We denote the linear span of $f\alpha_{k-i}(L)$ by V_i, for $0 \leqslant i < k$; $V_k = \{0\}$. This is a sequence of ideals in L. Let $\Phi(i, m, n)$ denote the nilpotency class of the factor-algebra L/V_i. Then $\Phi(0, m, n) = 1$, and $\Phi(k, m, n)$ is the nilpotency class of L.

If we put $d = (n!)^2$, $N = \Phi(i, m, n)$ in Lemma 3, we get that the Lie algebra V_i has a generating set consisting of less than

$$t = m^{n(n!)^2 \cdot \Phi(i, m, n)}((n!)^2)!$$

elements of $f_{\alpha_{k-1}}(L)$. Then $V_i^{\psi(t,n)} \subseteq V_{i+1}$. Therefore, L/V_{i+1} is soluble of length less than $\Phi(i, m, n) + \psi(t, n)$.

Lemma 4. *If the pair* $(\bar{L}, \bar{A})$ *is such that* $u^n = 0$ *for all* u *in* $\bar{L}$, *and* $\bar{L}$ *is soluble of length* l, *then* A *is nilpotent of class less than* n^l.

Proof. Let J be the $(l-1)$st term of the derived series of $\bar{L}$. Then $I_{\bar{A}}(J)^n \subseteq \bar{A}\underbrace{J \ldots J}_{n} = 0$. It is now enough to go over to the factor-algebra $\bar{A}/I_{\bar{A}}(J)$ and use induction on l (see also the proof of Theorem 6.2.2). □

It follows from the above and Lemma 4 that

$$\phi(i+1, m, n) < n^{\Phi(i,m,n)+\psi(t,n)} ,$$

We define a function $\tilde{\Phi}(i, m, n)$ by recursion, setting

$$\tilde{\Phi}(0, m, n) = 1 ;$$

$$\tilde{\Phi}(i+1, m, n) = n^{\tilde{\Phi}(i,m,n)} + \tilde{\psi}(m^{n(n!)^2\tilde{\Phi}(i,m,n)}((n!)^2!, n) ,$$

where $\tilde{\psi}$ is the function in A.1. Clearly, $\tilde{\Phi}((n!)^2, m, n)$ can be taken for the desired function. In other words, *every m-generator Lie algebra with* E_n *has nilpotency class*

$$\phi(m, n) < \tilde{\phi}(m, n) = \tilde{\phi}((n!)^2, m, n) . \quad \square$$

A1. Commentary

The impression might be created that the primitive recursive bound obtained here contradicts the statement in the preamble to Chap. 1, § 3 about the absolutely non-effective nature of the approach based on a consideration of the locally nilpotent radical. However, if effectiveness is given a meaning that is at all sensible, the latter will rise up in rebellion against the monstrous roughness of the explicit bounds that can be extracted when the primitive recursive functions are removed. This is clear also because the arguments in Appendix 1 and in the main text (in any of the variants) are virtually identical. Meanwhile, as has been emphasised repeatedly, to obtain realistic bounds it will be necessary to employ combinatorics of a completely different sort. No such exist for $p > 5$.

This appendix, which was written in January 1987 and has been processed slightly for the English translation, is perhaps remarkable not only for the construction of a bounding recursive function, but also for its conciseness and for the additional touches that have illuminated the main canvas of the book. The qualified reader will have no difficulty in providing some of the missing details for himself.

Currently, there exists another primitive recursive upper bound for $\phi(m, n)$. It appears in the long paper [8] by S. I. Adian and A. A. Razborov, where nearly all the calculations of a good half of this book (from a manuscript that was available to those authors) are reproduced in an excessively rectilinear fashion. It scarcely makes sense to compare the Zel'manov and Adian-Razborov bounds for $\phi(m, n)$. It is best to say simply that they exist. This trivial but sufficiently exact statement really does express the essence of the matter, since it is actually more difficult to devise an algebraical proof that could possibly fail to be "effectivizable" in the sense of recursive function theory.[1]

---

[1] A new approach to a constructive proof of the main theorem has been suggested quite recently by Vaughan-Lee [263].

Appendix II
A Short Biography of William Burnside (after A. R. Forsyth [52])

William Burnside was born on 2nd July 1852 in London, into the family of a merchant of Scottish ancestry. Although he was left an orphan at the age of six, Burnside received a good education. The happy student years at Cambridge, where he was recommended to study after a highly successful performance in the mathematical school at Christ's Hospital, gave way to ten years of teaching mathematics and hydrodynamics in various Cambridge colleges. In 1885, Burnside was appointed professor of mathematics in the Royal Naval College at Greenwich. The work of training naval officers was to his liking. Moreover, he was not burdened by administrative responsibility, and thus it left him plenty of leisure time, which he filled almost entirely with his scientific researches.

Burnside's natural endowments also showed up in fields other than mathematics. In his youth, there sprang up and flourished a long enthusiasm for rowing; his light weight notwithstanding, he enjoyed a deserved reputation as a brilliant number 7. In later years he developed a remarkable aptitude for fishing. Burnside retired in 1919 and settled with his family (wife, two sons and three daughters) in Kent. He died on 21st August 1927.

Burnside's life was not spoiled by the honours and rewards that would have been commensurate with his mathematical achievements. Nevertheless, he was awarded the honorary degree of Doctor of Science by the Universities of Dublin and Edinburgh, and elected Fellow of the Royal Society in 1893: he was awarded one of the Society's two medals in 1904, elected member of the Council of the London Mathematical Society from 1899 to 1917, awarded the de Morgan medal in 1899, was President of the Society from 1906 to 1908 and in 1900 he was elected to an honorary fellowship of his old college, Pembroke. Burnside was occasionally called upon to participate in the work of external examination in British universities, and also acted as a constant (and exemplary) referee for the Royal Society and the London Mathematical Society. In his declining years, Burnside expressed his deep conviction that the happy and successful pursuit of science was an honour in itself. His interests as a "pure" as well as an applied mathematician were extremely diverse. Burnside published over 150 articles, and also the widely-read tract "The Theory of Groups", the second, extended, version of which (1911) received universal acclaim.

Influenced as he was by the Cambridge school of natural philosophy, Burnside was slow to become aware of the existence of the fascinating world of pure mathematics. Despite this fact, and fully a decade devoted to problems in applied mathematics and to teaching hydrodynamics, a man of his intellect could not fail to take up the challenge of the burning questions of complex function theory. It is no accident that Burnside's first scientific publication was a paper on elliptic functions (1883). In subsequent articles he worked on: the kinetic theory of gases (1887–1888), deep-water waves (in connection with the eruption of the volcano Krakatoa), the two-dimensional potential problem with various boundary conditions (1891), and the kinematics of non-Euclidean space (1895). Right up to 1918, he maintained an interest in problems in the theory of functions of real variables and in probability theory.

However, at the beginning of the nineties, Burnside's imagination was taken by the theory of transformation groups and its applications in geometry and the theory of automorphic functions, which was being vigorously developed at exactly that time in the works of Poincaré. The greater part of Burnside's publications are small notes, and his only long paper (which is in two parts) is devoted to automorphic functions, and is a very substantial contribution to the discipline. In these same years, the theory of continuous groups was being developed powerfully in the works of Sophus Lie and his successors. As regards the theory of discrete groups (second half of the XIXth Century), after those of C. Jordan and L. Sylow, the first names to be quoted must be those of W. Burnside, O. Hölder, von Dyck, F. Klein and G. Frobenius[1].

Burnside devoted himself to the systematic development of finite group theory. His papers appeared year-by-year, dealing with different Chapters of the theory and marked by a magnificently wide and erudite mind. The first landmark in this creativity was the book "The Theory of Groups", which was published in 1897, and has been an influence on the state of the subject right up to the present time. The second, and greatly extended, edition of the book (1911) underpinned the original results of many years of Burnside's indefatigable activity (more than fifty papers!) on describing the diverse properties of finite groups and their representations.

*

* *

And now, after sixty years have elapsed since Burnside's death, it is fully clear how finely he penetrated to the heart of the objects of his study. Two of the problems he posed (B1 on the local finiteness of periodic groups, and B2 on the solubility of groups of odd order) have played an exceptionally important part in the development of group theory. At a certain stage, their fates are intertwined; the reduction of RBP for general exponent n in the paper of Hall and Higman [91] to RBP for prime-power exponents was a conditional result that rested on facts that could only be decided after the solution of problem B2. This solution, given in the fundamental paper of Feit and Thompson [1963] turned out to be not only unusually complicated but astonishingly rich in its ideological planning. Problem

[1] And, we add, F. E. Molin, a Russian algebraist who obtained independently, and somewhat differently, a number of the basic results in the representation theory of finite groups.

B1 has also been the source of a fruitful cycle of investigations differing very considerably in the variety of technical apparatus used.

One cannot but mention another contemporary of Burnside in connection with the solution of RBP for prime exponent p presented in this book. I have in mind Friedrich Engel (1860–1941), a remarkable mathematician who was a close friend and devoted collaborator of Sophus Lie; as was stated figuratively in [215], he was a gift of fate. This fine personality is associated not only with transformation groups and Engel Lie algebras: for us, Engel is a vital link in a long chain connecting such widely distant mathematical disciplines as Lie groups, Lie algebras, the condition E_{p-1}, and finite groups of given exponent.

Epilogue

The method of sandwiches lies at the heart of this book. Unfortunately, the book itself is written in the form of a sandwich, with layers of similar material in different sections and even in different chapters. The explanation for this phenomenon is that the plan of the book, apart from the very stable Chapters 1–4, changed during writing under the influence of new results. Perhaps it would be worthwhile to produce a slight rearrangement of the material, or even to equip it with constructive details. However, this is fairly difficult to accomplish technically, if one is not to sacrifice publishers' deadlines and the aims stated in the preface.

Following the survey [144], we add that work "Around Burnside", which is obviously pretty wide, is also very diverse. As is the case in any discipline distinguished by significant achievements, there remain unsolved problems, and new questions have been posed. Many of them are mentioned in the preceding pages, but the most important is old and obvious: RPB for arbitrary prime-power exponent p^k. The road to the truth here will scarcely be easy, unless the answer turns out to be negative and obtained *via* the construction of an elegant example. If the present book should inspire in some way the solution of the existence problem for the universal finite group $B_0(d, p^k)$ for $k > 1$, the author will account himself as being involved (albeit indirectly) in that exciting activity, even though his personal participation is ruled out as a possibility.

Note Added in Proof

This book is a very fortunate one: it was written at a turning-point in the history of the restricted Burnside problem. The latest achievements far outstrip the most daring of prognoses, and to attempt to give some idea of what they are like, even in commentary form, would be rather in the nature of newspaper reporting. At the very beginning of 1989, after the completion of the English translation, E. I. Zel'manov solved RPB for all prime-power exponents p^k. The case $p > 2$ will appear in *Izvestiya Akad. Nauk SSSR Ser. Mat.*, and the more complicated case $p = 2$ in *Mat. Sbornik.* The proof is based on the reduction, expounded in the commentary to § 4 of Chap. 7, to the theorem on sandwich algebras—in its improved version due to E. I. Zel'manov and A. I. Kostrikin—and on the new idea of "separated powers". The technical embodiment of the proof is too complex even to mention here; however, the reader who masters the methods in this book will find it comparatively easy to dig into Zel'manov's proof, at least in the odd exponent case.

The universal finite group $B_0(d, n)$ thus makes its debut. What are its structural properties like? Clearly, the order of $B_0(d, n)$ is not that interesting a parameter.

Finally, we add that A. D. Chanyshev and Yu. A. Medvedev have pushed the method of obtaining thick sandwiches to its logical conclusion.

A. I. Kostrikin,
15.09.89

References

1. Adian, S. I.: Subgroups of free groups of odd exponent. Trudy Mat. Inst. Steklov. *172*, 64–72 (1971) (in Russian).
2. Adian, S. I.: The Burnside Problem and Identities in Groups. Ergebnisse der Math. Bd. 95. Berlin-Heidelberg-New York: Springer 1979 (English translation). Rūss. Qūdle? Roshan: Nauka 1975.
3. Adian, S. I.: Periodic products of groups. Trudy Mat. Inst. Steklov. *142*, 3–21 (1976) (in Russian).
4. Adian, S. I.: An axiomatic method for constructing groups with prescribed properties. Uspekhi Mat. Nauk *32*, 3–15 (1977) (in Russian).
5. Adian, S. I.: The simplicity of periodic products of groups. Dokl. Akad. Nauk SSSR *241*, 745–748 (1978) (in Russian).
6. Adian, S. I.: Normal subgroups of free groups of finite exponent. Izv. Akad. Nauk SSSR Ser. Mat. *45*, 931–947 (1981) (in Russian).
7. Adian, S. I.: Investigations on the Burnside problem and associated questions. Trudy Mat. Inst. Steklov. *168*, 171–196 (1984) (In Russian).
8. Adian, S. I. and Razborov, A. A.: Periodic groups and Lie algebras. Uspekhi Mat. Nauk *42*, 3–68 (1987) (in Russian).
9. Adian, S. I. and Repin, N. N.: An exponential lower bound for the nilpotency classes of Engel Lie algebras. Mat. Zametki *39*, 444–452 (1986) (in Russian).
10. Adian, S. I. and Repin, N. N.: Lower bounds for the orders of maximal groups of prime exponent. Mat. Zametki *44*, 161–169 (1988) (in Russian).
11. Adian, S. I.: Identités dans les groupes. Acta Congres. Int. Math. *1*, 263–267 (1971).
12. Adian, S. I.: Periodic groups of odd exponent. Lecture Notes in Math. *372*, 8–12 (1974).
13. Adian, S. I.: Classifications of periodic words and their application in group theory. Lecture Notes in Math *806*, 1–40 (1980).
14. Aleshin, S. V.: Finite automata and the Burnside problem on periodic groups. Mat. Zametki *11*, 319–328 (1971) (in Russian).
15. Alford, W. A. and Pietsch, B.: An application of the nilpotent quotient program. Lecture Notes in Math. *806*, 47–48 (1980).
16. Amayc, R. and Stewart, I. N.: Infinite-dimensional Lie algebras. Leyden 1984.
17. Bachmuth, S., Heilbronn, H. A. and Mochizuki, H. Y.: Burnside metabelian groups. Proc. Roy. Soc. London Ser. A *307*, 235–250 (1968).
18. Bachmuth, S. and Mochizuki, H. Y.: The class of the free metabelian group with exponent p^2. Comm. Pure Appl. Math. *21*, 385–399 (1968).
19. Bachmuth, S. and Mochizuki, H. Y.: Third Engel groups and the Macdonald-Neumann conjecture. Bull. Austral. Math. Soc. *5*, 379–386 (1971).
20. Bachmuth, S. and Mochizuki, H. Y.: A criterion for non-solvability of exponent 4 groups. Comm. Pure Appl. Math. *26*, 601–608 (1973).

21. Bachmuth, S., Mochizuki, H. Y. and Walkup, D.: A nonsolvable group of exponent 5. Bull. Amer. Math. Soc. *76*, 638–640 (1970).
22. Bachmuth, S., Mochizuki, H. Y. and Walkup, D. W.: Construction of a non-solvable group of exponent 5. In: Word Problems. Decision problems and the Burnside problem in group theory. Amsterdam: North-Holland 39–66 (1973).
23. Bachmuth, S., Mochizuki, H. Y. and Weston, K.: A group of exponent 4 with derived length at least 4. Proc. Amer. Math. Soc. *39*, 228–234 (1973).
24. Baer, R.: The higher commutator subgroups of a group. Bull. Amer. Math. Soc. *50*, 143–160 (1944).
25. Baker, H. F.: Alternants and continuous groups. Proc. London Math. Soc. *3*, No. 2, 24–47 (1905).
26. Bakhturin, Yu. A.: Identities in Lie Algebras. Moscow: Nauka 1984 (in Russian).
27. Bayes, A. J., Kautsky, K. and Walmsley, J. W.: Computation in nilpotent groups (application). Lecture Notes in Math. *372*, 82–89 (1974).
28. Blackburn, N. and Espuelas, Alberto: The power structure of metabelian p-groups. Proc. Amer. Math. Soc. *92*, 478–484 (1984).
29. Block, R. E. and Wilson, R. L.: The restricted simple Lie algebras are of classical or Cartan type. Proc. Nat. Acad. Sci. USA *81*, 5271–5274 (1984).
30. Bolker, E. D.: Groups whose elements are of order two or three. Amer. Math. Monthly *79*, 1007–1010 (1972).
31. Bourbaki, N.: Éléments de Mathématique. Fasc. XXVI. Chapitre 1: Algèbres de Lie, 1960. Chapitre 2–3: Algèbres de Lie libres, Groupes de Lie, 1972. Chapitre 4–6: Groupes de Coxeter et systèmes de Tits. Groupes engendres par des refléxions. Systèmes de racines, 1968. Chapitre 7–8: Sous-algèbres de Cartan. Éléments reguliers. Algèbres de Lie semisimples déployées, 1975.
32. Braun, A.: Lie rings and the Engel condition. J. Algebra *31*, 287–292 (1974).
33. Bruck, R. H.: On the restricted Burnside problem. Arch. Math. *13*, 179–186 (1962).
34. Bruck, R. H.: Engel conditions in groups and related questions. Lecture Notes, Third Summer Research Institute, Australian Math. Soc., Canberra 1963.
35. Bryce, R. A.: On metabelian groups of prime-power exponent. Proc. Roy. Soc. London Ser. A *130*, 393–399 (1969).
36. Burnside, W.: On an unsettled question in the theory of discontinuous groups. Quart. J. Appl. Math. *33*, 230–238 (1902).
37. Burnside, W.: On criteria for the finiteness of the order of linear substitutions. Proc. London Math. Soc. *3*, 435–440 (1905).
38. Burnside, W.: The theory of Groups of Finite Order. Cambridge University Press, 1911. Reprint: New York: Dover 1955.
39. Campbell, J. E.: On a law of combination of operators. Proc. London Math. Soc. *29*, 14–32 (1898).
40. Caranti, A.: Automorphism groups of p-groups of class 2 and exponent p^2: a classification on 4 generators. Annali. Mat., Ser. 4 *134*, 93–146 (1983).
41. Chang, B.: On Engel rings of exponent p-1. Proc. London Math. Soc. *11*, 203–212 (1961).
42. Chanyshev, A. D.: On 2-generator Engel Lie algebras over a field. Vestnik Moskov. Univ. Ser. I. Mat. Mekh. (in press) (in Russian).
43. Cohn, P. M.: A non-nilpotent Lie ring satisfying the Engel condition and a non-nilpotent Engel group. Proc. Cambridge Philos. Soc. *51*, 401–405 (1955).
44. Coxeter, H. S. M. and Moser, W. O.: Generators and relations for discrete groups. Berlin-Gottingen-Heidelberg: Springer 1957.

45. Dixon, M. R. and Fournelle, T. A.: The construction of infinite finitely generated periodic groups using wreath products. J. Algebra *115*, 150–163 (1988).
46. Doyle, J. K., Mandelberg, K. I. and Vaughan-Lee, M. R.: On solvability of groups of exponent 4. J. London Mathematical *18*, 234–242 (1978).
47. Dubnov, Ya. S. and Ivanov, V. K.: Sur l'abaissement du degré des polynômes en affineurs. C. R. (Doklady) Acad. Sci. URSS (N. S.) *41*, 95–98 (1943).
48. Edmunds, C. C. and Gupta, N. D.: On groups of exponent 4, IV. Conference on Group Theory (Univ. Wisconsin-Parkside, Kenosha, Wis., 1972), pp. 57–70. Lecture Notes in Math. *319*, 1973.
49. Filippov, V.: On Engel Mal'tsev algebras. Algebra i Logika *15*, 89–109 (1976).
50. Filippov, V. T.: On nil-elements of index 2 in Mal'tsev algebras. VINITI, No. 249$\bar{9}$80. (Sibirsk. Mat. Zh. *22*, (1981)) (in Russian).
51. Fischer, I. and Struik, R. R.: Nil algebras and periodic groups. Amer. Math. Monthly *75*, 611–623 (1968).
52. Forsyth, A. R.: William Burnside. J. London Math. Soc. *3*, 64–80 (1928).
53. Glauberman, G., Krause, E. F. and Struik, R. R.: Engel congruences in groups of prime-power exponent. Canad. J. Math. *18*, 579–588 (1966).
54. Golod, E. S.: Nil-algebras and residually finite groups. Izv. Akad. Nauk SSSR Ser. Mat. *28*, 273–276 (1964) (in Russian).
55. Golod, E. S.: Some problems of Burnside type, in: Proc. Internat. Congr. Math. (Moscow 1966), 284–289 (in Russian).
56. Green, J. A.: On groups with odd prime-power exponent. J. London Math. Soc. *27*, 476–485 (1952).
57. Green, J. A. and Rees, D.: On semigroups in which $x^r = x$. Proc. Cambridge Philos. Soc. *48*, 35–40 (1952).
58. Grigorchuk, R. I.: The Burnside problem on periodic groups. Funktsional. Anal. i Prilozhen. *14*, 53–54 (1980) (in Russian).
59. Grigorchuk, R. I.: On Milnor's problem concerning growth in groups. Dokl. Akad. Nauk SSSR *271*, 30–33 (1983) (in Russian).
60. Grishkov, A. N.: The local nilpotency of the ideal of a Lie algebra generated by an element of order 2. Sibirsk. Mat. Zh. *23*, 181–183 (1982) (in Russian).
61. Grishkov, A. N.: The restricted Burnside problem for nilpotent Moufang loops. Dokl. Akad. Nauk *285*, 534–536 (1985) (in Russian).
62. Gross, F.: On finite groups of exponent $p^m q^n$. J. Algebra *7*, 238–253 (1967).
63. Gruenberg, K. W.: Two theorems on Engel groups. Proc. Cambridge Philos. Soc. *49*, 377–380 (1953).
64. Grün, O.: Zusammenhang zwischen Potenzbildung und Kommutatorbildung. J. reine angew. Math. *182*, 158–177 (1940).
65. Grün, O.: Eine obere Grenze für die Klasse eine *h*-stufigen *p*-Gruppe. Abh. Math. Sem. Univ. Hamburg *21*, 90–91 (1957).
66. Grunewald, F. J., Havas, G., Mennicke, J. L. and Newman, M. F.: Groups of exponent 8. Bull. Austral. Math. Soc. *20*, 7–16 (1979).
67. Grunewald, F. J., Havas, G., Mennicke, J. L. and Newman, M. F.: Groups of exponent 8. Lecture Notes in Math. *806*, 49–188 (1980).
68. Grunewald, F. J. and Mennicke, J. L.: Finiteness proofs for groups of exponent 8. Lecture Notes in Math. *806*, 189–210 (1980).
69. Gupta, C. K. and Gupta, N. D.: Some groups of prime exponent. J. Combinatorial Theorm *5*, 379–407 (1968).
70. Gupta, C. K. and Gupta, N. D.: On groups of exponent four, II. Proc. Amer, Math. Soc. *31*, 360–362 (1972).

71. Gupta, N. D.: Polynilpotent groups of prime exponent. Bull. Amer. Math. Soc. *74*, 559–561 (1968).
72. Gupta, N. D.: The free metabelian group of exponent p^2. Proc. Amer. Math. Soc. *22*, 375–376 (1969).
73. Gupta, N. D., Mochizuki, H. Y. and Weston, K. W.: On groups of exponent four with generators of exponent 2. Bull. Austral. Math. Soc. *10*, 135–142 (1974).
74. Gupta, N. D. and Newman, M. F.: Engel congruences in groups of prime-power exponent. Canad. J. Math. *20*, 1321–1323 (1968).
75. Gupta, N. D. and Newman, M. F.: The nilpotency class of finitely generated groups of exponent 4. Lecture Notes in Math. *372*, 330–332 (1974).
76. Gupta, N. D. and Newman, M. F.: Groups of finite exponent. Bull. Austral. Math. Soc. *12*, 99 (1975).
77. Gupta, N. D., Newman, M. F. and Tobin, S. J.: On metabelian groups of prime-power exponent. Proc. Roy. Soc. Ser. A *302*, 237–242 (1968).
78. Gupta, N. D. and Quintana, R. B.: On groups of exponent four, III. Proc. Amer. Math. Soc. *33*, 15–19 (1972).
79. Gupta, N. D. and Sidki, S.: On Burnside's problem for periodic groups. Math. Z. *182*, 385–388 (1983).
80. Gupta, N. D. and Sidki, S.: Some infinite *p*-groups. Algebra i Logika *22*, 584–589 (1983).
81. Gupta, N. D. and Tobin, S. J.: On certain groups with exponent four. Math. Z. *102*, 216–226 (1967).
82. Gupta, N. D. and Weston, K. W.: On groups of exponent four. J. Algebra *17*, 59–66 (1974).
83. Hall, M. Jr.: The Theory of Groups. New York: Macmillan 1959.
84. Hall, M. Jr.: A basis for free Lie rings and higher commutators in free groups. Proc. Amer. Math. Soc. *1*, 575–581 (1950).
85. Hall, M. Jr.: Solution of the Burnside problem for exponent 6. Proc. Nat. Acad. Sci. USA *43*, 751–753 (1957).
86. Hall, M. Jr.: Solution of the Burnside problem for exponent six, III. J. Math. *2*, 764–786 (1958).
87. Hall, M. Jr.: Generators and relations in groups—the Burnside problem. Lectures in Modern Math. II, 42–92. New York (1964).
88. Hall, M. Jr.: Notes on groups of exponent four. Conference on Group Theory (Univ. Wisconsin–Parkside, Kenosha, Wis., 1972), pp. 91–118. Lecture Notes in Math. *319*, 1973.
89. Hall, M. Jr.: Computers in group theory. Topics in group theory and computation (Proc. Summer School, University Coll., Galway, 1973), pp. 1–37. Academic Press, London 1977.
90. Hall, M. Jr. and Sims, C. C.: The Burnside group of exponent 5 with two generators. London Math. Soc. Lecture Note Ser. *71*, 207–220 (1982).
91. Hall, P. and Higman, G.: On the *p*-length of *p*-soluble groups and reduction theorems for Burnside's problem. Proc. London Math. Soc. *6*, 1–42 (1956).
92. Hausdorff, F.: Die symbolische Exponentialformel in der Gruppentheorie. Bericht. Sachs. Akad. Wiss. Leipzig *58*, 19–48 (1906).
93. Havas, G.: Computational approaches to combinational group theory. Bull. Austral. Math. Soc. *11*, 475–476 (1974).
94. Havas, G.: Commutators in groups expressed as products of powers. Commun. Algebra *9*, 115–129 (1981).
95. Havas, G. and Newman, M. F.: Application of computers to questions like those of Burnside. Lecture Notes in Math. *806*, 211–230 (1980).

96. Havas, G., Newman, M. F. and Vaughan-Lee, M. R.: A nilpotent quotient algorithm for graded Lie rings (to appear).
97. Havas, G., Newman, M. F. and Vaughan-Lee, M. R.: Some finite groups of prime-power exponent (to appear).
98. Havas, G., Wall, G. E. and Wamsley, J. W.: The two-generator restricted Burnside group of exponent 5. Bull. Austral. Math. Soc. *10*, 459–470 (1974).
99. Heineken, H.: Liesche Ringe mit Engelbedingung. Math. Ann. *149*, 232–236 (1963).
100. Hermanns, F.-J.: Eine metabelsche Gruppe vom Exponenten 8. Arch. Math. *29*, 375–382 (1977).
101. Hermanns, F. J.: On certain groups of exponent eight generated by three involutions. Lecture Notes in Math. *806*, 231–245 (1980).
102. Herstein, I.: Non-commutative rings. The Carus Mathematical Monographs. New York: Wiley 1968.
103. Herzog, M. and Praeger, C. E.: On the order of linear groups of fixed finite exponent. J. Algebra *43*, 216–220 (1976).
104. Hickin, K. K. and Phillips, R. E.: Non-isomorphic Burnside groups of exponent p^2. Canad. J. Math. *80*, 180–189 (1978).
105. Hickin, K. K. and Phillips, R. E.: Joins of periodic groups. Proc. London Math. Soc. *39*, 176–192 (1979).
106. Higgins, P. J.: Lie rings satisfying the Engel condition. Proc. Cambridge Philos, Soc. *50*, 8–15 (1954).
107. Higman, G.: On a conjecture of Nagata. Proc. Cambridge Philos. Soc. *52*, 1–4 (1956).
108. Higman, G.: On finite groups of exponent 5. Proc. Cambridge Philos. Soc. *52*, 381–390 (1956).
109. Higman, G.: Le problème de Burnside. Colloque d'algèbre supérieure, 123–128 (1957).
110. Higman, G.: Finite groups in which every element has prime-power order. J. London Math. Soc. *32*, 335–342 (1957).
111. Higman, G.: Lie ring methods in the theory of finite nilpotent groups. Proc. Internat. Congr. Math. Edinburgh 307–312 (1958).
112. Higman, G.: *p*-length theorems. Proc. Sympos. Pure Math., Vol VI, pp. 1–16. American Mathematical Society, Providence R. I., 1962.
113. Holenweg, W.: Die Dimensiondefekte der Burnside-Gruppen mit zwei Erzeugenden. Comm. Math. Helv. *35*, 169–200 (1961).
114. Holenweg, W.: Über die Ordnung von Burnside-Gruppen mit endlich vielen Erzeugenden. Comm. Math. Helv. *36*, 83–90 (1962).
115. Huppert, B.: Endliche Gruppen 1. Berlin: Springer-Verlag 1967.
116. Huppert, B. and Blackburn, N.: Finite groups II, III. Berlin, Springer-Verlag 1982.
117. Ivanyuta, I. D.: On groups of exponent 4. Dokl. Akad. Nauk. Ukr. SSR Ser. A., 787–789 (1969) (in Ukrainian).
118. Jacobson, N.: Abstract derivations and Lie algebras. Trans. Amer. Math. Soc. *42*, 206–224 (1937).
119. Jacobson, N.: Structure theory for algebraic algebras of bounded degree. Ann. Math. *46*, 695–707 (1945).
120. Jacobson, N.: Lie algebras. Interscience tracts in pure and applied mathematics. New York-London: Interscience 1962.
121. Jacobson, N.: Some recent developments in the theory of algebras with polynomial identities. Lecture Notes in Math. *697*, 8–46 (1978).

122. Jennings, S. A.: The structure of the group ring of a p-group over a modular field. Trans. Amer. Math. Soc. *50*, 175–185 (1941).
123. Kaplansky, U.: On a problem of Kurosch and Jacobson. Bull. Amer. Math. Soc. *52*, 496–500 (1946).
124. Kaplansky, I.: Lie algebras. In: Lectures in modern math, I. 115–132 (1963).
125. Kargapolov, M. I.. and Merzlyakov, Yu. I.: Foundations of Group Theory, 3rd Edition. Moscow: Nauka 1982 (in Russian).
126. Khukhro, E. I.: Finite groups of exponent $p^\alpha q^\beta$. Algebra i Logika *17*, 727–740 (1978) (in Russian).
127. Khukhro, E. I.: On the connection between the Hughes conjecture and relations in finite groups of prime exponent. Mat. Sb *116*(*158*), 352–364 (1981) (in Russian).
128. Khukhro, E. I.: The associated Lie ring of the free 2-generator group of prime exponent and the Hughes conjecture for 2-generator p-groups. Mat. Sb. *118*(*160*), 567–575 (1981) (in Russian).
129. Khukhro, E. I.: A new identity in the Lie ring of a free group of prime exponent and groups without the Hughes property. Izv. Akad. Nauk SSSR Ser. Mat. *50*, (1980) (in Russian).
130. Khukhro, E. I.: The local nilpotency of groups having splitting automorphisms of prime order. Mat. Sb. *130*(*172*), 121–127 (1986) (in Russian).
131. Kostrikin, A. I.: Solution of the restricted Burnside problem for exponent 5. Izv. Akad. Nauk SSSR Ser. Mat. *19*, 233–244 (1955) (in Russian).
132. Kostrikin, A. I.: Lie rings satisfying an Engel condition. Dokl. Akad. Nauk SSSR *108*, 580–582 (1956) (in Russian).
133. Kostrikin, A. I.: Lie rings satisfying an Engel condition. Izv. Akad. Nauk SSSR Ser. Mat. *21*, 595–540 (1957) (in Russian).
134. Kostrikin, A. I.: On the connection between periodic groups and Lie rings. Izv. Akad. Nauk SSSR Ser. Mat. *21*, 289–310 (1957) (in Russian).
135. Kostrikin, A. I.: On the local nilpotency of Lie rings satisfying an Engel condition. Dokl. Akad. Nauk SSSR *118*, 1074–1077 (1958) (in Russian).
136. Kostrikin, A. I.: On Burnside's problem. Dokl. Akad. Nauk SSSR *119*, 1081–1084 (1958) (in Russian).
137. Kostrikin, A. I.: On Burnside's problem. Izv. Akad. Nauk SSSR *23*, 3–34 (1959) (in Russian).
138. Kostrikin, A. I.: Engel properties of groups satisfying the identical relation $x^p = 1$. Dokl. Akad. Nauk SSSR *135*, 524–526 (1960) (in Russian).
139. Kostrikin, A. I.: On simple Lie p-algebras. Trudy Mat. Inst. Steklov. *64*, 79–89 (1961) (in Russian).
140. Kostrikin, A. I.: Lie algebras and finite groups. Proc. Internat. Congr. Math. Stockholm, 264–269 (1962) (in Russian).
141. Kostrikin, A. I.: Variations modulaires sur un thème de Cartan. Actes. Congres. Internat. Math. 1, 285–92 (1970).
142. Kostrikin, A. I.: Some related questions in the theory of groups and Lie algebras. In: Proc. Second. Internat. Conf. Theory of Groups, Canberra 1973. Lecture Notes in Math. *372*, 409–416 (1974).
143. Kostrikin, A. I.: Sandwich Lie algebras. Mat. Sb. *110*(*152*), 3–12 (1979).
144. Kostrikin, A. I.: Lie algebras and finite groups. Trudy Mat. Inst. Steklov. *168*, 132–154 (1984).
145. Kostrikin, A. I.: The nilpotency class of n-Engel Lie algebras. Sb. Algebra, MSU 1988 (to appear: in Russian).

146. Kostrikin, A. I. and Shafarevich, I. R.: Cartan pseudogroups and Lie ap-algebras. Dokl. Akad. Nauk SSSR *168*, 740–742 (1966) (in Russian).
147. Kovács, L. G.: Varieties of groups and Burnside's problem. Bull. Amer. Math. Soc. *74*, 599–601 (1968).
148. Kovács, L. G.: Varieties and finite groups. J. Austral. Math. Soc. *10*, 5–19 (1969).
149. Krause, E.: On the collection process. Groups of exponent 8 satisfy the 14th Engel congruence. Proc. Amer. Math. Soc. *15*, 491–496, 497–504 (1964).
150. Krause, E. and Weston, K.: An algorithm related to the restricted Burnside group of prime exponent. In: Computational Problems in Abstract Algebra. New York: Pergamon Press 1970, pp. 185–187.
151. Krause, E. and Weston, K. W.: On the Lie algebra of a Burnside group of exponent 5. Proc. Amer. Math. Soc. *27*, 463–470 (1971).
152. Kurosh, A. G.: Problems in ring theory connected with Burnside's problem on periodic groups. Izv. Akad. Nauk SSSR Ser. Mat. *5*, 233–240 (1941) (in Russian).
153. Kurosh, A. G.: The theory of groups, Vols 1 and 2. Moscow: Gostekh-izdat 1944 New York: Chelsea 1955 (English translation).
154. Kuz'min, E. N.: On the Nagata-Higman theorem. In: Mathematical Structures, 101–107, Sofia 1975 (in Russian).
155. Lazard, M.: Sur les groupes nilpotents et les anneaux de Lie Ann. Sci. École Norm. Sup. *71*, 101–190 (1954).
156. Lazard, M.: Lois de groupes et analyzeurs. Ann. Sci. École Norm. Sup. *72*, 299–400 (1955).
157. Lazard, M.: Groupes, anneaux de Lie et probleme de Burnside. Istituto Matematico dell'Universita, Roma 1960. pp. 1–60.
158. Leech, J.: Coset enumeration on digital computers. Proc. Cambridge Philos. Soc. *59*, 257–267 (1963).
159. Levi, F.: Über die Untergruppen der freien Gruppen II. Math. Zeit. *37*, 90–97 (1933).
160. Levi, F. und van der Waerden, B. L.: Über eine besondere Klasse von Gruppen. Abh. Math. Sem. Univ. Hamburg *9*, 154–158 (1933).
161. Loewy, A.: Zur Theorie der Gruppen linearer Substitutionen. Math. Ann. *53*, 225–242 (1900).
162. Lobych, V. P. and Skopin, A. I.: Relations in groups of exponent 8. Zap. Nauchn. Sem. Leningrad. Otdel. Mat. Inst. Steklov. *464*, 92–94 (1976) (in Russian).
163. Lyndon, R. C.: On Burnside's problem I. Trans. Amer. Math. Soc. *77*, 202–215 (1954); II, Trans. Amer. Math. Soc. *78*, 329–332 (1955).
164. Lyndon, R. C.: Burnside groups and Engel rings. Proc. Sympos. Pure Math. Vol 1, pp. 4–14. American Mathematical Society, Providence R. I. 1959.
165. Macdonald, I. D.: The Hughes problem and others. J. Austral. Math. Soc. *10*, 475–479 (1969).
166. Macdonald, I. D.: Solution of the Hughes problem for finite *p*-groups of class 2p. Proc. Amer. Math. Soc. *27*, 39–42 (1971).
167. Macdonald, I. D.: Computer results on Burnside groups. Bull. Austral. Math. Soc. *9*, 433–438 (1973).
168. Macdonald, I. D.: A computer application to finite *p*-groups. Bull. Austral. Math. Soc. *17*, 102–112 (1974).
169. MacLane, S.: Some recent advances in algebra. Studies in Modern Algebra, pp. 9–34, 1963.
170. McMullen, J. R.: Compact torsion groups. Lecture Notes in Math *372* (1974).
171. Magnus, W.: Beziehungen zwischen Gruppen und Idealen in einem spezieillen Ring. Math. Ann *111*, 259–280 (1935).

172. Magnus, W.: Neuere Ergebnisse über auflösbare Gruppen. Jber. Deutsch. Math.-Verein *47*, 69–78 (1937).
173. Magnus, W.: Über Beziehungen zwischen höhere Kommutatoren. J. reine angew. Math. *177*, 105–115 (1937).
174. Magnus, W.: Über Gruppen und Zugeordnete Liesche Ringe. J. reine angew. Math. *182*, 142–149 (1940).
175. Magnus, W.: A connection between the Baker-Hausdorff formula and a problem of Burnside. Ann. Math. *52*, 11–26 (1950); Errata, Ann. Math. *57*, 606 (1953).
176. Magnus, W. and Chandler, B.: A History of Combinatorial Group Theory: A Case Study in the History of Ideas. Berlin, Springer 1982.
177. Magnus, W., Karras, A. and Solitar, D.: Combinatorial Group Theory. New York: Wiley 1966.
178. Malyshev, F. M.: The nilpotency class of Engel Lie algebras. Vestnik Moskov. Univ. Ser. I Mat. Mekh. No. 2, 55–58 (1980) (in Russian).
179. Malyshev, F. M.: Simplicial systems of linear equations. In: Algebra. Moscow: Moscow University Press (1980), pp. 53–56 (in Russian).
180. Mann, A. J. S.: On the orders of groups of exponent 4. J. London Math. Soc. *28*, 64–76 (1982).
181. Mazurov, V. D.: The restricted Burnside problem for exponent 30. Algebra i Logika *8*, 460–477 (1968) (in Russian).
182. Meier-Wunderli, H.: Über endliche p-Gruppen, deren Elemente der Gleichung x^p: = 1 genügen. Comm. Math. Helv. *24*, 18–45 (1950).
183. Meier-Wunderli, H.: Metabelsche Gruppen. Comm. Math. Helv. *35*, 1–10 (1951).
184. Meier-Wunderli, H.: Über die Struktur der Burnsidegruppen mit zwei Erzeugenden und vom Primzahlexponenten $p > 3$. Comm. Math. Helv. *30*, 144–174 (1956).
185. Meixner, T.: Eine Bemerkung zu p-Gruppen vom Exponenten p. Arch. Math. *29*, 561–563 (1977).
186. Michel, Jean: Calculs dans algèbres de Lie libres: la serie de Hausdorff et le problème de Burnside. Asterisque No. 38–39, 139–148 (1976).
187. Miller, J. I.: Center-by-metabelian groups of prime exponent. Trans. Amer Math. Soc. *249*, 217–224 (1979).
188. Mishchenko, S. P.: The Engel identity and its applications. Mat. Sb. *121*, 423–430 (1983) (in Russian).
189. Mishchenko, S. P.: On the problem of the Engel property. Mat. Sb. *124*, 56–67 (1984) (in Russian).
190. Mochizuki, Horace, Y.: On groups of exponent four: a criterion for nonsolvability. In: Proc. Second Internat. Conf. Theory of Groups, Canberra 1973. Lecture Notes in Math. *372*, 499–503 (1974).
191. Monarkh, E. I. and Skopin, A. I.: An interactive system of symbolic computations in groups of Burnside type. Zap. Nauchn. Sem. Leningrad. Otdel. Inst. Mat. Steklov. *114*, 164–173 (1982) (in Russian).
192. Moran, S.: The product of powers in a finite p-group. Arch. Math. *17*, 112–120 (1966).
193. Muzalewski, M.: Burnside's problems, residual finiteness and finite reducibleness. Bull. Acad. Polon. Sci. Ser. Sci. Math. Astronom. Phys. *24*, 1067–1068 (1976).
194. Nagata, M.: On the nilpotency of nil-algebras. J. Math. Soc. Japan *4*, 296–301 (1952).
195. Neumann, B. H.: Identical relations in groups I. Math. Ann. *114*, 506–525 (1937).
196. Neumann, B. H.: Groups whose elements have bounded orders. J. London Math. Soc. *12*, 195–198 (1937).
197. Neumann, Hanna: Varieties of Groups. Ergebnisse der Math. Bd. 37. Berlin-Heidelberg-New York: Springer-Verlag 1967.

198. Newman, M. F.: A computer-aided study of a group defined by fourth powers. Bull. Austral. Math. Soc. *14*, 293–294 (1976); Addendum: Bull. Austral. Math. Soc. *15*, 477–479 (1976).
199. Newman, M. F.: Determination of groups of prime-power order. Lecture Notes in Math. *573* (1977). (Proc. Miniconf. Theory of Groups, Canberra (1975), 73–84).
200. Newman, M. F.: Groups of exponent dividing seventy. Math. Scientist *4*, 149–157 (1979).
201. Newman, M. F.: Problems (of Burnside's type). Lecture Notes in Math. *806*, 249–254 (1980).
202. Newman, M. F.: Bibliography compiled for the Proceedings of the Burnside Workshop. Lecture Notes in Math. *806*, 255–274 (1980).
203. Newman, M. F.: Groups of exponent six. Computational Group Theory, 39–41. London, Academic Press (1984).
204. Newman, M. F., Weston, K. W. and Tah-Zen Yau.: Polynomials associated with groups of exponent four. Bull. Austral. Math. Soc. *12*, 81–87 (1975).
205. Novikov, P. S.: Solution of the Burnside problem on periodic groups. Uspekhi Mat. Nauk *14*, 236–237 (1959) (in Russian).
206. Novikov, P. S.: On periodic groups. Dokl. Akad. Nauk SSSR *127*, 749–752 (1959) (in Russian).
207. Novikov, P. S. and Adian, S. I.: On infinite periodic groups I, II, III. Izv. Akad. Nauk SSSR *32*, No. 1, 212–244 (1968); *32*, No. 2, 251–524 (1968); *32*, No. 3, 709–731 (1968) (in Russian).
208. Novikov, P. S. and Adian, S. I.: Defining relations and the word problem for free groups of finite exponent. Izv. Akad. Nauk. SSSR Mat. *32*, 971–979 (1968) (in Russian).
209. Novikov, P. S. and Adian, S. I.: Commutative subgroups and the conjugacy problem for free groups of odd exponent. Izv. Akad. Nauk. SSSR Ser. Mat. *32*, 1176–1190 (1968) (in Russian).
210. Ol'shanskij, Yu. A.: On a theorem of Novikov and Adian. Mat. Sb. *118*, 203–235 (1982) (in Russian).
211. Ol'shanskij, Yu. A.: Groups of finite exponent whose proper subgroups are of prime order. Algebra i Logika *21*, 555–618 (1982) (in Russian).
212. Ol'shanskij, Yu. A.: The goemetry of defining relations in groups. Moscow: Mir 1989 (in press: in Russian).
213. Panella, Gianfranco: Un teorema di Golod-Šafarevič e alcune sue conseguenze. Conf. Sem. Math. Univ. Bari No. 104, 17 (1966).
214. Plonka, J.: On direct products of some Burnside groups. Acta. Math. Acad. Sci. Hung. *27*, 43–45 (1976).
215. Polishchuk, E. M.: Sophus Lie. Moscow: Nauka 1983 (in Russian).
216. Premet, A. A.: Lie algebras with strong degeneracy. Mat. Sb. *129*(*171*), 140–153 (1986) (in Russian).
217. Procesi, C.: The Burnside problem. J. Algebra *4*, 421–425 (1966).
218. Procesi, C.: Rings with polynomial identities. New York: Marcel Dekker 1973.
219. Quintana, R. B.: An attack on the restricted Burnside problem for groups of exponent 8 on 2 generators. Lecture Notes in Math. *319*, 140–147 (1973).
220. Razmyslov, Yu. P.: On Engel Lie algebras. Algebra i Logika *10*, 33–44 (1971) (in Russian).
221. Razmyslov, Yu. P.: On a problem of Hall and Higman. Izv. Akad. Nauk SSSR Ser. Mat. *42*, 833–847 (1978) (in Russian).
222. Razmyslov, Yu. P.: The nonsolvability of the variety of groups of exponent 4. In: Ring

Theory: Proc. 1978 Antwerp Conf., 219–231 (1978); Lecture Notes in Pure and Applied Math. *51*.
223. Razmyslov, Yu. P.: Identities in Lie algebras and their representations. Candidate's Dissertation, Moscow 1985 (in Russian).
224. Razmyslov, Yu. P.: Identities in algebras and their representations. Moscow: Nauka 1989 (in press: in Russian).
225. Rowen, L. H.: Polynomial identities in ring theory. New York: Academic Press 1980.
226. Rozhkov, A. V.: Subgroups of groups of Aleshin type. Mat. Sb. *129*(*171*), 422–433 (1986) (in Russian).
227. Sanov, I. N.: Solution of the Burnside problem for exponent 4. Uchen. Zap. Leningrad. Univ. *10*, 166–170 (1940).
228. Sanov, I. N.: Periodic groups of small exponent. Candidate's Dissertation, Leningrad 1946 (in Russian).
229. Sanov, I. N.: A property of a certain presentation of a free group. Dokl. Akad. Nauk SSSR *57*, 657–659 (1947) (in Russian).
230. Sanov, I. N.: On the Burnside problem. Dokl. Akad. Nauk SSSR *57*, 759–761 (1947) (in Russian).
231. Sanov, I. N.: On a certain system of relations in periodic groups of prime-power exponent. Izv. Akad. Nauk SSSR Ser. Mat. *15*, 477–502 (1951) (in Russian).
232. Sanov, I. N.: A connection between periodic groups of prime-power exponent and Lie rings. Izv. Akad. Nauk SSSR Ser. Mat. *16*, 23–58 (1952) (in Russian).
233. Schenkman, E.: Two theorems on finitely generated groups. Proc. Amer. Math. Soc. *5*, 497–498 (1954).
234. Schur, I.: Über Gruppen periodischer linearer Substitutionen. Sitzungsber. Preuss. Akad. 619–67 (1911).
235. Séguier, J. A. de: Théorie des groupes finis. Éléments de la théorie des groupes abstrait. Paris: Gauthier-Villars 1904.
236. Seligman, G.: Modular Lie Algebras. New York: Springer-Verlag 1967.
237. Shield, D.: The class of a nilpotent wreath product. Bull. Austral. Math. Soc. *17*, 53–89 (1977).
238. Shirvanyan, V. L.: The embedding of $B(\infty, n)$ in $B(2, n)$. Izv. Akad. Nauk SSSR Ser. Mat. *40*, 190–208 (1976) (in Russian).
239. Skopin, A. I.: The collecting formula. Zap. Nauchn. Sem. Leningrad. Otdel. Mat. Inst. Steklov. *46*, 59–63 (1974) (in Russian).
240. Skopin, A. I.: Relations in groups of exponent 8. Zap. Nauchn. Sem. Leningrad. Otdel. Mat. Inst. Steklov. *57*, 129–170 (1976) (in Russian).
241. Skopin, A. I.: Transmetabelian groups. Zap. Nauchn. Sem. Leningrad. Otdel. Mat. Inst. Steklov, *75*, 159–163 (1978) (in Russian).
242. Skopin, A. I.: On a group of exponent 8. Zap. Nauchn. Sem. Leningrad. Otdel. Mat. Inst. Steklov, *75*, 164–165 (1978) (in Russian).
243. Skopin, A. I.: A metabelian group of exponent 9 on two generators. Zap. Nauchn. Sem. Leningrad. Otdel. Mat. Inst. Steklov. *103*, 124–131 (1980) (in Russian).
244. Skopin, A. I.: The factors of the nilpotent series of some metabelian groups of prime-power exponent. Zap. Nauchn. Sem. Leningrad. Otdel. Mat. Inst. Steklov. *132*, 129–163 (1983) (in Russian).
245. Soublin, J.-P.: Problèmes de Burnside. C. R. des Journeés Mathématique de la Societé Mathématique de France (Univ. Sci. Tech. Languedoc, Montpelier, 1974) pp. 151–156.
246. Struik, R. R.: Notes on a paper by Sanov I. Proc. Amer. Math. Soc. *8*, 638–641 (1957); II, Proc. Amer. Math. Soc. *12*, 758–763 (1961).

247. Strunkov, S. P.: Subgroups of periodic groups. Dokl. Akad. Nauk SSSR *170*, 279–281 (1966) (in Russian).
248. Sushchanskij, V. I.: p-groups of permutations and the unrestricted Burnside problem. Dokl. Akad. Nauk SSSR *247*, 557–560 (1979) (in Russian).
249. Sushchanskij, V. I.: Wreath products over a sequence of permutation groups and residually finite groups. Dokl. Akad. Nauk SSSR Ser. A No. 2, 19–22 (1984) (in Russian).
250. Sushchanskij, V. I.: Groups of isometries of Baer p-spaces. Dokl. Akad. Nauk SSSR Ser. A. No. 8, 28–30 (1984) (in Russian).
251. Sushchanskij, V. I.: The Lie ring of the Sylow p-subgroup of the group of isometries of the space of p-adic integers. XVIII-th All-Union Algebraical Conference, p. 192 (1985) (in Russian).
252. Sushchanskij, V. I.: The representation of residually finite p-groups by isometries of the ring of p-adic integers. Dokl. Akad. Nauk Ukr. SSSR Ser. A (1986) (in press: in Russian).
253. Suzuki, M.: Group Theory II. Grundlehren Math. Wiss. *248*. New York: Springer-Verlag 1985.
254. Timofeenko, A. V.: On the 2-generator Golod p-groups. Algebra i Logika *24*, 211–225 (1985) (in Russian).
255. Tobin, S. J.: On groups with exponent 4. Ph. D. Thesis, Univ. of Manchester 1954.
256. Tobin, S. J.: On a theorem of Baer and Higman. Canad. J. Math. *8*, 263–270 (1956).
257. Tobin, S. J.: Simple bounds for Burnside p-groups. Proc. Amer. Math. Soc. *11*, 704–706 (1960).
258. Tobin, S. J.: On groups with exponent four. Proc. Roy. Ir. Acad. Sect. A *75*, 115–120 (1975).
259. Tokarenko, A. I.: Linear groups over rings. Sibirsk. Mat. Zh. *9*, 951–959 (1968) (in Russian).
260. Tritter, A. L.: A module-theoretic computation related to the Burnside problem. In: Computational Problems in Abstract Algebra, 189–198 (1970).
261. Vaughan-Lee, M. R.: Derived lengths of Burnside groups of exponent 4. Quart. J. Math. Oxford *30*, 495–504 (1979).
262. Vaughan-Lee, M. R.: The restricted Burnside problem. Bull. London Math. Soc. *17*, 113–133 (1985).
263. Vaughan-Lee, M. R.: Towards a Constructive Proof of Kostrikin's Theorem (in press).
264. Venkov, B. B.: Some homological properties of Burnside groups. Zap. Nauchn. Sem. Leningrad. Otdel. Mat. Inst. Steklov. *31*, 38–54 (1973) (in Russian).
265. Volichenko, I. B.: Some connections between Engel, two-term and standard identities in Lie algebras. Preprint No. 8, Int. Mat. Akad. Nauk BSSR (1977).
266. Wagner, A. and Mosenthal, V. A.: A bibliography of William Burnside. Historia Math. *5*, 307–312 (1978).
267. Wall, G. E.: On Hughes' H_p-problem. In: Proc. Internat. Conf. Theory of Groups, Canberra 1965. New York: Gordon and Breach 1967, pp. 357–362.
268. Wall, G. E.: On the Lie ring of a group of prime exponent. In: Proc. Second Internat. Conf. Theory of Groups, Canberra 1973. Lecture Notes in Math. *372*, 667–690 (1974).
269. Wall, G. E.: On the Lie ring of a group of prime exponent II. Bull. Austral. Math. Soc. *19*, 11–29 (1978).
270. Wall, G. E.: Lie Methods in Group Theory. Lecture Notes in Math. *697*, 137–173 (1978).
271. Wall, G. E.: On the multilinear identities which hold in the Lie ring of group of prime-power exponent. Preprint, Dept. Pure Math., University of Sydney.

272. Wall, G. E.: On the multilinear Lie relators for varieties of groups (to appear).
273. Wiegold, J.: Kostrikin's proof of the restricted Burnside conjecture for prime exponent. Lectures delivered at the Institute of Advanced Studies, Australian National Univ. 1965.
274. Wright, C. R. B.: On groups of exponent 4 with generators of order 2. Pacific J. Math. *10*, 1097–1105 (1960).
275. Wright, C. R. B.: On the nilpotency class of a group of exponent four. Pacific J. Math. *11*, 387–394 (1961).
276. Yager, R. I.: The Burnside Problem. Honours Essay, Univ. of Sydney, 1977.
277. Zassenhaus, H.: Über Liesche Ringe mit Primzahlcharakteristik. Abh. Math. Sem. Univ. Hamburg *13*, 1–100 (1939).
278. Zassenhaus, H.: Ein Verfahren, jeder endlichen *p*-Gruppe einem Lie-Ring mit der Charakteristik *p* zu zuordnen. Abh. Math. Sem. Univ. Hamburg *13*, 200–207 (1940).
279. Zel'manov, E. I.: Jordan nil-algebras of bounded index. Dokl. Akad. Nauk SSSR *249*, 30–33 (1979) (in Russian).
280. Zel'manov, E. I.: Absolute zero divisors in Jordan pairs and Lie algebras. Mat. Sb. *112*, 611–629 (1980) (in Russian).
281. Zel'manov, E. I.: Absolute zero divisors and algebraic Jordan algebras. Sibirsk. Mat. Zh. *23*, 100–116 (1982) (in Russian).
282. Zel'manov, E. I.: Lie algebras with algebraic adjoint representation. Mat. Sb. *12*, 545–561 (1983) (in Russian).
283. Zel'manov, E. I.: Lie algebras with a finite grading. Mat. Sb. *124*, 353–392 (1984) (in Russian).
284. Zel'manov, E. I.: Lie algebras with the Engel property. Dokl. Akad. Nauk SSSR *292*, 265–268 (1987) (in Russian).
285. Zel'manov, E. I.: On the nilpotency of nil-algebras. Lecture Notes in Math. 1988 (in press).
286. Zel'manov, E. I.: Some problems in the theory of groups of Lie algebras. Mat. Sb. No. 2 (1989) (in press; in Russian).
287. Zel'manov, E. I. and Kostrikin, A. I.: A theorem on sandwich algebras. Trudy Mat. Moskov. Obshch. (in press).
288. Zolotykh, A. A.: On 4-Engel Lie algebras. Vestnik Moskov. Univ. Ser. I. Mat. Mekh, No. 2, 79–81 (1986) (in Russian).

Author Index

Subject Index

Notation